LABORATORY MANUAL FOR

STARR'S
BIOLOGY
CONCEPTS AND APPLICATIONS

JAMES W. PERRY and DAVID MORTON
Frostburg State University

Contributions to Animal Diversity
by Ronald E. Barry, Jr.
and Evolutionary Agents
by James H. Howard

Wadsworth Publishing Company
Belmont, California
A Division of Wadsworth, Inc.

BIOLOGY EDITOR Jack C. Carey
MANAGING DESIGNER Andrew H. Ogus
ART EDITOR Donna Kalal
ILLUSTRATORS Creative Circle, Cecile Duray-Bito,
Tasa Graphic Arts, Inc.

Printed in the United States of America   49

2 3 4 5 6 7 8 9 10—95 94 93 92

ISBN 0-534-13373-8

CONTENTS

This laboratory manual is designed for students at the college entry level and assumes that the student has had no previous college biology or chemistry.

In preparing this manual we paid particular attention to **pedagogy, flexibility of format, clarity of instructions, illustrations, and practicality of the materials** to be used.

Page References to Starr's
Biology: Concepts and Applications

The manual has been written to conform with the terminology in *Biology: Concepts and Applications*, by Cecie Starr. Page references to the text follow the main headings. This material provides useful background reading for the exercise. You will find, however, that the exercises support virtually any biology text used in an introductory course.

Length of Lab and Demonstration Options

We realize there is wide variation in the amount of time each instructor devotes to laboratory activities. To provide additional *flexibility* for the instructor, some of the more experimental exercises include alternative activities. For example, in Exercise 4, Diffusion, two options are provided for experimentation with osmosis.

In the diversity exercises, once students have completed the introductory portion, instructors may choose which of the experimental procedures to assign based upon the length of time available. For example, after Phase I of Exercise 5 has been completed, options II through V offer instructors several independent ways to demonstrate the properties of enzymes.

If your course has only two-hour labs, we suggest two alternatives: Portions of the exercise may be deleted, or portions of the exercise may be put on demonstration.

Logical Organization

Pedagogically, we made certain that each portion of the exercise is part of a continuous flow of thought. Thus we do not wait until the post-lab questions to ask students to record conclusions when it is more appropriate to do so within the body of the procedure.

Clarity in Procedures and Terminology

The exercises have been written so that the conscientious student can accomplish the objectives of each exercise with minimal input from an instructor. The procedure section of each exercise is much more detailed and step-by-step than in other manuals. Each step is numbered. Instructions follow a natural progression of thought so the instructor need not conduct every movement. The format of each exercise includes:

1. a list of **objectives** to be accomplished;
2. an **introduction** to pique student interest and indicate relevance;
3. a list of **materials** for each portion of the exercise so that a student can gather quickly the supplies necessary;
4. the **procedure**, including safety notes, illustrations of apparatus, figures to be labeled, drawings to be made, tables for recording data, graphs to be drawn, and questions that lead to conclusions to be drawn; the procedure is often in list form with each step numbered, and terms required to accomplish objectives are **boldfaced**;
5. ten **pre-lab questions** about the introduction and procedure, which the student should be able to answer after reading the exercise and prior to entering the laboratory; and
6. **post-lab questions** that draw upon knowledge gained from *doing* the exercise, which the student should be able to answer *after* finishing.

To further increase *clarity*, the use of scientific names within the procedure has been deemphasized when it is not relevant to understanding of the subject. Generally, these names appear within parentheses because the label on many prepared microscope slides bears only the scientific name.

Flexible Quiz Options

Each exercise has a set of *pre-lab questions*. We have found, through 20-odd years of experience, that students typically walk into the laboratory unprepared to do the exercise. They may or may not have had a lecture presentation on the subject, but few read the exercise before coming to lab. The obvious answer was to incorporate some sort of pre-exercise activity. However, we recognize that grading a large number of lab papers each week may put an unreasonable burden on instructors. Consequently, we decided on a multiple-choice format (which is easy to grade and still accomplishes the pedagogical goal). In the **Instructor's Manual** described below, the pre-lab questions have been duplicated with questions and answers in a different order from that in the lab manual to avoid memorization. In our course, we have students take a pre-lab quiz, consisting of pre-lab questions that have been scrambled (both sequence of questions and sequence of answers for each question). This quiz takes literally about two minutes of lab time and counts as part of the lab grade, thus rewarding students for preparation.

Post-lab questions are included as well. To answer these questions the student generally has to apply the knowledge gained from the exercise. In some cases the questions may have no "right or wrong" answer. In others, consultation with the textbook may be necessary. In short, these questions search out understanding, rather than simple recall.

Post-lab questions are not universally intended to be answered simply on the basis of information gained directly from the lab; if they were, they would be a restatement of the objectives in question form. Rather, they seek to extend and apply knowledge by encouraging students to consult their textbook, information in the library, or their own creative thought. We have provided answers to the post-lab questions in the **Instructor's Manual** with the realization that you and your students may find other, equally acceptable answers to the questions.

There is also a great deal of *flexibility* in how these pre- and post-lab questions may be used. For example, some believe the pre-lab questions could be used to measure understanding after the exercise has been completed. If you think this is the case, by all means use them in this manner.

Illustrations

The charge from our editor, Jack Carey, was to provide as many *illustrations* as possible. This stemmed from the recognition that many portions of the biological experience are visual as well as cognitive. Moreover, most textbooks are unable to provide enough illustrations to accompany each laboratory activity. Our survey administered by Wadsworth Publishing Company indicated that there is a diversity of opinion on how illustrations should appear and/or be used. Thus, some illustrations of microscopic specimens are labeled to provide orientation and clarity. Most micrographs are unlabeled but provide leaders to which students can attach labels. In other cases, more can be gained by requesting the student to do simple drawings. Space has been included in the manual for them, with boxes for macroscopic and circles for microscopic specimens.

Materials

The *materials* used in the exercises are readily available from biological and laboratory supply houses. Within the exercises a materials list is broken into categories of "Per Student, Per Pair, Per Group, Per Lab Room" so that required supplies are obvious to students.

Dissection Options

We have included a mouse dissection (Exercise 14) as an abbreviated option to the longer fetal pig dissection (Exercises 31–33). Mice are significantly less expensive than rats, are procreative enough to raise in relatively large numbers over the course of several months, and offer an alternative to those instructors who wish for their students to study the anatomy of a mammal, without going into the detail provided by the fetal pig exercises. There are also the added advantages of using an animal that has been recently alive (and shows it internally) and avoiding the necessity of chemical fixatives.

As anyone who lives in a temperate climate is aware, it may be necessary to adjust the sequence of the laboratory exercises to accommodate seasonal availability of certain materials. However, we have suggested the use of preserved specimens wherever possible to avoid this problem.

Instructor's Manual

A separate Instructor's Manual includes:
- approximate quantities of materials needed;
- procedures for preparing reagents, materials, and equipment;
- vendors and item numbers for supplies;
- answers to pre-lab questions;
- answers to post-lab questions;
- answers to Mendelian genetics problems (Exercise 10); and
- tear-out sheets of pre-lab questions (which appear in different order from students' Laboratory Manual) for those who wish to duplicate them for quizzes.

There are very few things in life that are perfect. We don't suppose that this lab manual is one of them. We think your students will enjoy the exercises. Perhaps you and they will find places where rephrasing will make the activity better. Please, let us know your concerns, and encourage your students to do likewise.

We encourage your criticism, comments, and suggestions. Call us!

James W. Perry
(301) 689-4173
David Morton
(301) 689-4355
Frostburg, Maryland
1990

REVIEWERS

Our thanks to the following colleagues for their assistance during the extensive review and class-testing processes.

Louis Avosso, Nassau Community College

J. Wesley Bahorik, Kutztown University of Pennsylvania

Virginia Buckner, Johnson County Community College

Tommi Lou Carosella, California State University–Stanislaus

Christine L. Case, Skyline College

Jerry D. Davis, University of Wisconsin–La Crosse

Jean DeSaix, University of North Carolina–Chapel Hill

Lowell V. Diller, Frostburg State University

Roland R. Duke, Auburn University

Barbara Hanson, Canisius College

Amy Harman, Frostburg State University

James H. Howard, Frostburg State University

John D. Jackson, North Hennepin Community College

Robert Kull, U.S. Air Force Academy

Earl L. Nollenberger, Shippensburg University

Joy B. Perry, Savage River Tissue Culture

Pat Pietropaolo, Community College of Finger Lakes

Mary L. Powell, Frostburg State University

Thomas A. Redick, Frostburg State University

Marian Reeve, Merritt College

Rosemary Richardson, Bellevue Community College

Robert K. Riley, Frostburg State University

Jerry Skinner, Keystone Junior College

John Tiftickjian, Delta State University

Carol A. Wilson, Portland State University

Wayne A. Yoder, Frostburg State University

Welcome! You are about to embark on a journey through the cosmos of life. You will learn things about yourself and your surroundings that will broaden and enrich your life. You will have the opportunity to marvel at the microscopic world, to be fascinated by the cellular events occurring in your body at this very moment, and to gain an appreciation for the environment, including the marvelous diversity of the plant and animal world.

We can offer a number of suggestions to make your collegiate experience in biology a pleasant one. The first step toward that goal has been taken by us; we have written a laboratory guide that is "user friendly." You will be able to hear the authors speaking with you as though we were there to share your experience. Both of us share a personal belief that the more we make you feel comfortable with us, the more likely you are to share our enthusiasm for biology. It would be naive for us to suppose that each and every one of you will be biology majors at graduation. But one thing we all must realize is that we are citizens of "spaceship Earth." The fate of our spaceship is largely in your hands because you are the decision makers of the future. As has been so aptly stated, "We inherited the earth from our parents and grandparents, but we are only the caretakers for our children and grandchildren."

As caretakers we need to be informed about the world about us. That's why we enroll in colleges and universities with the hope of gaining a liberal education. In doing so, we establish a basis on which to make educated decisions about the future of the planet. Each exercise in this manual contains a lesson in life that is of a more global nature than the surroundings of your biology laboratory.

In order to enhance your biology education, take the initiative to put yourself at the best possible advantage: Don't miss class; read your text assignment routinely; read the laboratory exercise before you come to the lab.

Each exercise in the manual is organized in the same way:

1. **Objectives** tell exactly what you should learn from the exercise. If you wish to know what will be on the exam, consult the objectives for each exercise.
2. The **introduction** provides you with background information for the exercise and stimulates your interest.
3. The **materials** list for each portion of the exercise allows you to determine at a glance whether you have all the necessary supplies needed to do the activity.
4. The **procedure** for each section, often in easy-to-follow step-by-step fashion, describes the activity. Within the procedure, spaces are provided to make drawings. Questions, with space for answers, are posed asking you to draw conclusions about an activity you are engaged in. You'll find a lot of illustrations, some of which are labeled and others which are not but have leaders for you to attach labels. The terms to be used as labels are found in the procedure and in a list accompanying the illustration. Sometimes we believe it best for you to make a simple drawing, and have inserted boxes or circles for your sketches. Where appropriate, tables and graphs are present for recording your data.
5. **Pre-lab questions** can be answered easily by simply reading the exercise. They're meant to "set the stage" for the lab period by emphasizing some of the more salient points.
6. **Post-lab questions** are intended to be done after the laboratory is completed. Some are straightforward interpretations of what you have done, while others require additional thought and perhaps some research in your textbook. In fact, some have no "right" or "wrong" answer at all!

In our experience, students are much too reluctant to ask questions for fear of appearing stupid. Remember, there is no such thing as a stupid question. *Speak up!* Think of yourselves as "basic learners" and your instructors as "advanced learners." Interact and ask questions so that you and your instructors can further your/their respective educations.

MATERIALS AND SUPPLIES KEPT IN THE LAB AT ALL TIMES

The materials listed below will be always available in the lab room. Familiarize yourself with their location prior to beginning the exercises.

- compound light microscopes
- dissection microscopes
- glass microscope slides
- coverslips
- lens paper
- tissue wipes
- plastic 15-cm rulers
- dissecting needles
- razor blades
- assorted glassware-cleaning brushes
- detergent for washing glassware
- distilled water
- hand soap
- paper toweling
- safety equipment (see separate list)

LABORATORY SAFETY

None of the exercises in this manual are inherently dangerous. Some of the chemicals are corrosive (causing burns to the skin), others are poisonous if ingested or inhaled in large amounts. Contact with your eyes by otherwise innocuous substances may result in permanent eye injury; remember, once your sight is lost, it's probably lost forever. **Locate the safety items described below and then study the list of basic safety rules.**

1. Eyewash bottle or eye bath
Should any substance be splashed in your eyes, wash them for 15 minutes.

2. Fire extinguisher
Read the directions for use of the fire extinguisher.

3. Fire blanket
Should someone's clothing catch fire, wrap the blanket around the individual and roll the person on the floor to smother the flames.

4. First-aid kit
Minor injuries such as small cuts can be treated effectively in the lab. Open the first-aid kit to determine its contents.

5. Safety goggles
Eye protection should be worn during the more experimental exercises.

Safety Rules

1. Do not eat, drink, or smoke in the laboratory.

2. Wash hands with soap and warm water before leaving the laboratory.

3. When heating a test tube, point the mouth of the tube away from yourself and other people.

4. Always wear shoes in the laboratory.

5. Keep extra books and clothing in designated places so that your work area is as uncluttered as possible.

6. If you have long hair, tie it back when in the laboratory.

7. Read labels carefully before removing substances from a container. **Never return a substance to a container.**

8. Discard used chemicals and materials into appropriately labeled containers. Certain chemicals should not be washed down the sink; these will be indicated by your instructor.

Report all accidents and spills to your instructor immediately!

INSTRUCTIONS FOR WASHING LABORATORY GLASSWARE

1. Place contents to be discarded in proper waste container as described in exercise.

2. Rinse glassware with tap water.

3. Add a small amount of glassware cleaning detergent.

4. Scrub with an appropriately sized brush.

5. Rinse with tap water until detergent disappears.

6. Rinse three times with distilled water (dH_2O).

7. Allow to dry in inverted position on drying rack (if available).

When glassware is clean, dH_2O sheets off rather than remaining on the surface in droplets.

MEASUREMENT AND MICROSCOPY

OBJECTIVES After completing this exercise you will be able to:

1. define meniscus, magnification, resolving power, contrast, field of view, parfocal, parcentral, depth of field, working distance;

2. recognize graduated cylinders, beakers, Erlenmeyer flasks, different types of pipets, and a triple beam balance;

3. explain the concepts of length, volume, and mass in metric units;

4. measure and estimate length, volume, and mass in metric units;

5. describe how to care for a compound light microscope;

6. recognize and give the function of the parts of a compound light microscope;

7. accurately align a compound light microscope;

8. correctly use a compound light microscope;

9. describe the usefulness of the dissecting microscope;

10. use your skills to enjoy a fascinating world unavailable to the unaided eye.

INTRODUCTION Biology relies on observation and experimentation followed by further observation to gather knowledge about life. This process is called the *scientific method.*

Biologists use scientific tools or instruments to enhance their ability to observe living phenomena. Some instruments measure characteristics that could otherwise only be described in words. Other instruments expand our senses, such as when a light microscope magnifies small organisms and cells.

I. MEASUREMENT

(Starr, p. 40)

We examine our world in two ways, qualitatively and quantitatively. A *qualitative observation* of something describes a characteristic important to understanding what it is and is not numerical. A *quantitative observation* of the same characteristic would involve a measurement or count. The height of a person can be described as short, average, or tall (qualitative), or as a number of a unit of length (quantitative).

One of the rules of the scientific method is that results be repeatable. Therefore, scientific observations are usually made as quantitative as possible.

Logically, units in the ideal system of measurement should be easy to convert from one to another (e.g., feet to miles) and from one related measurement to another (e.g., length to area, and area to volume). The metric system meets these requirements and is used by the majority of people and countries in the world, and universally is preferred by science educators and researchers. In most nonmetric countries, governments have launched programs to hasten the conversion to metrics. Any country that fails to do so will be at a serious economic and scientific disadvantage. Let us examine the metric system and compare it to the American Standard, or English, system of measurement with these ideas in mind.

The metric system is a decimal system of measurement. Metric units are 10, 100, 1,000 and sometimes 1 million or more times larger or smaller than the reference unit. The metric reference units are the **meter** for length, the **liter** for volume, and the **gram** for mass. Differences between units are always in 10s or multiples of 10. Regardless of the type of measurement, the same prefixes are used to designate the relationship of a unit to the reference unit. Table 1-1 lists the prefixes we will introduce in this and subsequent exercises.

Table 1-1	Prefixes for Metric System Units	
Prefix of Unit	Part of Reference Unit	
nano	1/1,000,000,000 or 0.000000001	or 10^{-9}
micro	1/1,000,000 or 0.000001	or 10^{-6}
milli	1/1,000 or 0.001	or 10^{-3}
centi	1/100 or 0.01	or 10^{-2}
kilo	1,000	or 10^3

MATERIALS

Per student pair:

- 30-cm ruler, with metric and American Standard (English) units on opposite edges
- 250-mL beaker
- 250-mL Erlenmeyer flask
- 3 graduated cylinders: 10 mL, 25 mL, 100 mL
- one-piece plastic dropping pipet (not graduated) or Pasteur pipet and bulb
- graduated pipet and safety bulb or filling device (optional)
- Styrofoam coffee cup
- 1 gallon milk bottle
- 15 cm ruler
- metric tape measure
- 1-liter measuring cup

Per student group (table):

- triple beam balance

Per lab room:

- source of distilled water
- metric bathroom scale

PROCEDURE

A. Length

Length is the measurement of a line, end to end. The standard unit is the *meter*, and the most commonly used related units of length are:

```
1,000 millimeters (mm) = 1 meter (m)
100 centimeters (cm) = 1 m
              1,000 m = 1 kilometer (km)
```

1. How many centimeters are there in 1 km?

$$\frac{100 \text{ cm}}{1 \text{ m}} \times \frac{1,000 \text{ m}}{1 \text{ km}} \quad \text{or}$$
$$(100 \text{ cm/m})(1,000 \text{ m/km}) = 100,000 \text{ cm/km}$$

If you wish to convert any number of kilometers to centimeters, simply multiply by 100,000 cm/km. Convert 1.7 km to centimeters.

_____ cm

How many meters are there in 1 mm?

```
1/1,000 m/mm    or    0.001 m/mm
```

If you wish to convert any number of millimeters to meters, multiply by 0.001 m/mm. Convert 17 mm to meters.

_____ m

In a similar fashion, complete the construction of a conversion table for length (Table 1-2). Construct it so that if you wish to convert a measurement in a unit listed at the top of a column to a unit at the front of a row, multiply it by the number at the intersection of the column and row.

Table 1-2 Conversion Table for Metric Units of Length				
	mm	cm	m	km
mm	1	_____	0.001	_____
cm	_____	1	_____	_____
m	1,000	100	1	_____
km	_____	100,000	1,000	1

Of course, you could just as easily convert a measurement in a unit listed in the row to one in a unit listed in the column by dividing it by the appropriate number from the table. Either way, the answer can be obtained by simply shifting the decimal point in the original measurement.

2. Measure the length of this page in centimeters to the nearest tenth of a centimeter with the metric edge of a ruler. Note that the space between each centimeter is divided by nine lines into 10 millimeters.

The page is _____ cm long.

Use the conversion table you just completed to calculate the length of this page in millimeters, meters, and kilometers.

_____ mm   _____ m   _____ km

Now repeat the above measurement using the American Standard edge of the ruler. Measure the length of this page in inches to the nearest eighth of an inch.

_____ in.

Convert your answer to feet and yards.

_____ ft   _____ yd

Write a statement in the space below, telling how easy it is to convert *units of length* in the metric system compared to converting units of length in the American Standard system.

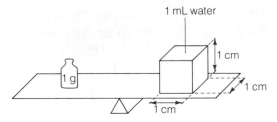

Figure 1-1 Illustration of the relationship between the units of length, volume, and mass in the metric system

B. Volume

Volume is the space a given object occupies. The standard unit of volume is the *liter* (L); and the most commonly used subunit, the *milliliter* (mL). There are 1,000 mL in 1 liter.

The volume of a box is the height multiplied by the width multiplied by the depth. The amount of water contained in a cube with sides of 1 cm in length, or 1 cubic centimeter (cc), for all practical purposes equals 1 mL (Fig. 1-1).

1. How many milliliters are there in 1.7 liters?

_____ mL

How many liters are there in 1.7 mL?

_____ L

2. Use the illustrations in Fig. 1-2 to recognize **graduated cylinders, beakers, Erlenmeyer flasks,** and the different types of **pipets.** Some of this apparatus may be made of glass; some may be plastic. Some will be calibrated in milliliters and liters; some may not.

3. Pour some water into a 100-mL graduated cylinder and observe the boundary between fluid and air, the **meniscus.** Due to surface tension, the meniscus is curved, not flat. Draw the meniscus in the plain cylinder outlined in Fig. 1-3. The correct reading of the volume is at the lowest point of the meniscus.

4. Pipets are used to transfer small volumes from one vessel to another. Some pipets are not graduated (e.g., Pasteur pipets and most one-piece plastic dropping pipets).

Half fill the 250-mL Erlenmeyer flask with distilled water. Use either a plastic dropping pipet or a Pasteur pipet with a bulb to find out how many drops there are in 1 mL of distilled water. Count the number of drops needed to fill a 10-mL graduated cylinder to the 1-mL mark.

There are _____ drops/mL.

The following section is optional. Other pipets are graduated. Note that some graduated pipets deliver a specific maximum volume from between the two lines that are farthest apart (measuring type). Others do so by allowing as much fluid as possible to drain out

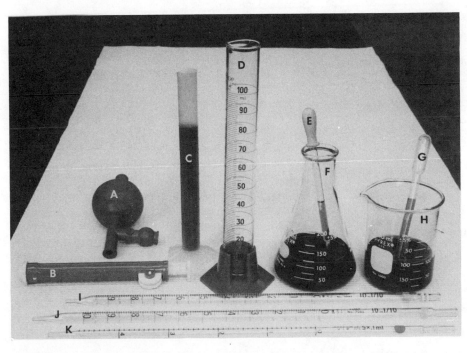

Figure 1-2 Apparatus commonly used to measure volume: pipet safety bulb (A), pipet filling device (B), plastic graduated cylinder (C), glass graduated cylinder (D), Pasteur pipet and bulb (E), Erlenmeyer flask (F), one-piece plastic dropping pipet (G), beaker (H), graduated pipets (I to K). (Photo by David Morton.)

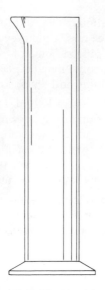

Figure 1-3 Draw a meniscus in this plain cylinder.

from the highest line (serological type). The latter type usually is marked TD for "To Deliver." If it is so marked and has one frosted or painted ring, do not force out the fluid that remains in the tip. If there are two frosted or painted rings, the fluid that remains in the tip is forced out to deliver the measured volume.

CAUTION *Never Mouth Pipet!*

Always use the safety bulbs and filling devices provided with pipets. Your instructor will demonstrate the use of specific devices. If you never mouth pipet, there is no chance of accidentally drinking any caustic or toxic material.

Do you have a measuring or serological type graduated pipet?

If it is a serological pipet, does it have one or two frosted or painted rings?

_____ ring(s)

What is the maximum amount of fluid that your pipet can transfer?

_____ mL

What is the minimum amount of fluid that can be accurately transferred with your pipet?

_____ mL

Pipet half the maximum volume into the smallest graduated cylinder that will hold this volume. Graduated pipets are the most accurate apparatus introduced in this lab for measuring fluid volume.

C. Mass

Mass is the quantity of matter in a given object. The standard unit is the **kilogram** (kg), and other commonly used units are the *milligram* (mg) and *gram* (g). There are 1,000,000 mg in 1 kg and 1,000 g in 1 kg.

1. How many milligrams are there in 1 g?

_____ mg

Convert 1.7 g to milligrams and kilograms.

_____ mg _____ kg

2. Our 1-cc cube, if filled with 1 mL of water, would have a mass of 1 g (Fig. 1-1). The mass of other materials depends on their **density** (i.e., water has a density of 1).

density = mass/volume

Approximately how many liters are present in 1 cubic meter of water?

_____ L

In kg, what is its mass?

_____ kg

In your "mind's eye," think about how many pounds there are in a cubic yard of water. Write a statement in the space below, telling how easy it is to convert *between different units* of the metric system compared to converting between different units of the American Standard system.

3. Determine the mass of an unknown volume of water.* Mass may be measured with a **triple beam balance,** which gets its name from its three beams (Fig. 1-4). A movable mass hangs from each beam.

Slide all of the movable masses to zero. The middle and back masses each click into the left-most notch, and the front mass is moved to the far left. Clear the pan of all objects and make sure it is clean. The balance marks should line up, indicating that the beam is level and that the pan is empty. If the balance marks do not line up, rotate the zero adjust knob until they do. Place a 250-mL beaker on the pan. The

---

*Modified from C. M. Wynn and G. A. Joppich, *Laboratory Experiments for Chemistry: A Basic Introduction*, 3rd ed. Wadsworth, 1984.

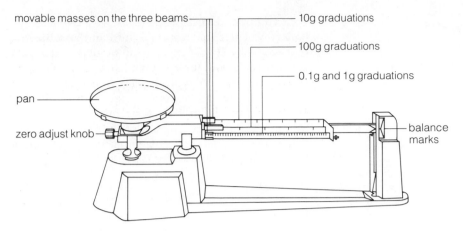

Figure 1-4 Triple beam balance. (After Wynn and Joppich, 1984.)

right side of the beam should rise. Slide the mass on the middle beam until it clicks into the notch at the 100-g mark. If the right end of the beam tilts down below the stationary balance mark, you have added too much mass. Move the mass back a notch. If the right end remains tilted up, additional mass is needed. Keep adding 100-g increments until the beam tilts down and then move the mass back one notch. Repeat this procedure on the back beam, adding 10 g at a time until the beam tilts down, and then backing up one notch. Next slide the front movable mass until the balance marks line up. The sum of the masses indicated on the three beams gives the mass of the beaker. The space between the numbered gram markings on the front beam is divided by nine unnumbered lines into 10 sections, each representing 0.1 g. The mass of the beaker to the nearest tenth of a gram is

_____ g.

Add an unknown amount of water and repeat the above procedure. The mass of the water will equal the combined masses of the beaker and water minus that of the beaker alone.

mass of the beaker plus the water _____ g

minus the mass of the beaker _____ g

equals the mass of the water _____ g

Now measure the volume of the water in milliliters with a graduated cylinder. What was the volume?

_____ mL

You may have wondered why we have avoided the term *weight* in the above discussion. That is because mass is a quantity of matter, while weight depends on the gravitational field in which the matter is lo-

cated. Thus, if you were on the moon you would weigh less but your mass would be the same as on earth.

D. Estimation of Measurements

1. Estimate the length of your index finger in centimeters.

_____ cm

2. Estimate your lab partner's height in meters.

_____ m

3. How many milliliters will it take to fill a Styrofoam coffee cup?

_____ mL

4. How many liters will it take to fill a 1-gallon milk bottle?

_____ L

5. Estimate the weight of some small personal item (e.g., loose change) in grams.

_____ g

6. Estimate your or your lab partner's weight in kilograms.

_____ kg

7. Check each of your results using one of the following: 15-cm metric ruler, metric tape measure, 100-mL graduated cylinder, 1-liter measuring cup, triple beam balance, metric bathroom scale.

8. Today, many packaged items have the volume or weight listed in both American Standard and metric units. Before your next lab period, find and list 10 such items in Table 1-3.

Table 1-3   Measure of Everyday Packaged Items

Item	American Standard	Metric
_____	_____	_____
_____	_____	_____
_____	_____	_____
_____	_____	_____
_____	_____	_____
_____	_____	_____
_____	_____	_____
_____	_____	_____
_____	_____	_____
_____	_____	_____

II. THE COMPOUND LIGHT MICROSCOPE

(Starr, pp. 38, 39)

A microscope is an instrument that contains at least one lens and is used to view a specimen, or the detail in a specimen, that cannot be seen with the unaided eye. A magnifying glass is a simple light microscope.

Microscopy involves three basic concepts: magnification, resolving power, and contrast. **Magnification** is the degree to which the image of a specimen is enlarged. **Resolving power** determines how well specimen detail is preserved during the magnifying process. **Contrast** is the ability to see a particular detail against its background.

Why can't you simply bring a specimen closer to the eye to see its detail, much as you might do to read the fine print in a contract? The lens of the eye focuses an image of what you view onto the light-sensing surface of the eye, the retina. Unfortunately, the normal eye lens cannot focus on an object closer than about 10 cm. At this distance you can see two specimen details separated by 0.1 mm. Because most cells are between 0.01 mm and 0.1 mm in diameter, they cannot be seen without a microscope.

The compound light microscope brings an image of a specimen very close to the eye and helps the lens focus its image on the retina. The greater the proportion of the retina covered by the image of a specimen, the greater its magnification. Magnification without detail is empty, and with a light microscope, the maximum useful magnification is about 1,000 times the diameter of the specimen (1,000×). Above this value, detail is missing.

To see detail, there has to be contrast. *Dyes* are usually added to sections of biological specimens to increase contrast.

A compound microscope has at least two lens systems: an **ocular** that you look into and an **objective** that scans the specimen. Like automobiles, there are many models, and for a given microscope, many possible accessories. A typical example is diagramed in Fig. 1-5. If your microscope is significantly different from the one illustrated, your instructor will distribute an unlabeled diagram of the microscope you will be using. *Before removing the microscope assigned to you from the cabinet, read Section A.*

MATERIALS

Per student:

- compound light microscope
- lens paper
- lint-free cloth (optional)
- unlabeled diagram of the compound light microscope model used in your course (optional)
- prepared slide with mounted letter *e*
- prepared slide with crossed colored threads coded for thread order
- prepared slide with unstained fibers
- prepared slide with Wright-stained smear of mammalian blood
- 15-cm transparent plastic ruler
- directions on how to calibrate and use an ocular micrometer (optional)

Per student group (4):

- bottle of lens-cleaning solution (optional)
- dropper bottle of immersion oil (optional)

Per lab room:

- labeled chart of a compound light microscope

PROCEDURE

A. Care of a Compound Light Microscope

1. To carry a microscope to and from your lab bench, grasp the **arm** with your dominant hand and support the **base** with the other hand, always keeping the microscope upright. *Do not try to carry anything else at the same time.* Label the arm and the base on Fig. 1-5.

2. Remove the dust cover and clean the exposed parts of the optical system. First blow off any loose dust that may be on the ocular and then gently brush off any remaining dust with a piece of lens paper.

CAUTION   *Never Wipe a Glass Lens with Anything Other Than Lens Paper.*

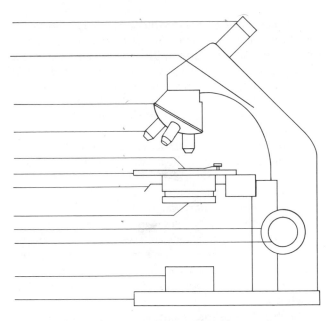

Labels: ocular, objective, arm, base, illuminator, condenser, lever for iris diaphragm of condenser, stage, stage clip, coarse adjustment knob, fine adjustment knob, nosepiece

Figure 1-5 Compound light microscope

If the part is still dirty, breathe on the lens and gently polish with a rotary motion using a fresh piece of lens paper. If the part is still dirty, *and with your instructor's approval*, clean the lens with a piece of tissue paper moistened with lens cleaning solution.

3. Always remember that your microscope is a precision instrument. Never force any of its moving parts.

4. It is just as difficult to see clearly through a dirty slide as through a dirty microscope. Clean dirty slides with a lint-free cloth or with lens paper before using.

5. At the end of an exercise, make sure the last slide has been removed from the stage and *rotate the nosepiece so that the low power objective is in the light path*. If your instrument focuses by moving the body tube, turn the coarse adjustment so that it is racked all the way down. If your microscope has an electric cord, neatly fold it up on itself and tie it with a plastic strap or rubber band. Otherwise, wind the cord around the base of the arm of the microscope.

6. Replace the dust cover before returning your microscope to the cabinet.

B. Parts of the Compound Light Microscope

Now that you know how to care for it, remove the microscope assigned to you from the cabinet and place it on your lab bench. Use the chart on the wall of your lab room and the narrative in this section to identify the various parts of your microscope. Label Fig. 1-5 or the diagram given you by your instructor. Read each

step below and manipulate the parts *only where indicated.*

1. *Light source.* The compound microscope uses transmitted light to illuminate a transparent specimen usually mounted on a glass slide. Older microscopes may have a *mirror* to reflect light through the specimen. Newer instruments have a built-in illuminator.

If your microscope has a double-sided mirror, use the flat surface. Angle the mirror so that the light path is directed through the hole in the stage (platform where the specimen sits). Use the curved surface of the mirror only when illumination is inadequate.

If your microscope has a built-in illuminator, it will have a cord, an *off/on switch*, and perhaps also a *rheostat* to vary the intensity of the light. On some models the switch and the rheostat are combined. Turn on the light source. If the illuminator has a rheostat, adjust the intensity so that the light is not too bright.

Label the illuminator in Fig. 1-5.

2. *Condenser.* For maximum resolving power, a **condenser**—with a *condenser lens* and **condenser iris diaphragm**—focuses the light source on the specimen so that each of its points is evenly illuminated. The lever for the iris diaphragm of the condenser is used to limit the cone of light arising from the central point of the field of view until it almost fills the back lens element of the objective (Fig. 1-8). The **field of view** is the circle of light you see when looking into the microscope. Determine if there is a *condenser adjustment knob* to set the height of the condenser. *Do not turn the knob; you will learn how to use it later.*

Simpler models have a revolving disk with a series of holes of different sizes to regulate the illumination of the specimen. As long as the specimen has adequate contrast, use the smallest hole that gives maximum illumination. The correct hole is usually marked to match it with the objective or total magnification in use at that time.

There may be a *filter holder* under the condenser with a blue or frosted glass disk. Many manufacturers of microscopes believe that blue light is more pleasing to the eye, and theoretically at least, blue light gives better resolving power because of its shorter wavelength. The frosted glass disk scatters light and may be useful in producing even illumination at low magnifications.

Label the condenser and lever for iris diaphragm on Fig. 1-5.

3. *Stage.* A specimen mounted on a glass slide is held in place on the **stage** by a pair of **stage clips** so that the specimen is suspended over a central hole. *Do not remove the stage clips, as they make it easier to move a slide in small increments.*

If your microscope is especially well equipped, it has a *mechanical stage* instead of clips. In this case, position a slide on the mechanical stage by releasing

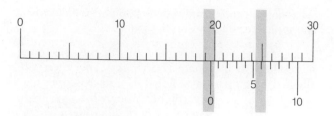

Figure 1-6 A vernier scale

the tension on the spring-loaded movable arm. There are two knobs to the right or left of the stage: one to move the specimen forward and backward, and the other for its lateral movement.

On most mechanical stages, each direction has a vernier scale so that you can easily locate interesting fields again and again. A *vernier scale* consists of two scales running side by side, a long one in millimeters and a short one, 9 mm in length and divided into 10 equal subdivisions.

To take a reading, note the whole number on the long scale coinciding with or just below the zero line of the short scale. If the whole number of the long scale and the zero of the short scale coincide, the first place after the decimal point is zero. Otherwise, the first place after the decimal point is the value of the line on the short scale which coincides with one of the next nine lines after the whole number on the long scale. For example, the correct reading of the vernier scale in Fig. 1-6 is 19.6 mm. Label the stage and stage clips on Fig. 1-5.

4. *Focusing knobs.* The **coarse adjustment knob** is for use with low and medium power objectives, while the **fine adjustment knob** is for critical focusing, especially with the high-dry and oil immersion objectives. On older microscopes, you move the *body tube* of the instrument up and down to focus the specimen; in modern models you move the stage.

Modern microscopes also have a *preset focus lock*, which stops the stage at a particular height. After setting this lock, you can lower the stage with the coarse adjustment knob to facilitate changing of the specimen, and then raise it to focusing height without fear of colliding the specimen against the objective.

There may also be a *focus tension adjustment knob,* usually located inside of the left-hand coarse adjustment knob.

Turn the coarse adjustment knob. Do you turn the knob toward you or away from you to bring the stage and objective closer together?

Label the coarse and fine adjustment knobs on Fig. 1-5.

5. *Objectives.* An objective lens projects a real, inverted, and magnified image of the specimen just beneath the ocular (Fig. 1-7a). Real means the image

exists and inverted means the image is upside-down. Most microscopes have several objectives mounted on a revolving **nosepiece.** The magnifying power of each objective is labeled on its side. Usually included are these objectives: a *low power* or scanning (4×), a *medium power* (10×), a *high-dry* (about 40×), and perhaps an *oil immersion objective* (about 100×). A few models have one *zoom objective*.

The other number often labeled on to the side of nosepiece objectives is the **numerical aperture (N.A.).** The larger the numerical aperture, the greater the resolving power and useful magnification.

Objectives are **parfocal.** That is, once one objective has been focused, you can rotate to another one and the image will remain in coarse focus, requiring only slight movement of the fine adjustment knob. Objectives are also **parcentral,** meaning that the center of the field of view remains about the same for each objective.

Label the nosepiece and objective on Fig. 1-5.

Now follow these steps to use each objective:

a. Rotate the low power objective into the light path.

b. Place a prepared slide of mammalian blood on the stage, securing it with either the stage clips or the movable arm of the mechanical stage.

c. Look through the ocular. Bring the blood cells into focus. At this magnification the blood cells are very small. Center a specimen detail by moving the slide.

d. Rotate the nosepiece so that the medium power objective is in the light path. Focus the specimen detail.

e. Rotate the nosepiece so that the high-dry objective is in the light path. Focus the specimen detail using the *fine adjustment knob.*

To use the oil immersion objective:

f. Rotate the nosepiece so that the light path is midway between high-dry and oil immersion objectives.

g. Place a small drop of immersion oil onto the coverslip above the specimen.

h. Rotate the nosepiece so that the oil immersion objective is in the oil above the specimen detail.

i. Focus on the specimen detail with the fine adjustment knob.

j. After examining the specimen detail, rotate the oil immersion objective out of the light path. Carefully wipe the oil from the oil immersion objective with lens paper.

k. Remove the slide from the stage and wipe the oil from the coverslip with lens paper.

Immersion oil, because it has an index of refraction similar to that of glass, increases resolving power and useful magnification.

6. *Ocular.* The ocular, or eyepiece, lens magnifies the image produced by the objective and, in conjunction with the eye lens, projects a second real, inverted,

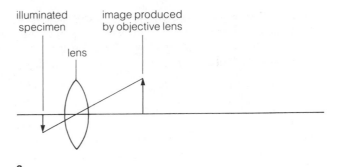

a

image produced
by ocular lens

image produced
by objective lens

image produced
by lens of eye

ocular lens

eye lens

b

Figure 1-7 Images formed by a compound light microscope: (a) objective lens, (b) ocular and eye lenses. The image produced by the ocular lens is represented by a dashed line because it is virtual (not real) and can only be seen when viewed by another lens such as the lens of the eye.

but demagnified image onto the retina (Fig. 1-7b). Label the ocular in Fig. 1-5.

It should not surprise you that the retinal image is demagnified, especially if you consider the size of the retina in comparison to the world around you. That is, all retinal images are demagnified.

Does your conscious mind perceive retinal images as inverted?

(yes or no) _____

The brain inverts all retinal images.

(true or false) _____

Oculars are generally 10×. Since each objective has a different magnifying power, the total magnification is calculated by multiplying the magnifying power of the ocular by that of the objective in use. What is the total magnification with a 10× ocular and a 40× objective?

_____ ×

Models with zoom objectives usually have the total magnification printed on the *zoom control*.

Your microscope will have one or two oculars mounted on a *monocular* or *binocular head*, respectively. There may be a pointer mounted in an ocular so that you can show a specimen detail to the instructor or another student. For a monocular microscope, it is best to use your dominant eye to look down the ocular, keeping your other eye open.

To determine your dominant eye:

a. Look at a small object on the far wall of your room with both eyes open.

b. Form the thumb and index finger of one hand into a circle and place this circle in your line of sight so that it surrounds the object.

c. Close your right eye. If the object shifts out of the circle to your left, your right eye is probably dominant. If the object remains in the circle, your left eye is probably dominant.

d. Repeat a and b. This time close your left eye. If the object shifts to the right, your left eye is dominant. If the object remains within the circle, your right eye is dominant. The more pronounced the shift, the greater the dominance. If there is no shift, neither eye is dominant.

Binocular microscopes need to be adjusted to the distance between your pupils (*interpupillary distance*) and for any difference in power (*diopter*) between the lenses of each eye. These adjustments are described in Section C.

C. Aligning a Compound Light Microscope with In-base Illumination, a Condenser with an Iris Diaphragm, and No Lamp Iris Diaphragm

Aligning your microscope properly will not only help you see specimen detail clearly but also will protect your eyes from strain.

1. Rotate the nosepiece until the medium power objective is in the light path and open the condenser iris diaphragm.

2. Place the prepared slide with the letter *e* on the stage; center and carefully focus on it. *Skip steps 3 and 4 if your microscope is monocular. Skip step 5 if your microscope does not have a control to adjust the height of the condenser.*

3. If your microscope is binocular, adjust the interpupillary distance. Hold a different ocular tube with each hand and, while looking at the specimen, pull the tubes apart or push them together until you see one field of view. After making this adjustment, read the number off the scale. Record your personal interpupillary distance here:

_____

From now on you can set the interpupillary distance at this number.

4. Now compensate for any difference in diopter between the lenses of each eye:

   a. Usually there is one diopter adjustment ring around the left ocular tube. In this case, with your left eye closed focus the microscope using the fine adjustment knob. Now open the left eye and close the right one. Use the diopter adjustment ring to bring the specimen into focus.

   b. Sometimes both ocular tubes have a diopter adjustment ring. In this case set the left one to the same number as the interpupillary distance, close the right eye, and focus on the specimen. Then close the left eye and open the right one. Rotate the diopter adjustment ring on the right ocular tube to bring the specimen into focus.

5. Place a sharp point (pencil, dissecting needle, etc.) on top of the illuminator and bring the silhouette into sharp focus by adjusting the height of the condenser.

6. Use the lever to set the iris diaphragm of the condenser:

   a. If the ocular on your microscope is removable and *with the permission of the instructor,* carefully slide it out and put the ocular open end down upon a piece of lens paper in a safe place. Then, while looking down the ocular tube, adjust the condenser iris diaphragm until the edge of the aperture lies just inside the margin of the back lens element of the objective (Fig. 1-8). Replace the ocular.

   b. If the ocular cannot be removed, close the condenser diaphragm and then open it until there is no further increase in brightness. Now, close it again, stopping when you see the brightness begin to diminish.

7. If your microscope has a rheostat, adjust the illumination to a level that lets you see specimen detail and that is comfortable for your eyes. To maintain the same illumination at higher magnifications, you will have to increase its intensity.

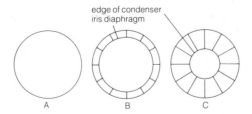

edge of condenser
iris diaphragm

A   B   C

Figure 1-8 Correct setting for condenser iris diaphragm; drawing B is correct. In A you cannot see the edge of the iris diaphragm. In C the diaphragm has been closed too much.

8. Repeat steps 6 and 7 each time you use a different objective.

D. Depth of Field

The **depth of field** is the distance through which you can move the specimen and have it remain in focus.

1. Obtain a prepared slide of three crossed colored threads. *This exercise requires care since you are probably not yet adept at focusing on a specimen.* Once you have a slide in focus with the low power objective, you need only use the fine adjustment knob to focus with the medium power and high-dry objectives. After switching to the higher power objectives, try rotating the fine focus knob a half turn away from you, then a full turn toward you. If you have not found the plane of focus, next try one and a half turns away from you and two full turns toward you, and so on. If you work deliberately, you will find the plane of focus and will not crack the coverslip.

   Working distance, the space between the objective lens and the coverslip, decreases with increasing magnifying power. Therefore, the higher the power of the objective in use, the closer the objective is to the slide and the more careful you must be. How many threads are in focus using the

   low power objective? _____

   medium power objective? _____

   high-dry objective? _____

   With which objective is it easiest to focus a specimen?

   _____

   At which magnification is it most difficult to focus a specimen?

   _____

2. Specimens have depth. Using the high-dry objective, determine which of the three threads is closest to the slide. (Each slide label has a code on it. When you believe that you have determined which thread is closest to the slide, check with your instructor to find out if you are correct.) Which thread is on the bottom?

   _____

   Which of the three threads is closest to the objective?

   _____

   Focusing carefully with the fine adjustment knob, move from the bottom to the upper thread. Did you move the knob away from you or toward you?

   _____

Figure 1-9 Drawing of the letter *e* as seen through the ocular (_____×)

E. Orientation of the Image

1. Place the prepared slide with the letter *e* right-side up on the stage with the medium power objective in the light path, center the letter in the field of view, and carefully bring it into focus.

2. Draw in Fig. 1-9 the *e* as you see it through the ocular. Record the total magnification used in the line at the end of the legend.

3. Move the specimen to the right while watching it through the microscope. In which direction does the image move?

_____

4. Move the specimen away from you. In which direction does the image move?

_____

F. Using Different Magnifications

1. Place a prepared slide of a Wright-stained smear of mammalian blood on the stage. Focus on the slide with the low, then the medium power objectives. Most of the cells are red blood cells stained pink. A few cells are white blood cells with prominent blue-stained nuclei.

2. Using the medium power objective locate a lymphocyte, a small white blood cell with a round nucleus and very little cytoplasm (Fig. 1-10). The other white blood cells are about twice the size of the red blood cells and have larger or more complicated nuclei. Center the lymphocyte in the field of view and switch to the high-dry objective. If your microscope has an oil immersion objective, now switch to it. Remember to use the oil immersion procedure (Section B.5).

3. Look for another lymphocyte, and then another. Use two methods: (a) continue using the high power objective and simply move the specimen; (b) switch to the medium power objective and repeat the above procedure. Which method is quicker?

_____

In general, it is easier to switch to a lower power objective to locate a specific specimen detail than to stay with a higher power one. This is why the low power objective is sometimes called the scanning objective. Because the magnification is in diameters, the area of the field of view decreases dramatically with increasing magnification (Fig. 1-11). Figure 1-11 also illustrates why it is so easy to lose a specimen detail at higher magnifications. Once a specimen detail is lost, it is always easier to find it again if you switch to a lower power objective.

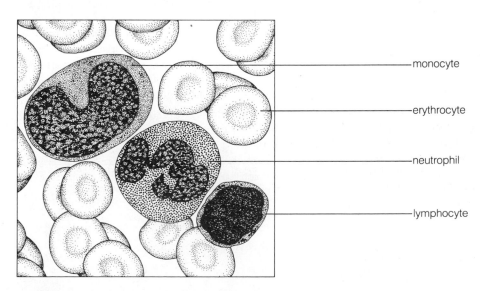

monocyte

erythrocyte

neutrophil

lymphocyte

Figure 1-10 Several blood cells. (After Fowler, 1984.)

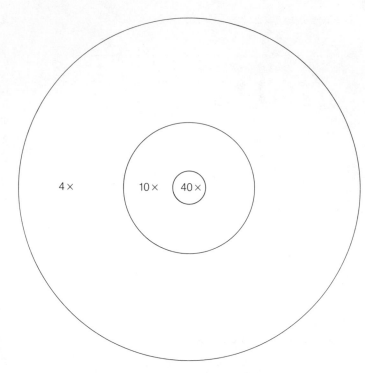

Figure 1-11 Illustration of the decreasing area of the field of view when a 4×, 10×, and 40× objective is used with a 10× ocular. The actual area of each circle has been enlarged 10×.

G. Using the Condenser Iris Diaphragm or Revolving Disk with Different Diameter Holes

Make sure the condenser and condenser iris diaphragm, or revolving disk, are correctly set. Place a specimen of unstained fibers on the stage. Locate and focus on these fibers using the medium power objective. Close the condenser iris diaphragm or move the revolving disk to smaller holes. Does this procedure increase or decrease contrast?

This procedure should be used as a last resort to see a specimen detail because resolving power is decreased.

H. Units of Measurement

The basic metric unit of length at the light microscopic level is the micrometer (μm). There are 1,000 micrometers in 1 mm. Even smaller units, the nanometer (nm) and Angstrom (Å), are often used at the electron microscopic level. There are 1,000 nm and 10,000 Å in 1 μm. How many nanometers are there in 1 mm?

_____ nm

How many Angstroms are there in 1 nm?

_____ Å

I. Determining the Diameter of the Field of View

1. Rotate the low power objective into the light path. What is the total magnification?

_____ ×

2. Place a transparent 15-cm ruler on the stage.

3. Focus on the ruler. What is the diameter of the field of view?

_____ mm

4. Repeat step 3 with the medium power objective in the light path.

The total magnification is _____ ×.

The diameter of the field of view is _____ mm.

5. Use the following formula to estimate the diameter of the field of view when the high-dry objective is in the light path.

$$\frac{\begin{array}{cc} \text{total magnification} & \text{mm counted} \\ \text{using low power} \times & \text{with that} \\ \text{objective} & \text{objective} \end{array}}{\text{total magnification of high-dry objective}} = \underline{\quad\quad} \text{ mm}$$

Convert this value to micrometers.

_____ μm

6. Complete Table 1-4.

Table 1-4 Diameter of Field of View

Objective (×)	Total Magnification (×)	Diameter (mm)	Diameter (μm)
———	———	———	
———	———	———	
———	———		———
———	———		———

7. You will use this information to estimate size by observing the percentage of the diameter of the field of view taken up by the specimen or part of the specimen. Using the medium power objective, estimate the percentage of the diameter of the field of view covered by the letter *e*. The approximate diameter of the letter *e* is:

$$\frac{\text{percent} \times \text{diameter of field of view (mm)}}{100\%} = \underline{\qquad} \text{ mm}$$

8. If your microscope is equipped with an ocular micrometer, your instructor will provide directions on how to accurately measure specimen details.

III. DISSECTING MICROSCOPE

Dissecting microscopes (Fig. 1-12) are compound microscopes with a large working distance between the specimen and the objective lens. They are especially useful when viewing larger specimens, including thicker slide-mounted specimens, and when manipulating the specimen is necessary (when dissection is required, for example).

The large working distance also allows for illumination of the specimen from above (reflected light) as well as from below (transmitted light). Reflected light shows up surface features on the specimen better than transmitted light does.

MATERIALS

Per student group (4):

- dissecting microscope
- specimens appropriate for viewing with the dissecting microscope (e.g., a prepared slide with a whole mount of a small organism, bread mold, an insect mounted on a pin stuck in a cork, a small flower)

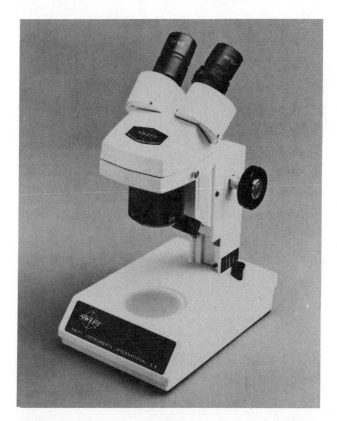

Figure 1-12 Dissecting microscopes. (Photos courtesy Swift Instruments and Bausch & Lomb.)

PROCEDURE

1. View under a dissecting microscope one or more of the specimens provided by your instructor. What is the magnification range of this microscope?

_____ × to _____ ×

2. Is the image of the specimen inverted as in the compound microscope?

(yes or no) _____

3. Describe the type of illumination used by your dissecting microscope. Is there a choice?

_____

_____

_____

_____

_____

PRE-LAB QUESTIONS

____ **1.** The observation that a person is tall is (a) qualitative, (b) quantitative, (c) a and b, (d) none of the above.

____ **2.** Length is the measurement of (a) a line, end to end, (b) the space a given object occupies, (c) the quantity of matter present in an object, (d) none of the above.

____ **3.** Volume is the measurement of (a) a line, end to end, (b) the space a given object occupies, (c) the quantity of matter present in an object, (d) none of the above.

____ **4.** Mass is the measurement of (a) a line, end to end, (b) the space a given object occupies, (c) the quantity of matter present in an object, (d) none of the above.

____ **5.** A millicurie, a unit of radioactivity, is (a) a tenth of a curie, (b) a hundredth of a curie, (c) a thousandth of a curie, (d) a millionth of a curie.

____ **6.** A magnifying glass is an example of (a) a simple light microscope, (b) a compound light microscope, (c) a dissecting microscope, (d) an electron microscope.

____ **7.** The degree to which the image of a specimen is enlarged by a microscope is called (a) contrast, (b) magnification, (c) resolving power, (d) none of the above.

____ **8.** The degree to which detail is preserved in the image of a specimen is called (a) contrast, (b) magnification, (c) resolving power, (d) none of the above.

____ **9.** The two image-forming lenses of a compound light microscope are (a) the condenser and objective, (b) the condenser and ocular, (c) the objective and ocular, (d) none of the above.

____ **10.** The image of the specimen projected by the objective lens is (a) magnified, (b) inverted, (c) a and b, (d) none of the above.

TAXONOMY:
CLASSIFYING AND NAMING ORGANISMS

OBJECTIVES After completing this exercise you will be able to:

1. define common name, scientific name, binomial, genus, specific epithet, species, taxonomy, phylogenetic system, dichotomous key;

2. distinguish common names from scientific names;

3. describe why scientific names are preferred over common names in biology;

4. identify the genus and specific epithet in a scientific binomial;

5. write out scientific binomials in the form appropriate to the Linnean system;

6. construct a dichotomous key;

7. use a dichotomous key to identify plants, animals, or other organisms as provided by your instructor.

INTRODUCTION We are all great classifiers. Every day, we consciously or unconsciously classify and categorize the objects around us. We recognize an organism as a cat or a dog, a pine tree or an oak tree. But there are numerous kinds of oaks, so we refine our classification, giving the trees distinguishing names such as "red oak," "white oak," or "bur oak." These are examples of **common names,** names with which you are probably most familiar.

Scientists are continually exchanging information about living organisms. But not all scientists speak the same language. The common name "white oak," familiar to an American, would likely be unfamiliar to a Spanish biologist, even though the tree we know as white oak may exist in Spain as well as in our own backyard. Moreover, even within our own language, the same organism may have several common names. For example, within North America a "gopher" may also be called a "ground squirrel," a "pocket mole," or a "groundhog." On the other hand, the same common name may actually describe many different organisms; there are more than 300 different trees called "mahogany"! To circumvent the problems associated with common names, biologists use **scientific names** that are unique to each kind of organism and that are used throughout the world.

An eighteenth-century Swedish naturalist, Carl von Linné (now most frequently known by the latinized form of his name, Linnaeus), is largely responsible for creating the system of scientific names that we use today. Linnaeus undertook the formidable task of naming and classifying all plants and animals, assigning each organism a two-part name called a **binomial.** The first word of the binomial designates the group to which the organism belongs; this is the **genus** name (the plural of *genus* is *genera*). All oak trees belong to the genus *Quercus,* a word derived from Latin, the universal scholarly language of Linnaeus's time. Each kind of organism within a genus is given a **specific epithet,** always an adjective. Thus, the scientific name in the Linnean system for white oak is *Quercus alba* (specific epithet is *alba*), while that of bur oak is *Quercus macrocarpa* (specific epithet is *macrocarpa*).

Notice that the genus name is always capitalized; the specific epithet usually is not capitalized (although it may be if it is the proper name of a person or place). The binomial is written in *italics* (since these are Latin names); if italics are not available, the genus name and specific epithet are underlined.

In this course you will hear discussion of species of organisms. For example, if you are on a field trip you may be asked "What species is this tree?" Assuming you are looking at a white oak, your reply would be "*Quercus alba.*" Note that the name of the **species** includes *both* the genus name and specific epithet.

If a species is named more than once within textual material, it is accepted convention to write out the full genus name and specific epithet the first time and to abbreviate the genus name every time thereafter. For example, if white oak is being described, the first use would be written *Quercus alba,* and each subsequent naming would appear as *Q. alba.*

Similarly, when a number of species, all of the same genus, are being listed, the accepted convention is to write both the genus name and specific epithet for the first species and to abbreviate the genus name for each species listed thereafter. Thus, it would be acceptable to list the scientific names for white oak and bur oak as *Quercus alba* and *Q. macrocarpa*, respectively.

Taxonomy is the science of classification (categorizing) and nomenclature (naming). Biologists prefer a system that indicates the evolutionary relationships between organisms. To this end, classification became a **phylogenetic system;** that is, one indicating the presumed evolutionary ancestry among organisms.

Most current taxonomic thought separates all living organisms into five kingdoms:

- Kingdom Monera (bacteria and blue-green algae)
- Kingdom Protista (euglenoids, chrysophytes, diatoms, dinoflagellates, slime molds, and protozoans)
- Kingdom Fungi (fungi)
- Kingdom Plantae (plants)
- Kingdom Animalia (animals)

Let's consider the scientific system of classification, using ourselves as examples. All members of our species belong to:

- Kingdom Animalia (animals)
- Phylum Chordata (animals with a notochord)
- Class Mammalia (animals with mammary glands)
- Order Primates (mammals that walk upright on two legs)
- Family Hominidae (human forms)
- Genus *Homo* (mankind)
- Specific epithet *sapiens* (wise)
- Species: *Homo sapiens*

The more closely related evolutionarily two organisms are, the more categories they share. You and I are different individuals of the same species. We share the same genus and specific epithet, *Homo* and *sapiens*. 1.5 million years ago a creature believed to be our closest extinct ancestor walked the earth. That creature shared our genus name but had a different specific epithet, *erectus*. Thus, *Homo sapiens* and *Homo erectus* are *different* species.

Unfortunately, there is one bit of confusion in the classification system we currently use: While animal biologists recognize the category called a "phylum," plant biologists use the term "division" instead.

Like all science, taxonomy is subject to change as new information becomes available. Modifications are made to reflect revised interpretations. This is particularly true in tropical biology, where our knowledge is exceptionally limited.

I. CONSTRUCTING A DICHOTOMOUS KEY

MATERIALS

Per lab room:

- several meter sticks or metric height charts taped to a wall

PROCEDURE   To classify organisms, you must first identify them. A taxonomic key is a device for identifying an object unknown to you but that someone else has described. The user makes choices between a set of alternative characteristics of the unknown object, and by making the correct choices arrives at the name of the object.

Keys that are based upon successive choices between two alternatives are known as **dichotomous keys** (*dichotomous* means to fork into two equal parts). When using a key, always read both choices even though the first appears to be the logical one. Don't guess at measurements, use a scale. Since living organisms vary in their characteristics, don't base your conclusion on a single specimen if more are available.

Suppose the geometric shapes below have unfamiliar names. Look at the dichotomous key below the figures. Notice there is a 1a and a 1b. Start with 1a. If the description in 1a fits the figure you are observing, then proceed to the choices listed under number 2, as shown at the end of line 1a. If 1a does not describe the figure in question, 1b does. Looking at the end of line 1b, you see that the figure would be called an Elcric.

Using the key provided, determine the hypothetical name for each object. Write the name beneath the object and then check with your instructor to see if you have made the correct choices.

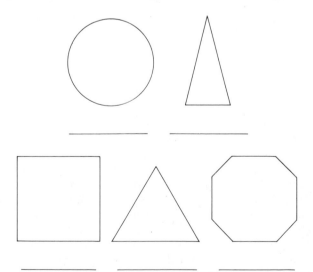

Key

1a. Figure with distinct corners 2

1b. Figure without distinct corners . . . . Elcric

2a. Figure with three sides 3

2b. Figure with four or more sides 4

3a. All sides of equal length . . . . Legnairtosi

3b. Only two sides equal . . . . Legnairt

4a. Figure with only right angles . . . . Eraqus

4b. Figure with other than right angles . . . Nogatco

Now you will construct a dichotomous key, using your classmates as "objects." The class should divide up into groups of eight (or as evenly as the class size will allow). Working with the individuals in your group, fill in Table 2-1, measuring height with a metric ruler or the scale attached to the wall.

Let's use a branch diagram to see how we might plan a dichotomous key. If there are both men and women in a group, the most obvious first split is male/female (although other possibilities for the split could be chosen as well). Let's follow the course of splits for two of the men in the group.

Note that each choice has *only* two alternatives. Thus we split into "under 1.75 m" and "1.75 m or taller." Likewise, our next split is into "blue eyes" and "nonblue eyes" rather than all the possibilities.

On a separate sheet of paper, construct a branch diagram for your group using the characteristics in Table 2-1, and then condense it into the dichotomous key below. When you have finished, exchange your key with that of an individual in another group. Key out the individuals in the other group without speaking until you believe you know the name of the individual you are examining. Ask that individual if you

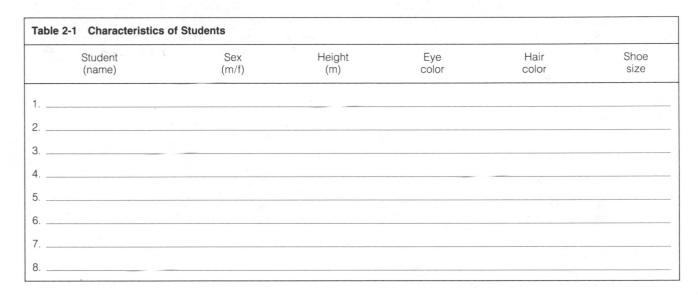

Table 2-1 Characteristics of Students

Student (name)	Sex (m/f)	Height (m)	Eye color	Hair color	Shoe size
1.					
2.					
3.					
4.					
5.					
6.					
7.					
8.					

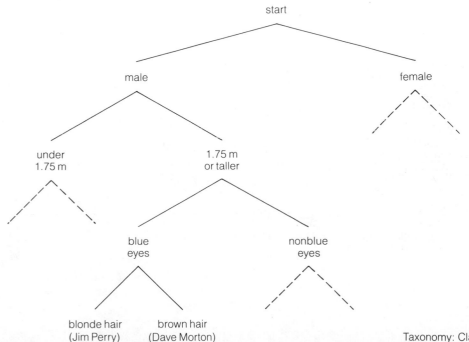

are correct. If not, ⟍ back to find out where you made a mistake, o⟍ssibly where the key was misleading. (Dependi⟍ n how you construct your key, you may need m⟍ or fewer lines than have been provided below.⟍

Key to Students i⟍ oup _____

1a. _____

1b. _____

2a. _____

2b. _____

3a. _____

3b. _____

4a. _____

4b. _____

5a. _____

5b. _____

6a. _____

6b. _____

7a. _____

7b. _____

8a. _____

8b. _____

II. ⟍SE OF A TAXONOMIC KEY

A. ⟍OMMON TREES AND SHRUBS

M⟍ERIALS

P⟍ tudent group (table):

- ⟍ t of eight tree twigs with leaves (fresh or herb-⟍ium specimens) or

- ⟍ees and shrubs in leafy condition

⟍ROCEDURE Use the key that follows to identify ⟍e tree and shrub specimens that have been pro-⟍ided in the lab or that you find on your campus. Re-⟍er to the Glossary to Accompany Tree Key (page 23) and Figs. 2-1 through 2-8, Illustrations of Plant Parts (pages 20–22) when you encounter an unfamiliar term.

Note: Some descriptions within the key have more characteristics than your specimen will exhibit. For example, the key may describe a fruit type when the specimen does not have a fruit on it. However, other specimen characteristics are described, and these should allow you to identify the specimen.

Common names within parentheses follow the scientific name.

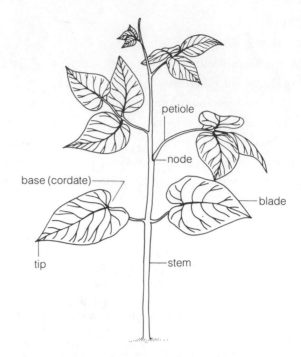

Figure 2-1 Structure of a typical plant (bean)

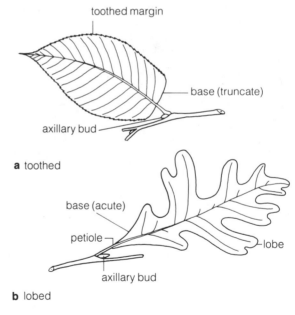

a toothed

b lobed

Figure 2-2 Simple leaves

Key to Some Common Genera of Trees

1a. Leaves broad and flat; plants producing flowers and fruits (angiosperms) 2

1b. Leaves needlelike or scalelike; plants producing cones, but no flowers or fruits (gymnosperms) . . 22

2a. Leaves compound 3

2b. Leaves simple 9

3a. Leaves alternate 4

3b. Leaves opposite 7

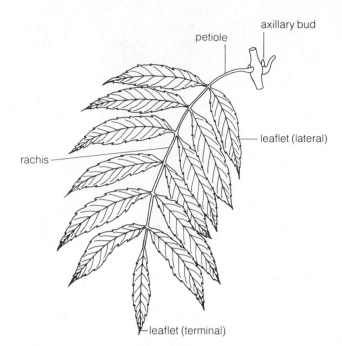

Figure 2-3 Pinnately compound leaf

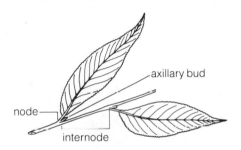

Figure 2-4 Simple leaves—alternate

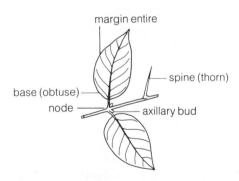

Figure 2-5 Simple leaves—opposite

4a. Leaflets short and stubby, less than twice as long as broad; branches armed with spines or thorns; fruit a beanlike pod 5

4b. Leaflets long and narrow, more than twice as long as broad; trunk and branches unarmed; fruit a nut 6

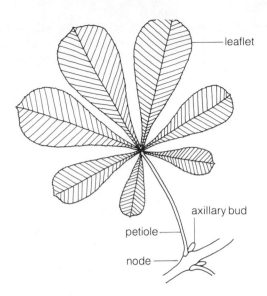

Figure 2-6 Palmately compound leaf

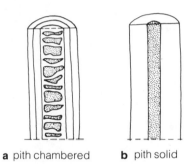

a pith chambered   **b** pith solid

Figure 2-7 Pith types

5a. Leaflet margin without teeth; terminal leaflet present; small deciduous spines at leaf base . . . .
. . . . *Robinia* (black locust)

5b. Leaflet margin with fine teeth; terminal leaflet absent; large permanent thorns on trunk and branches
. . . . *Gleditsia* (honey locust)

6a. Leaflets usually numbering less than 11; pith of twigs solid . . . . *Carya* (hickory)

6b. Leaflets numbering one or more, pith of twigs divided into chambers . . . . *Juglans* (walnut, butternut)

7a. Leaflets pinnately arranged; fruit a light-winged samara 8

7b. Leaflets palmately arranged; fruit a heavy leathery spherical capsule . . . . *Aesculus* (buckeye)

8a. Leaflets numbering mostly three to five; samaras borne in pairs, with curved wings . . *Acer* (box elder)

8b. Leaflets numbering mostly more than five; samaras borne singly, with straight wings . . *Fraxinus* (ash)

9a. Leaves alternate 10

9b. Leaves opposite 21

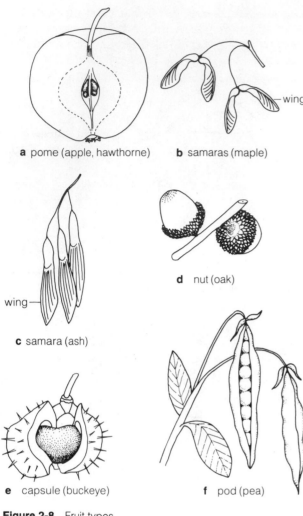

a pome (apple, hawthorne)   **b** samaras (maple)

wing

c samara (ash)

d nut (oak)

e capsule (buckeye)   **f** pod (pea)

Figure 2-8 Fruit types

10a. Leaves very narrow, at least three times as long as broad; axillary buds flattened against stem. . . . . *Salix* (willow)

10b. Leaves broader, less than three times as long as broad 11

11a. Leaf margin without small, regular teeth . . . 12

11b. Leaf margin with small, regular teeth 13

12a. Fruit a pod with downy seeds; leaf blade obtuse at base; petioles flattened, or if rounded, bark smooth . . . . *Populus* (poplar, popple, aspen)

12b. Fruit an acorn; leaf blade acute at the base; petioles rounded; bark rough . . . . *Quercus* (oaks)

13a. Leaves (at least some of them) with lobes or other indentations in addition to small, regular teeth 14

13b. Leaves without lobes or other indentations except for small, regular teeth 16

14a. Lobes asymmetrical, leaves often mitten-shaped . . . . *Morus* (mulberry)

14b. Lobes or other indentations fairly symmetrical 15

15a. Branches thorny (armed); fruit a small applelike pome . . . . *Crataegus* (hawthorne)

15b. Branches unarmed 17

16a. Bark smooth and waxy, often separating into thin layers; leaf base symmetrical . . . . . *Betula* (birch)

16b. Bark rough and furrowed, leaf base asymmetrical . . . . *Ulmus* (elm)

17a. Leaf base asymmetrical, strongly heart-shaped, at least on one side . . . . . . *Tilia* (basswood or linden)

17b. Leaf base acute, truncate, or slightly cordate. 18

18a. Leaf base asymmetrical; bark on older stems (trunk) often warty . . . . *Celtis* (hackberry)

18b. Leaf base symmetrical 19

19a. Leaf blade usually about twice as long as broad, generally acute at the base; fruit fleshy 20

19b. Leaf not much longer than broad, generally truncate at base; fruit a dry pod. . . . . *Populus* (poplar, popple, aspen)

20a. Leaf tapering to a pointed tip, glandular at base . . . . *Prunus* (cherry)

20b. Leaf spoon-shaped with a rounded tip, no glands at base . . . . *Crataegus* (hawthorne)

21a. Leaf margins with lobes and points, fruit a samara . . . . *Acer* (maple)

21b. Leaf margins without lobes or points; fruit a long capsule . . . . *Catalpa* (catalpa)

22a. Leaves needlelike, with two or more needles in a cluster 23

22b. Leaves needlelike or scalelike, occurring singly 24

23a. Leaves more than five in a cluster, soft, deciduous, borne at the ends of conspicuous stubby branches . . . . *Larix* (larch, tamarack)

23b. Leaves two to five in a cluster . . . *Pinus* (pines)

24a. Leaves soft, not sharp to the touch 25

24b. Leaves stiff, sharp, often unpleasant to the touch 27

25a. Leaves rounded at tip, whitened beneath . . . . *Tsuga* (hemlock)

25b. Leaves pointed at tip 26

26a. Tree; without distinct petioles, with two white lines on undersurface; bases circular . . . . . *Abies* (firs)

26b. Low shrub; leaves without petioles that follow down twigs, lighter green beneath but without distinct white lines . . . . *Taxus* (yew)

27a. Leaves more or less four-sided, neither in opposing pairs nor in whorls of three . . *Picea* (spruces)

27b. Leaves three-sided, either in opposing pairs or in whorls of three 28

28a. Twigs very strongly flattened; cones consisting of a few brown, dry scales. . . . .
. . . . *Thuja* (white cedar, arbor vitae)

28b. Twigs not flattened, easy to roll between the fingers; cones blue, spherical, berrylike 29

29a. Leaves about 1 cm long, all needlelike. . . . .
. . . . *Juniperus* (juniper, Eastern red cedar)

29b. Leaves mostly less than 0.5 cm long, often needle-like and scalelike on the same individual. . . . .
. . . . *Thuja* (Western red cedar)

Glossary to Accompany Tree Key

- *Acorn*—The fruit of an oak, consisting of a nut and its basally attached cup (Fig. 2-8d).
- *Acute*—Sharp-pointed (Fig. 2-2b).
- *Alternate*—Describing the arrangement of leaves or other structures that occur singly at successive nodes or levels; not opposite or whorled (Fig. 2-4).
- *Angiosperm*—A flowering seed plant (e.g., bean plant, maple tree, grass).
- *Armed*—Possessing thorns or spines.
- *Asymmetrical*—Not symmetrical.
- *Axil*—The upper angle between a branch or leaf and the stem from which it grows.
- *Axillary bud*—A bud occurring in the axil of a leaf (Figs. 2-2a, b, 2-4, 2-5).
- *Basal*—At the base.
- *Blade*—The expanded, more or less flat portion of a leaf (Fig. 2-1).
- *Capsule*—A dry fruit that splits open at maturity (e.g., buckeye; Fig. 2-8e).
- *Compound leaf*—Composed of two or more separate parts (leaflets) (Figs. 2-3, 2-6).
- *Cordate*—Heart-shaped (Fig. 2-1).
- *Deciduous*—Falling off at the end of a functional period (such as a growing season).
- *Fruit*—A ripened ovary, in some cases with associated floral parts (Figs. 2-8a–f).
- *Glandular*—Bearing secretory structures (glands).
- *Gymnosperm*—Seed plant lacking flowers and fruits (e.g., pine tree).
- *Lateral*—On or at the side (Fig. 2-3).
- *Leaflet*—One of the divisions of the blade of a compound leaf (Figs. 2-3, 2-6).
- *Lobed*—Separated by indentations (sinuses) into segments (lobes) larger than teeth (Fig. 2-2b).
- *Node*—Region on a stem where leaves or branches arise (Figs. 2-1, 2-4, 2-5, 2-6).
- *Nut*—A hard, one-seeded fruit that does not split open at maturity (e.g., acorn; Fig. 2-8d).
- *Obtuse*—Blunt (Fig. 2-5).

- *Opposite*—Describing the arrangement of leaves or other structures that occur two at a node, each separated from the other by half the circumference of the axis (Fig. 2-5).
- *Palmately compound*—With leaflets all arising at apex of petiole (Fig. 2-6).
- *Petiole*—Stalk of a leaf (Figs. 2-1, 2-2, 2-6).
- *Pinnately compound*—A leaf constructed somewhat like a feather, with the leaflets arranged on both sides of the rachis (Fig. 2-3).
- *Pith*—Internally, the central-most region of a stem (Figs. 2-7a,b).
- *Pod*—A dehiscent, dry fruit; a rather general term sometimes used when no other more specific term is applicable (Fig. 2-8f).
- *Pome*—Fleshy fruit containing several seeds (e.g., apple or pear; Fig. 2-8a).
- *Rachis*—Central axis of a compound leaf (Fig. 2-3).
- *Samara*—Winged, one-seeded, dry fruit (e.g., maple and ash fruits; Figs. 2-8b,c).
- *Simple leaf*—One with a single blade, not divided into leaflets (Figs. 2-2, 2-4, 2-5).
- *Spine*—Strong, stiff, sharp-pointed outgrowth on a stem or other organ (Fig. 2-5).
- *Symmetrical*—Capable of being divided longitudinally into similar halves.
- *Terminal*—Last in a series (Fig. 2-3).
- *Thorn*—Sharp, woody, spinelike outgrowth from the wood of a stem; usually a reduced, modified branch.
- *Tooth*—Small, sharp-pointed marginal lobe of a leaf (Fig. 2-2a).
- *Truncate*—Cut off squarely at end (Fig. 2-2a).
- *Unarmed*—Without thorns or spines.
- *Whorl*—A group of three or more leaves or other structures at a node.

B. Some Microscopic Members of the Kingdom Protista

MATERIALS

Per student:

- compound microscope
- microscope slide
- coverslip
- dissecting needle

Per student group (table):

- 1 culture of Volvocales mixture
- 4 disposable plastic pipets
- methyl cellulose in dropping bottle

PROCEDURE Suppose you wished to identify the specimens in some pond water. The easiest way would be to key them out with a dichotomous key, now that you know how to use one. Let's do just that.

1. Obtain a clean glass microscope slide and clean coverslip.

2. Using a disposable plastic pipet, withdraw a small amount of the culture provided.

3. Place *one* drop of the culture on the center of the slide.

4. Gently lower the coverslip onto the liquid.

5. Using your compound light microscope, observe your wet mount. Focus first with the low power objective and then with the medium or high-dry objective, depending on the size of the organism in the field of view.

6. Concentrate your observation on a single specimen, keying out the specimen using the "Key to Selected Protistans" below.

7. In the space below, write the scientific name of each organism you identify. After each identification, have your instructor verify your conclusion.

Key to Selected Protistans*

1a. Cells grouped into a colony 2

1b. Cells not grouped into a colony; single, round cells . . . . *Chlamydomonas*

2a. Colony one-cell thick; flat or cup-shaped 7

2b. Colony a round ball or sphere 3

3a. Colony composed of fewer than 100 cells 4

3b. Colony a hollow round ball of more than 500 cells; new colonies can be seen forming inside the mature colony . . . . *Volvox*

4a. All cells in a colony are the same size; seldom more than 32 cells 5

4b. Cells in a colony of two different sizes; colony large, composed of up to 100 cells; a hollow sphere . . . . *Eudorina californica*

5a. When viewed under high power, cells are round or spindle-shaped, never triangular- or wedge-shaped; cells separated from each other and not tightly packed 6

5b. When viewed under high power, cells are triangular- or wedge-shaped; cells are very tightly packed and close together . . . . *Pandorina*

6a. Cells round; 16 to 64 cells in a colony . . . . *Eudorina elegans*

6b. When viewed from the side, cells are spindle-shaped; usually four or eight cells in a colony; when viewed under low power, the colony looks like a doughnut or crown . . . . *Stephanosphaera*

7a. Colony like a flattened horseshoe with several projections from the posterior; 16 to 32 cells in a colony . . . . *Platydorina*

7b. Colony a loose square or rectangle; 2 to 16 cells in a colony . . . . *Gonium*

Organism 1 is _____ .

Organism 2 is _____ .

Organism 3 is _____ .

Organism 4 is _____ .

Organism 5 is _____ .

Organism 6 is _____ .

Organism 7 is _____ .

Organism 8 is _____ .

PRE-LAB QUESTIONS

____ **1.** The name "human" is an example of a (a) common name, (b) scientific name, (c) binomial, (d) polynomial.

____ **2.** The person primarily responsible for the scientific nomenclature used today is (a) Darwin, (b) Linnaeus, (c) Watson, (d) Hooke.

____ **3.** The scientific name for the ruffed grouse is *Bonasa umbellus*. *Bonasa* is (a) the family name, (b) the genus, (c) the specific epithet, (d) all of the above.

____ **4.** A binomial is always a (a) genus, (b) specific epithet, (c) scientific name, (d) two-part name.

____ **5.** The science of classifying and naming organisms is known as (a) taxonomy, (b) phylogeny, (c) morphology, (d) physiology.

____ **6.** Which scientific name for the wolf is presented correctly? (a) Canis lupus, (b) canis lupus, (c) *Canis lupus,* (d) Canis Lupus.

____ **7.** A road that dichotomizes is a (an) (a) intersection of two crossroads, (b) road that forks into two roads, (c) road that has numerous entrances and exits, (d) road that leads nowhere.

____ **8.** Most scientific names are derived from (a) English, (b) Latin, (c) Italian, (d) French.

____ **9.** One objection to common names is that (a) many organisms may have the same common name; (b) many common names may exist for the same organism; (c) the common name may not be familiar to an individual not speaking the language of the common name; (d) all of the above.

____ **10.** Phylogeny is the apparent (a) name of an organism, (b) ancestry of an organism, (c) nomenclature, (d) dichotomy of a system of classification.

---

*Adapted from "Dichotomous Key to the Volvocales Mixture," Carolina Biological Supply Company.

EXERCISE 2
TAXONOMY: CLASSIFYING AND NAMING ORGANISMS

POST-LAB QUESTIONS

1. If you were to use a binomial system to identify the members of your family (mother, father, sisters, brothers), how would you write their names so that your system would most closely approximate that used to designate species?

2. If you owned a large, varied record collection, how might you keep track of all your different kinds of music?

3. Describe several advantages of the use of scientific names over common names.

4. Based upon the following classification scheme, which two organisms are most closely phylogenetically related? Why?

	Organism 1	*Organism 2*	*Organism 3*	*Organism 4*
Kingdom	Animalia	Animalia	Animalia	Animalia
Phylum	Arthropoda	Arthropoda	Arthropoda	Arthropoda
Class	Insecta	Insecta	Insecta	Insecta
Order	Coleoptera	Coleoptera	Coleoptera	Coleoptera
Genus	*Caulophilus*	*Sitophilus*	*Latheticus*	*Sitophilus*
Specific epithet	*oryzae*	*oryzae*	*oryzae*	*zeamaize*
Common name	broadnosed grain weevil	rice weevil	longheaded flour beetle	maize weevil

Consider the drawing of Plants A and B in answering questions 5 to 7.

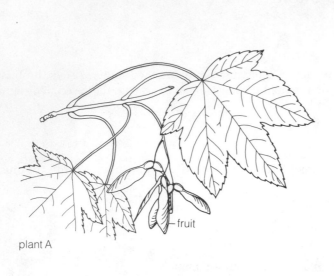

plant A

fruit

cone

plant B

5. Using the taxonomic key, identify the two plants as either angiosperms or gymnosperms.

Plant A is a (an) _____ .

Plant B is a (an) _____ .

6. To what genus does Plant A belong? What is its common name?

Genus: _____

Common name: _____

7. To what genus does Plant B belong? What is its common name?

Genus: _____

Common name: _____

Consider the drawing of Plants C and D in answering questions 8 to 10.

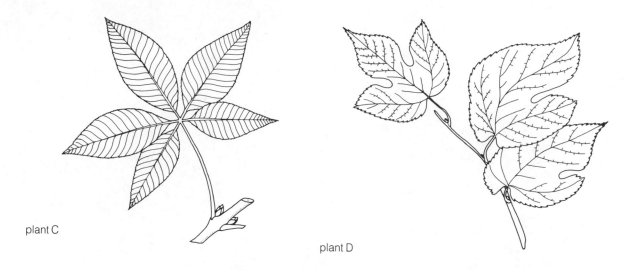

plant C

plant D

8. As completely as possible, describe the leaf of Plant C.

9. To what genus does Plant C belong? What is its common name?

Genus: _____

Common name: _____

10. Is the leaf of Plant D simple or compound?

_____

What is the genus of Plant D?

Genus: _____

STRUCTURE AND FUNCTION OF LIVING CELLS

OBJECTIVES After completing this exercise you will be able to:

1. define cell, cell theory, prokaryotic, eukaryotic, nucleoid, nucleus, cytomembrane system, organelle, *karyon*, multinucleate, cytoplasmic streaming, colloid, sol, gel, envelope;

2. describe the similarities and differences between prokaryotic and eukaryotic cells;

3. identify the cell parts described within this exercise;

4. state the function for each cell part;

5. distinguish between plant and animal cells.

INTRODUCTION Structurally and functionally, all life has one common feature: All living organisms are composed of **cells.** The development of this concept began with Robert Hooke's seventeenth-century observation that slices of cork were made up of small units. He called these units "cells" because their structure reminded him of the small cubicles that monks lived in. Over the next 100 years the **cell theory** emerged. This theory has three principles: (1) All organisms are composed of one or more cells; (2) the cell is the basic *living* unit of organization; and (3) all cells arise from preexisting cells.

Although cells vary in organization, size, and function, all share three structural features: (1) All possess a *plasma membrane* defining the boundary of the living material; (2) all contain a region of *DNA* (deoxyribonucleic acid), which stores genetic information; and (3) all contain *cytoplasm,* everything inside the plasma membrane that is not part of the DNA region.

With respect to internal organization, there are two basic types of cells, **prokaryotic** and **eukaryotic.** Study Table 3-1, comparing the more important differences between prokaryotic and eukaryotic cells.

Table 3-1 Comparison of Prokaryotic and Eukaryotic Cells

Characteristics	Cell Type	
	Prokaryotic	Eukaryotic
Genetic material	Located in **nucleoid,** an irregularly shaped region of cytoplasm not bounded by a special membrane	Located in **nucleus,** a double membrane-bounded compartment within the cytoplasm
	Consists of a single molecule of DNA	Numerous molecules of DNA combined with protein
	——	Organized into chromosomes
Cytoplasmic structures	Small ribosomes	Large ribosomes
	Photosynthetic membranes arising from the plasma membrane (in some representatives only)	**Cytomembrane system,** a system of connected membrane structures
	——	**Organelles,** membrane-bounded compartments specialized to perform specific functions
Kingdoms represented	Monera	Protista Fungi Plantae Animalia

The Greek word **karyon** means "kernel," referring to the nucleus. Thus, prokaryotic means "before a nucleus," while eukaryotic indicates the presence of a "true nucleus."

Prokaryotic cells typical of modern bacteria and cyanobacteria are believed to be similar to the first cells that arose on earth 3.5 billion years ago. Eukaryotic cells probably evolved from prokaryotes.

This exercise will familiarize you with the basics of cell structure and the function of prokaryotes (prokaryotic cells) and eukaryotes (eukaryotic cells).

I. PROKARYOTIC CELLS (Starr, pp. 45, 225)

MATERIALS

Per student:

- dissecting needle
- compound microscope
- microscope slide
- coverslip

Per student pair:

- distilled water (dH₂O) in dropping bottle

Per student group (table):

- culture of a cyanobacterium (either *Anabaena* or *Oscillatoria*)

Per lab room:

- 3 bacterium-containing nutrient agar plates (demonstration)
- 3 demonstration slides of bacteria (coccus, bacillus, spirillum)

PROCEDURE   Observe the culture plate containing bacteria growing on the surface of a nutrient medium. Can you see the individual cells with your naked eye?

NO

Observe the microscopic preparations of bacteria on *demonstration* next to the culture plate. The three slides represent the three different shapes of bacteria. Which objective lenses are being used to view the bacteria?

10 1.25

Would you say bacteria are large or small organisms?

Small

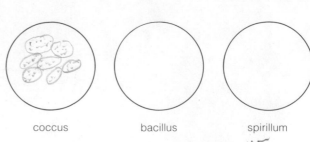

coccus   bacillus   spirillum

Figure 3-1   Drawing of several bacterial cells (___45___×).
Approximate size = _____ μm.

Can you discern any detail within the cytoplasm?

A GHIE

In the space provided for Fig. 3-1, make a sketch of what you see through the microscope. Record the magnification you are using to view the bacteria in the blank provided at the end of the legend. Next to your sketch record the approximate size of the bacterial cells. (Return to page 13 of Exercise 1 if you've forgotten how to estimate the size of an object being viewed through the microscope.)

Examine the electron micrograph of the bacterium *Escherichia coli* (Fig. 3-2). The cell is bound by the *cell wall*, a structure chemically distinct from the wall of plant cells but serving the same primary function. The *plasma membrane* is tightly appressed to (lying flat against) the internal surface of the cell wall and is difficult to distinguish. Look for two components of the *cytoplasm: Ribosomes* are electron-dense particles (they appear black) that give the cytoplasm its granular appearance; the *nucleoid* is a relatively electron-transparent region (appears light) containing fine threads of DNA.

Label Fig. 3-2, an electron micrograph of *E. coli*.

Another type of prokaryotic cell is exemplified by cyanobacteria, such as *Oscillatoria* and *Anabaena*. Cyanobacteria (sometimes called blue-green algae) are commonly found in water and damp soils. They obtain their nutrition by converting the sun's energy through photosynthesis.

With a dissecting needle, remove a few filaments from the cyanobacterial culture, placing them in a drop of water on a clean microscope slide. Place a coverslip over the material and examine it first with the low power objective and then using the high-dry objective (or oil immersion objective, if your microscope is so equipped). Make a sketch in Fig. 3-3 of the cells you see at high power. Estimate the size of a *single* cyanobacterial cell, and record the magnification you used in making your drawing.

LABCI

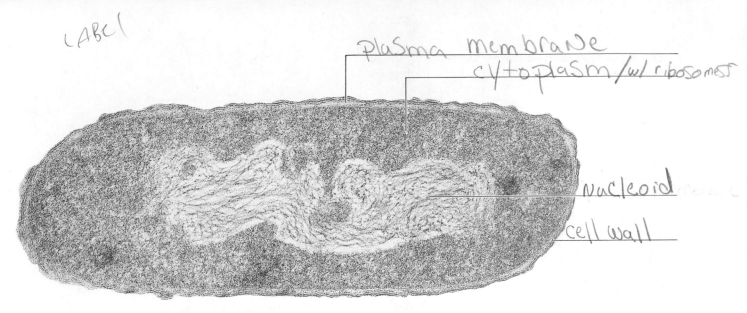

Plasma membrane
cytoplasm/w/ ribosomes
Nucleoid
cell wall

Labels: cell wall, cytoplasm with ribosomes, nucleoid, plasma membrane

Figure 3-2 Electron micrograph of the bacterium *Escherichia coli* (55,000×). (Photo courtesy G. Cohen-Bazire.)

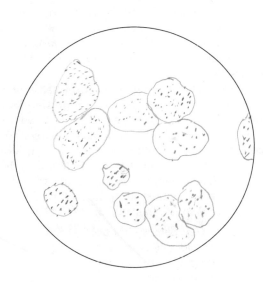

Figure 3-3 Drawing of several prokaryotic cells of a cyanobacterium (___45___×). Approximate size = _____ μm.

Now examine the electron micrograph of *Anabaena* (Fig. 3-4), identifying and labeling *cell wall, plasma membrane, cytoplasm, ribosomes,* and *nucleoid*. The cyanobacteria also possess membranes that function in photosynthesis. Identify these *photosynthetic membranes.*

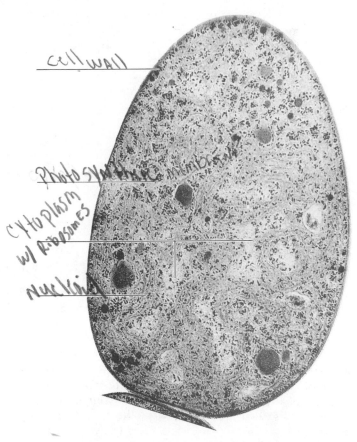

cell wall
Photosynthetic membrane
cytoplasm w/ ribosomes
Nucleoid

Labels: cell wall, cytoplasm with ribosomes, nucleoid, photosynthetic membranes.

Figure 3-4 Electron micrograph of *Anabaena* (13,000×). (Photo courtesy R. D. Warmbrodt.)

Se

II. EUKARYOTIC CELLS   (Starr, pp. 45–56)

MATERIALS

Per student:

- textbook
- toothpick
- microscope slide
- coverslip
- culture of *Physarum polycephalum*
- compound microscope

Per student pair:

- methylene blue in dropping bottle
- safranin O in dropping bottle
- distilled water (dH₂O) in dropping bottle

Per student group (table):

- *Elodea* in water-containing culture dish
- forceps
- onion bulb
- tissue paper

Per lab room:

- model of animal cell
- model of plant cell

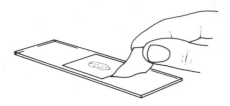

Figure 3-5   Method for staining specimen under coverslips of microscope slide.

PROCEDURE

A. Animal Cells Observed with the Light Microscope

1. *Human cheek cells.* Using the broad end of a clean toothpick, gently scrape the inside of your cheek. Stir the scrapings into a drop of distilled water on a clean microscope slide and add a coverslip. Because the cells are almost transparent, decrease the amount of light entering the objective lens to increase the contrast. (See Section G, Exercise 1, page 12.) Find the cells using the low power objective of your microscope, then switch to the high-dry objective for detailed study. Find the *nucleus*, a centrally located spherical body within the *cytoplasm* of each cell.

Now stain your cheek cells with a dilute solution of methylene blue, a dye that stains the nucleus darker than the surrounding cytoplasm. To stain your slide, follow the directions illustrated in Fig. 3-5.

First add a drop of the stain to one edge of the coverslip. Then draw the stain under the coverslip by touching a piece of tissue paper to the *opposite* side of the coverslip. It is not necessary to remove the coverslip.

In Fig. 3-6, make a sketch of the cheek cells, labeling *cytoplasm*, *nucleus*, and the location of the *plasma membrane*. (A light microscope cannot resolve the plasma membrane, but the boundary between the cytoplasm and the external medium indicates its location.) Many of the cells will be folded or wrinkled due to their thin, flexible nature. Estimate and record in your sketch the size of the cells. (The method for estimating the size is found in step 7 on page 13.)

2. *Physarum polycephalum.* Although, strictly speaking, the slime mold *Physarum* is not considered an animal (depending upon the authority, it's considered a protist or a fungus), its cellular structure is typical of animal cells. *Physarum* is a unicellular organism, so it contains all the metabolic machinery for independent existence.

Place a plain microscope slide on the stage of your compound microscope. This will serve as a platform on which you can place a culture dish. Now obtain a petri dish culture of *Physarum*, remove the lid, and place it on the platform. Observe initially with the low power objective, then with the medium power objective. Place a coverslip over part of the organism

Labels: cytoplasm, nucleus, plasma membrane

Figure 3-6   Drawing of human cheek cells (_____45_____ ×).
Approximate size = _____ mm.

Labels: cytoplasm (gel), cytoplasm (sol), plasma membrane

Figure 3-7   Drawing of a portion of *Physarum* (_____ ×)

before rotating the high-dry objective into place. This prevents the agar from getting on the lens.

Physarum is **multinucleate** (more than one nucleus occurs within the cytoplasm). Unfortunately, the nuclei are tiny; you will not be able to distinguish them from other granules in the cytoplasm. The outer boundary of the cytoplasm is the *plasma membrane*. Locate the boundary. Once again, the resolving power of your microscope probably is not sufficient to allow you to actually view the membrane.

Watch the cytoplasm of the organism move. This intracellular motion is known as **cytoplasmic streaming.** Although not visible with the light microscope without using special techniques, contractile proteins called *microfilaments* are believed responsible for cytoplasmic streaming.

Like all cells, the cytoplasm of *Physarum* is a semisolid **colloid,** a state of matter in which the particles are too large to be dissolved but too small to settle out. Note that the outer portion of the cytoplasm appears solid; this is the **gel** state of the cytoplasm. Notice that the contents closer to the interior are in motion within a fluid; this portion of the cytoplasm is in the **sol** state. Movement of the organism occurs as

the sol-state cytoplasm at the advancing tip pushes against the plasma membrane, causing the region to swell outward. The sol-state cytoplasm rushes into the region, becoming converted to the gel state along the margins.

In Fig. 3-7 make a labeled sketch of the portion of *Physarum* that you have been observing.

B. Animal Cells as Observed with the Electron Microscope

Studies with the electron microscope have yielded a wealth of information on the structure of eukaryotic cells. Structures too small to be seen with the light microscope have been identified. These include many **organelles**, structures in the cytoplasm that have been separated ("compartmentalized") by enclosure in membranes. Examples of organelles are the nucleus, mitochondria, endoplasmic reticulum, and Golgi bodies. Although the cells in each of the five kingdoms have some peculiarities unique to that kingdom, electron microscopy has revealed that all cells are fundamentally similar.

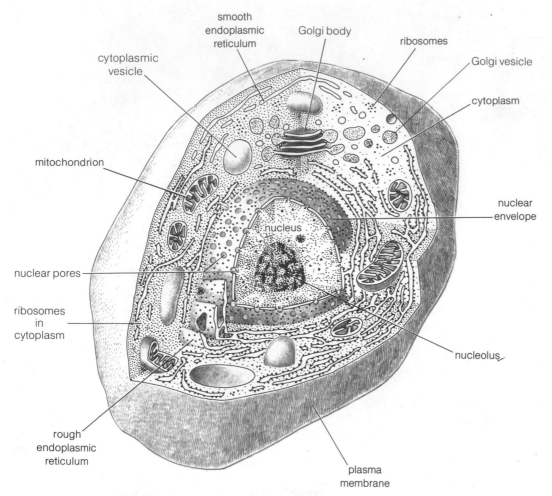

smooth
endoplasmic
reticulum

Golgi body

ribosomes

Golgi vesicle

cytoplasmic
vesicle

cytoplasm

mitochondrion

nuclear
envelope

nucleus

nuclear pores

ribosomes
in
cytoplasm

nucleolus

rough
endoplasmic
reticulum

plasma
membrane

Figure 3-8   Three-dimensional representation of an animal cell
as seen with the electron microscope. (After Wolfe, 1985.)

Study Fig. 3-8, a three-dimensional representation of an animal cell. With the aid of Fig. 3-8, identify the parts on the model of the animal cell that is on *demonstration*.

Figure 3-9 is an electron micrograph (EM) of an animal cell (kingdom Animalia). Study the electron micrograph and identify, with the aid of Fig. 3-8 and any electron micrographs in your textbook, each structure listed below.

Pay particular attention to the membranes surrounding the nucleus and mitochondria. Note that these two structures are each bounded by *two* membranes, which are commonly referred to collectively as an **envelope.**

Using your textbook as a reference, list the function for each of the following cellular components.

1. plasma membrane _____

_____

2. cytoplasm _____

_____

3. nucleus (the plural is *nuclei*) _____

_____

   a. nuclear envelope _____

   b. nuclear pores _____

4. chromatin _____

_____

5. nucleolus (the plural is *nucleoli*) _____

_____

6. endoplasmic reticulum (ER)

   a. rough ER _____

   b. smooth ER _____

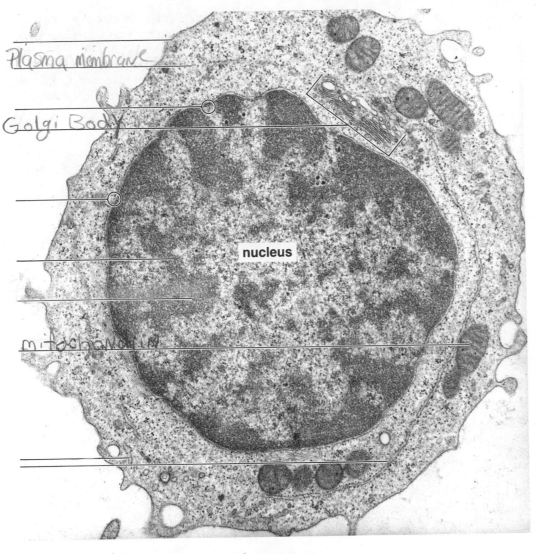

Plasma membrane (handwritten)

Golgi Body (handwritten)

nucleus

mitochondria (handwritten)

Plants will show a cell wall + membrane. Animals do NOT. (handwritten)

Labels: plasma membrane, cytoplasm, nuclear envelope, nuclear pore, chromatin, rough ER, smooth ER, ~~Golgi body~~, ~~mitochondrion~~

Figure 3-9 Electron micrograph of an animal cell (2000×). (Photo courtesy W. R. Hargreaves.)

7. Golgi body _____

8. mitochondrion (the plural is *mitochondria*) _____

C. Plant Cells Seen with the Light Microscope

1. *Elodea leaf cells.* Young leaves at the growing tip of *Elodea* are particularly well suited for studying cell structure because these leaves are only a few cell layers thick.

With a forceps, remove a single young leaf, mount it in a drop of distilled water, and cover with a coverslip. Examine the leaf first with the low-power objective. Then concentrate your study on several cells using the high-dry objective.

You will probably be struck by the abundance of green bodies in the cytoplasm. These are the *chloroplasts*, organelles that function in photosynthesis and that are typical of green plants. You may see numerous dark lines running parallel to the long axis of the leaf. These are the air-containing *intercellular spaces*. The *cell wall*, a structure distinguishing plant from animal cells, may be visible as a clear area surrounding the cytoplasm.

After the cells have warmed a bit, you should see **cytoplasmic streaming** taking place. Movement of the chloroplasts along the cell wall is the most obvious evidence of cytoplasmic streaming. Microfilaments are thought to be responsible for this intracellular motion.

Remember that you are looking at a three-dimensional object. In the middle portion of the cell, find the large, clear *central vacuole*, which can take up to 50 to 90% of the cell interior.

The chloroplasts occur in the cytoplasm surrounding the vacuole, so they will appear to be in different locations, depending on where you focus in the cell. If your focus is the upper or lower surface, the chloroplasts will appear to be scattered throughout the cell. But if you focus in the center of the cell (by raising or lowering the objective with the fine focus knob), you will see the chloroplasts in a thin layer of cytoplasm along the wall. The vacuole is the "empty" space in the center of the cell.

Locate the *nucleus* within the cytoplasm. It will appear as a clear or slightly amber body that is slightly larger than the chloroplasts. (You may need to examine several cells to find a clearly defined nucleus.)

How would you describe the three-dimensional shape of the *Elodea* leaf cell?

_____

_____

_____

The shape of the chloroplasts and nucleus?

_____

_____

_____

Now add a drop of safranin stain to make the cell wall more obvious. Add the stain the same way you stained your cheek cells with methylene blue (Fig. 3-5).

Sketch and label several *Elodea* cells in Fig. 3-10. Indicate where the plasma membrane would be located in the cells.

Labels: cell wall, plasma membrane, cytoplasm, nucleus, chloroplasts, central vacuole

Figure 3-10 Drawing of *Elodea* cells (_____ ×)

2. *Onion scale cells.* Make a wet mount of a colorless scale of an onion bulb, using the technique described in Fig. 3-11.

Observe your preparation with your microscope, focusing first with the low-power objective. Continue your study, switching to the medium power and finally the high-dry objective.

Identify the *cell wall* and *cytoplasm.* The *nucleus* should be a prominent sphere within the cytoplasm. Examine the nucleus more carefully at high magnification. Within it find one or more *nucleoli.* Nucleoli are rich in a nucleic acid known as RNA (ribonucleic acid) while the rest of the nucleus is largely DNA (deoxyribonucleic acid), the genetic material.

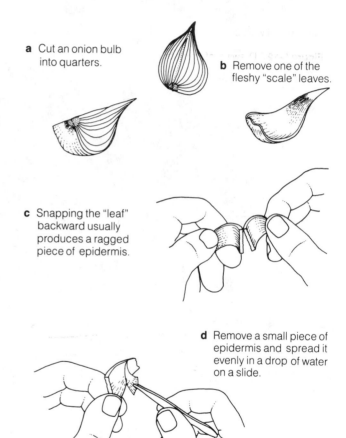

a Cut an onion bulb into quarters.

b Remove one of the fleshy "scale" leaves.

c Snapping the "leaf" backward usually produces a ragged piece of epidermis.

d Remove a small piece of epidermis and spread it evenly in a drop of water on a slide.

e Gently lower a coverslip to prevent trapping air bubbles. Examine with your microscope. Add more water to the edge of the coverslip with an eye dropper if the slide begins to dry.

Figure 3-11 Method for obtaining onion scale cells. (From Peter Abramoff and Robert G. Thomson, *Laboratory Outlines in Biology III.* Copyright © 1962, 1963 Peter Abramoff and Robert G. Thomson. Copyright © 1964, 1966, 1972, 1982 W. H. Freeman and Company. Used by permission.)

Labels: cell wall, cytoplasm, nucleus, nucleolus, oil droplets

Figure 3-12   Drawing of onion bulb leaf cells (_____×)

Numerous *oil droplets* should be visible in the form of granule-like bodies within the cytoplasm. These oil droplets are a form of stored food material. You may be surprised to learn that onion scales are actually leaves! Which cellular components present in *Elodea* leaf cells are absent in onion leaf cells?

_____

In Fig. 3-12 sketch and label several cells from onion scale leaves.

D. Plant Cells as Seen with the Electron Microscope

The electron microscope has made obvious some of the unique features of plant cells. Study Fig. 3-13, a three-dimensional representation of a typical plant cell.

With the aid of Fig. 3-13, identify the structures present on the model of a plant cell that is on *demonstration*.

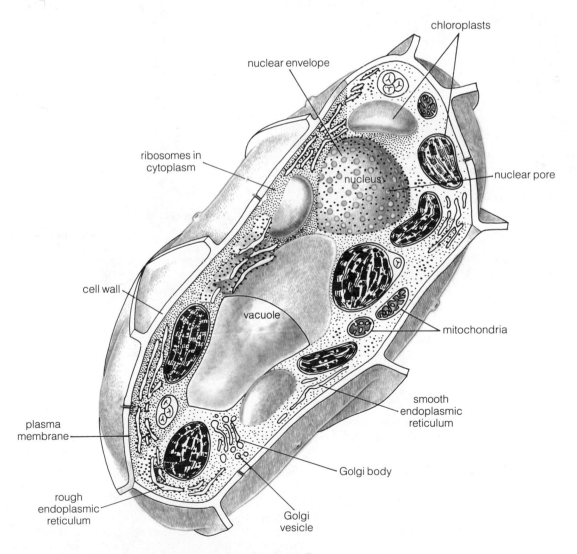

Figure 3-13   Three-dimensional representation of a plant cell as seen with the electron microscope. (After Wolfe, 1985.)

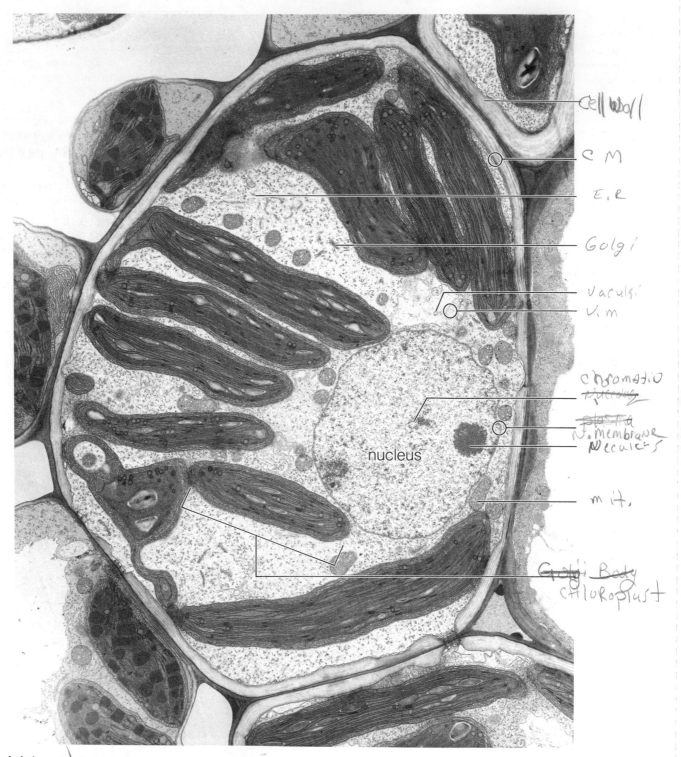

Handwritten annotations on the figure:
- Cell wall
- C M
- E.R
- Golgi
- Vaculi
- V.m
- chromatin ~~nucleus~~
- ~~plasm~~ N. Membrane
- Neculies
- mit.
- ~~Golgi Body~~ chloroplast
- nucleus

Labels: cell wall, chloroplast, vacuole, vacuolar membrane, plasma membrane, nuclear envelope, chromatin, nucleolus, endoplasmic reticulum (ER), Golgi body, mitochondrion

Figure 3-14 Electron micrograph of a corn leaf cell (3,000×). (Courtesy R. F. Evert and M. A. Walsh.)

DIFFUSION, OSMOSIS, AND THE FUNCTIONAL SIGNIFICANCE OF BIOLOGICAL MEMBRANES

OBJECTIVES After completing this exercise you will be able to:

1. define solvent, solute, solution, differentially permeable, diffusion, osmosis, concentration gradient, equilibrium, turgid, plasmolyzed, plasmolysis, turgor pressure, tonicity, hypertonic, isotonic, hypotonic;

2. describe the structure of cellular membranes;

3. distinguish between diffusion and osmosis;

4. determine the effects of concentration and temperature on diffusion;

5. describe the effects of hypertonic, isotonic, and hypotonic solutions on red blood cells and *Elodea* leaf cells.

INTRODUCTION Water is a great environment. Life is believed to have originated in the water. Without it, life as we know it would cease to exist. If, as suspected, life in our solar system is unique to earth, it is probably because ours is the only planet known to possess free water on its surface.

Living cells are made up of 75 to 85% water. Virtually all substances entering and leaving cells are dissolved in water, making it the **solvent** most important for life processes. The substances dissolved in water are called **solutes** and include such substances as salts and sugars. The combination of a solvent and dissolved solute is a **solution.**

All cells possess membranes composed of a phospholipid bilayer that contains embedded and surface proteins. Membranes are boundaries that solutes must cross to reach the cellular site where they will be utilized for the processes of life. These membranes regulate the passage of substances into and out of the cell. They are **differentially permeable,** allowing some substances to move easily while completely excluding others.

Although there are several methods by which solutes may enter cells, the most common is diffusion. Once within the cell, solutes move through the cytoplasm by diffusion, sometimes assisted by cyto-

plasmic streaming. Simply stated, **diffusion** is the movement of solute molecules from a region of high concentration to one of lower concentration. Diffusion occurs without requiring the expenditure of cellular energy.

So far, we've described the movement only of solute molecules across membranes. However, water (the solvent) also moves across the membrane. The movement of *water* across differentially permeable membranes is **osmosis.** Think of osmosis as a special form of diffusion, one occurring from a region of higher *water* concentration to one of lower *water* concentration.

The difference in concentration of like molecules in two regions is called a **concentration gradient.** Diffusion and osmosis take place *down* concentration gradients. Over time, the concentration of solvent and solute molecules becomes equally distributed, the gradient ceasing to exist. At this point the system is said to be at **equilibrium.**

Molecules are always in motion, even at equilibrium. Thus, solvent and solute molecules continue to move because of randomly colliding molecules. However, at equilibrium there is no *net change* in the concentration.

This exercise will introduce you to the principles of diffusion and osmosis.

NOTE: Start parts IIB and III before doing any other portion of this exercise.

RATE OF DIFFUSION OF SOLUTES

(Starr, pp. 41–42)

Solutes move within a cell's cytoplasm largely due to diffusion. However, the rate of diffusion (the distance diffused in a given amount of time) is affected by factors such as temperature and the size of the solute molecules. Recall from Exercise 3 that cytoplasm is a colloid. The following experiment demonstrates the effects of these two factors within a common colloid, gelatin (the substance of Jell-O®), to simulate cytoplasm.

MATERIALS

Per student:

- metric ruler

Per student group (table):

- 1 set of 4 screw-cap test tubes half-filled with 5% gelatin, to which the following dyes have been added: potassium dichromate, aniline blue, Bismarck brown, Janus green; labeled with each dye and marked "5°C"

- 1 set of 4 screw-cap test tubes as above but marked "Room Temperature"

Per lab room:

- 5°C refrigerator

PROCEDURE   Two sets of four screw-cap test tubes have been half-filled with 5% gelatin. One mL of a dye has been added to each test tube. On the line below, record the time at which the experiment was started.

_____

Set 1 is in a 5°C refrigerator; set 2 is at room temperature.

1. Remove set 1 from the refrigerator and compare the distance the dye has diffused in corresponding tubes of each set.

2. While holding each tube vertically in front of a white sheet of paper, use a metric ruler to measure how far each dye has diffused from the gelatin's surface. Record this distance in Table 4-1.

Roomtemp only.

3. Determine the *rate* of diffusion for each dye by using the formula:

rate = distance/elapsed time (hours)

Time experiment ended: _____

Time experiment started: _____

Elapsed time: _____ hours

Which of the solutes diffused the slowest (regardless of temperature)?

_____

Which diffused the fastest?

_____

What effect did temperature have on the rate of diffusion?

_____

Make a conclusion about diffusion of a solute in a gel, relating the rate of diffusion to the molecular weight of the solute.

_____

_____

_____

Return set 1 to the refrigerator.

_____

_____

Table 4-1   Effect of Temperature on Diffusion Rate of Various Solutes

Solute (dye)	Set 1 (5°C)		Set 2 (Room Temp.)	
	Distance (mm)	Rate	Distance (mm)	Rate
Potassium dichromate (MW* = 294)				
Bismarck brown (MW = 419)				
Janus green (MW = 511)				
Aniline blue (MW = 738)				

*MW = molecular weight, a reflection of the mass of a substance. (MW is determined by adding the atomic weights of all elements comprising a compound.)

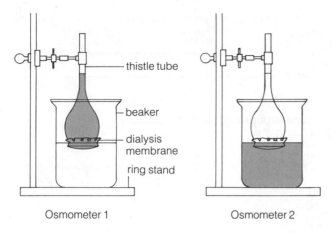

thistle tube

beaker

dialysis membrane

ring stand

Osmometer 1

Osmometer 2

Figure 4-1 A simple osmometer

II. OSMOSIS

(Starr, p. 42)

Osmosis occurs when different concentrations of water are separated by a differentially permeable membrane. One example of a differentially permeable membrane within a living cell is the plasma membrane. This experiment demonstrates osmosis by using dialysis membrane, a differentially permeable cellulose sheet that permits the passage of water but obstructs passage of larger molecules.

A. A Simple Osmometer Demonstration

MATERIALS

Per student:

■ metric ruler

Per lab room:

■ simple osmometer apparatus (Fig. 4-1)

PROCEDURE Observe the lab setup of two simple osmometers, illustrated in Fig. 4-1. The mouths of two thistle tubes have been covered with dialysis membrane. The thistle tube in osmometer 1 contains concentrated molasses, a substance consisting largely of sucrose dissolved in water. At the start of the experiment, your instructor will immerse the bowl of this thistle tube in a beaker of distilled water. (Distilled water, dH_2O, is virtually free of solutes.)

The thistle tube in osmometer 2 contains distilled water. Your instructor will immerse the bowl of this osmometer in a beaker containing molasses.

Note the mark on the stem of the thistle tube indicating the level of the solution at the start of the experiment. After 60 minutes, use a metric ruler to measure the distance that the solution has risen or fallen in each tube.

Record your observations in Table 4-2.

Which way did the water move?

_____

Relate the movement of water molecules to the *concentration gradient* of the water.

_____

_____

_____

Has the molasses in thistle tube 1 become more or less concentrated?

_____

B. A Weighing Method Demonstrating Osmosis

MATERIALS

Per student group (4):

■ 5 15-cm lengths of dialysis tubing, soaking in dH_2O

■ 10 10-cm pieces of string

■ ring stand and funnel apparatus (Fig. 4-2)

■ 25-mL graduated cylinder

■ 5 small string tags

■ china marker

Per student group (table):

■ dishpan half-filled with dH_2O

■ paper toweling

■ balance

Per lab room:

■ source of dH_2O (at each sink)

Table 4-2	Osmosis in Simple Osmometers		
	Contents		Distance and Direction Solution Has Moved After 60 Minutes
Osmometer	Thistle tube	Beaker	
1	Concentrated molasses	dH_2O	
2	dH_2O	Concentrated molasses	

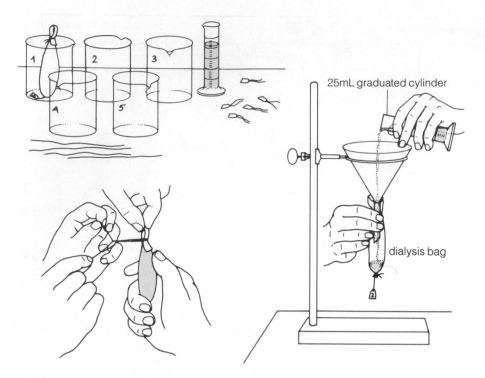

Figure 4-2 Method for filling dialysis bags

PROCEDURE Work in groups of four for this experiment.

1. Obtain five sections of dialysis tubing, each 15-cm long, that have been presoaked in distilled water. Recall that the dialysis tubing is permeable to water molecules but not to sucrose.

2. Fold over one end of each tube and tie it tightly with string.

3. Attach a string tag to the tied end of each bag and number them from 1 through 5.

4. Slip the open end of the bag over the stem of a funnel (Fig. 4-2). Using a graduated cylinder to measure volume, fill the bags as follows:

Bag 1. 15 mL of distilled water (*Note: Be sure to rinse the cylinder if it has been used for measuring sucrose.*)

Bag 2. 15 mL of 10% sucrose

Bag 3. 15 mL of 20% sucrose

Bag 4. 15 mL of 40% sucrose

Bag 5. 15 mL of distilled water

5. As each bag is filled, force out excess air by squeezing the bottom end of the tube.

6. Fold the end of the bag and tie it securely with another piece of string.

7. Rinse each filled bag in the dishpan containing distilled water (dH$_2$O); gently blot off the excess water with paper toweling.

8. Weigh each bag to the nearest 0.5 g.

9. Record the weights in the column marked "0 min" of Table 4-3.

10. Number five 600-mL beakers with a china marker.

11. Add 400 mL of dH$_2$O to beakers 1 through 4.

12. Add 400 mL of 40% sucrose solution to beaker 5.

13. Place bags 1 through 4 in the correspondingly numbered beakers.

14. Place bag 5 in the beaker containing 40% sucrose.

15. At intervals of 20 minutes, remove each bag from its beaker, blot off the excess fluid, and weigh each bag.

16. Record the weight of each bag in Table 4-3.

17. Return the bags to their respective beakers immediately after weighing.

18. Repeat steps 15 to 17 at 40, 60, and 80 minutes from time zero.

At the end of the experiment, take the bags to the sink, cut open, pour the contents down the drain, and discard the bags in the wastebasket. Pour the contents of the beakers down the drain and wash them according to the instructions given on page ix.

Make a *qualitative* statement about what you observed.

_____

_____

_____

Table 4-3   Change in Weight as a Consequence of Osmosis

No.	Bag Contents/ Beaker Contents	Bag Weight (grams)				
		0 Min	20 Min	40 Min	60 Min	80 Min
1	distilled water/ distilled water	32.58				
2	10% sucrose/ distilled water	38.11				
3	20% sucrose/ distilled water					
4	40% sucrose/ distilled water					
5	distilled water/ 40% sucrose	got lighter				

Was the direction of net movement of water in bags 1 to 4 into or out of the bags?

_____

Which bag gained the most weight? Why?

_____

_____

III. DIFFERENTIAL PERMEABILITY OF MEMBRANES

(Starr, pp. 41–42)

Dialysis tubing is a differentially permeable material that provides a means to demonstrate the movement of substances through cellular membranes.

MATERIALS

Per student group (4):

- 1 25-cm length of dialysis tubing, soaking in dH_2O
- 2 10-cm pieces of string
- bottle of 1% soluble starch in 1% sodium sulfate (Na_2SO_4)
- dishpan half-filled with dH_2O
- 600-mL graduated beaker
- ring stand and funnel apparatus (Fig. 4-2)
- bottle of 1% albumin in 1% sodium chloride (NaCl)
- 8 test tubes
- test tube rack
- china marker
- 25-mL graduated cylinder
- iodine (I_2KI) solution in dropping bottle
- 2% barium chloride ($BaCl_2$) in dropping bottle
- 2% silver nitrate ($AgNO_3$) in dropping bottle
- biuret reagent in dropping bottle
- albustix reagent strips (optional)
- scissors

Per lab room:

- series of 4 test tubes in test tube rack demonstrating positive tests for starch, sulfate ion, chloride ion, protein

PROCEDURE   Work in groups of four for this experiment.

1. Obtain a 25-cm section of dialysis tubing that has been soaked in distilled water (dH_2O).

2. Fold over one end of the tubing and tie it securely with string to form a leakproof bag (Fig. 4-2).

3. Slip the open end of the bag over the stem of a funnel and fill the bag approximately half full with 25 mL of a solution of 1% soluble starch in 1% sodium sulfate (Na_2SO_4).

4. Remove the bag from the funnel; fold and tie the open end of the bag.

5. Rinse the tied bag in a dishpan partially filled with dH_2O.

6. Pour 250 mL of a solution of 1% albumin (a protein) in 1% sodium chloride (NaCl) into a 600-mL beaker.

7. Place the bag into the fluid in the beaker.

8. Record the time in this space:

_____

9. With a china marker, label eight test tubes, numbering them 1 through 8.

Table 4-4 Results of Tests for Substances in Beaker		
Contents of Beaker: (+) = Presence, (−) = Absence		
	At Start of Experiment	After 75 Minutes
Starch		−
Sulfate ion		−
Chloride ion		+
Albumin		+

Table 4-5 Results of Tests for Substances in Dialysis Bag		
Contents of Dialysis Bag: (+) = Presence, (−) = Absence		
	At Start of Experiment	After 75 Minutes
Starch	+	
Sulfate ion	+	
Chloride ion		−
Albumin		−

10. Seventy-five minutes after the start of the experiment, pour 20 mL of the *beaker contents* into a *clean* 25-mL graduated cylinder.

11. Decant (pour out) 5 mL from the graduated cylinder into each of the first four test tubes.

12. Perform the following tests, recording the results in Table 4-4. Your instructor will have a series of test tubes showing positive tests for starch, sulfate and chloride ions, and proteins. You should compare your results with the known positives.

a. *Test for starch.* Add several drops of iodine solution (I_2KI) from the dropper bottle to test tube 1. If starch is present, the solution will turn blue-black.

b. *Test for sulfate ion.* Add several drops of 2% barium chloride ($BaCl_2$) from the dropper bottle to test tube 2. If sulfate ions (SO_4^-) are present, a white precipitate of barium sulfate ($BaSO_4$) will form.

c. *Test for chloride ion.* Add several drops of 2% silver nitrate ($AgNO_3$) from the dropper bottle to test tube 3. A milky-white precipitate of silver chloride ($AgCl$) indicates the presence of chloride ions (Cl^-).

d. *Test for protein.* Add several drops of biuret reagent from the dropper bottle to test tube 4. If protein is present, the solution will change from blue to pinkish-violet. The more intense the violet hue, the greater the quantity of the protein.

An alternative method for determining the presence of protein is the use of albustix reagent strips. Presence of protein is indicated by green or blue-green coloration of the paper.

13. Wash the graduated cylinder, using the technique described on page x.

14. Thoroughly rinse the bag in the dishpan of dH_2O.

15. Using a scissors, cut the bag open and empty the contents into the 25-mL graduated cylinder.

16. Decant 5-mL samples into each of the four remaining test tubes.

17. Perform the tests for starch, sulfate ions, chloride ions, and protein on tubes 5 through 8, respectively.

18. Record the results of this series of tests in Table 4-5.

To which substances was the dialysis tubing permeable?

_____

What physical property of the dialysis tubing might explain its differential permeability?

_____

19. Discard contents of test tubes and beaker down sink drain. Wash glassware by technique described on page x.

20. Discard dialysis tubing in wastebasket.

IV. PLASMOLYSIS IN PLANT CELLS

(Starr, pp. 41–43)

Plant cells are surrounded by a rigid cell wall, composed primarily of the glucose polymer, cellulose. Recall from Exercise 3 that many plant cells have a large central vacuole surrounded by the vacuolar membrane. The vacuolar membrane is differentially permeable. Normally the solute concentration within the cell's central vacuole is greater than that of the external environment. Consequently, water moves into the cell, creating **turgor pressure**, which presses the cytoplasm against the cell wall. Such cells are said to be **turgid.** Many non-woody plants (like beans and peas) rely on turgor pressure to maintain their rigidity and erect stance.

This experiment demonstrates the effects of external solute concentration on the structure of plant cells.

MATERIALS

Per student:

- forceps
- 2 microscope slides
- 2 coverslips
- compound microscope

Per student group (table):

- *Elodea* in tap water
- 2 dropping bottles of distilled water (dH$_2$O)
- 2 dropping bottles of 20% sucrose

PROCEDURE

1. With a forceps, remove two young leaves from the tip of an *Elodea* plant.

2. Mount one leaf in a drop of distilled water on a microscope slide and the other in 20% sucrose solution on a second microscope slide.

3. Place coverslips over both leaves.

4. Observe the leaf in distilled water with the compound microscope. Focus first with the medium-power objective and then switch to the high-dry objective.

5. Label the photomicrograph of turgid cells (Fig. 4-3).

6. Now observe the leaf mounted in 20% sucrose solution. After several minutes the cell will have lost water, causing it to become **plasmolyzed**. (This process is called **plasmolysis**.) Label the plasmolyzed cells shown in Fig. 4-4.

Tonicity is a description of one solution's solute concentration compared to that of another solution.

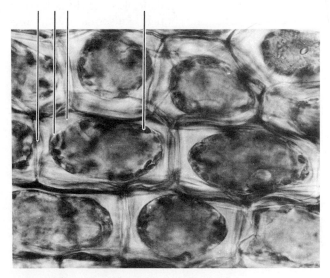

Labels: cell wall, chloroplasts in cytoplasm, plasma membrane, space (between cell wall and plasma membrane)

Figure 4-4 Plasmolyzed *Elodea* cells (650×). (Photo courtesy R. F. Evert.)

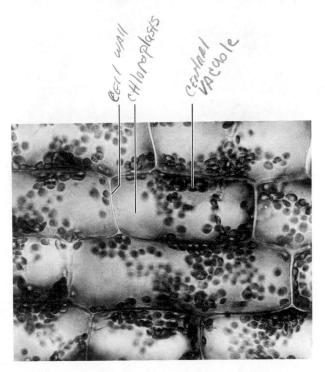

Labels: cell wall, chloroplasts in cytoplasm, central vacuole

Figure 4-3 Turgid *Elodea* cells (650×). (Photo courtesy R. F. Evert.)

A solution containing a lower concentration of solute molecules than another is **hypotonic** *relative to the second*. Solutions containing equal concentrations of solute are **isotonic** to each other, while one containing a greater concentration of solute relative to a second is **hypertonic**.

Were the contents of the vacuole in the *Elodea* leaf in distilled water hypotonic, isotonic, or hypertonic compared to the distilled water?

_____

Was the 20% sucrose solution hypertonic, isotonic, or hypotonic relative to the cytoplasm?

_____

If a hypotonic and a hypertonic solution are separated by a differentially permeable membrane, which direction will the water move?

_____

Name two differentially permeable membranes that are present within the *Elodea* cells and that were involved in the plasmolysis process.

1. _____

2. _____

V. OSMOTIC CHANGES IN RED BLOOD CELLS

(Starr, pp. 41–42)

Animal cells lack the rigid cell wall of a plant. The external boundary of an animal cell is the differentially permeable plasma membrane. Consequently, an animal cell increases in size as water enters the cell. However, since the plasma membrane is relatively fragile, it ruptures when too much water enters the cell. This is due to excessive pressure pushing out against the membrane. Conversely, if water moves out of the cell, it will become plasmolyzed and look spiny.

In this experiment, you will use red blood cells to demonstrate the effects of diffusion and osmosis in animal cells.

MATERIALS

Per student:

- compound microscope

Per student group (4):

- 3 clean screw-cap test tubes
- test tube rack
- metric ruler
- china marker
- bottle of 0.9% sodium chloride (NaCl)
- bottle of 10% NaCl
- bottle of distilled water (dH₂O)
- 3 disposable plastic pipets
- 3 clean microscope slides
- 3 coverslips

Per student group (table):

- bottle of sheep blood (in ice bath)

Per lab room:

- source of distilled water

PROCEDURE Work in groups of four for this experiment, but do the microscopic observations individually.

1. Observe the scanning electron micrographs in Fig. 4-5. Figure 4-5a illustrates the normal appearance of red blood cells. They are described as biconcave discs; that is, they are circular in outline with a depression in the center of both surfaces. Cells in an isotonic solution would appear like those in Fig. 4-5a.

Figure 4-5b shows cells that have been plasmolyzed. (In the case of red blood cells, plasmolysis is given a special term, *crenation*; the blood cell is said to be *crenate*.)

Figure 4-5c represents cells that have taken in water, but have not yet burst. (Burst red blood cells are said to be *hemolyzed*.) Note their swollen, spherical appearance.

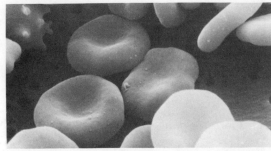

a Red blood cells in an isotonic solution ("normal")

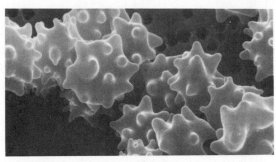

b Red blood cells in a hypertonic solution ("crenate")

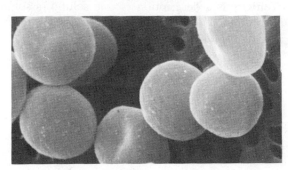

c Red blood cells in a hypotonic solution

Figure 4-5 Scanning electron micrographs of red blood cells. (Photos from M. Sheetz, R. Painter, and S. Singer. Reproduced from *The Journal of Cell Biology*, 1976, 70:193, by copyright permission of The Rockefeller University Press and M. Sheetz.)

2. Obtain three clean screw-cap test tubes.

3. Lay test tubes 1 and 2 against a metric ruler and mark lines indicating 5 cm *from the bottom of each tube.*

4. Fill each tube as follows:

Tube 1: 5 cm of 0.9% sodium chloride (NaCl)
       five drops of sheep blood

Tube 2: 5 cm of 10% NaCl
       five drops of sheep blood

5. Lay test tube 3 against a metric ruler, and mark lines indicating 0.5 cm and 5 cm *from the bottom of the tube.*

6. Fill tube 3 to the 0.5 cm mark with 0.9% NaCl, and to the 5 cm mark with distilled water. Then add five drops of sheep blood.

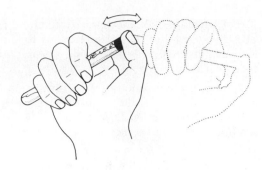

a

b

Figure 4-6 Methods for studying effects of different solute concentrations on red blood cells. (After Abramoff and Thomson, 1982.)

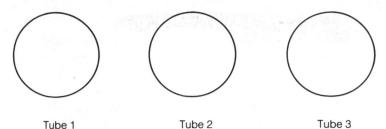

Tube 1   Tube 2   Tube 3

Figure 4-7 Microscopic appearance of red blood cells in different solute concentrations (_____×)

7. Replace the caps and mix the contents of each tube by inverting several times (Fig. 4-6a).

8. Hold each tube flat against the printed page of your lab manual (Fig. 4-6b). *Only if the blood cells are hemolyzed should you be able to read the print.*

9. Record your observations in the column marked "Print Visible?" of Table 4-6.

10. Number three clean microscope slides.

11. With three *separate* disposable pipets, remove a small amount of blood from each of the three tubes. Place one drop of blood from tube 1 on slide 1, one drop from tube 2 on slide 2, and one drop from tube 3 on slide 3.

12. Cover each drop of blood with a coverslip.

13. Observe the three slides with your compound microscope, focusing first with the medium power objective and finally with the high-dry objective. (Hemolyzed cells are virtually unrecognizable; all that remains are membranous "ghosts," which are difficult to see with the microscope.)

14. In Fig. 4-7 make a sketch of the cells from each tube. Label the sketches, indicating whether the cells are normal, plasmolyzed (crenate), or hemolyzed.

15. Record the microscopic appearance in Table 4-6.

16. Record in Table 4-6 the tonicity of the sodium chloride solutions you added to the test tubes.

Why do red blood cells burst when put in a hypotonic solution whereas *Elodea* leaf cells do not?

After completing all experiments, take your dirty glassware to the sink and wash it as directed on page x. Invert the test tubes in the test tube rack so they drain. Reorganize your work area, making certain all materials used in this exercise are present for the next class.

Table 4-6	Effect of Salt Solutions on Red Blood Cells				
Tube No.	Contents of Tube	Print Visible?	Microscopic Appearance of Cells	Tonicity of External Solution*	
1					
2					
3					

*With respect to that inside the red blood cell at the start of the experiment.

PRE-LAB QUESTIONS

_____ **1.** If one were to identify the most important compound for sustenance of life, it would probably be (a) salt, (b) $BaCl_2$, (c) water, (d) I_2KI.

_____ **2.** A solvent is (a) the substance in which solutes are dissolved, (b) a salt or sugar, (c) one component of a biological membrane, (d) differentially permeable.

_____ **3.** Diffusion (a) is an energy-requiring process, (b) is the movement of molecules from a region of higher concentration to one of lower concentration, (c) occurs only across differentially permeable membranes, (d) none of the above.

_____ **4.** Cellular membranes (a) consist of a phospholipid bilayer containing embedded proteins, (b) control the movement of substances into and out of cells, (c) are differentially permeable, (d) all of the above.

_____ **5.** An example of a solute would be (a) Janus green B, (b) water, (c) sucrose, (d) a and c above.

_____ **6.** Dialysis membrane (a) is differentially permeable, (b) is used in these experiments to simulate cellular membranes, (c) is permeable to water but not to sucrose, (d) all of the above.

_____ **7.** Specifically, osmosis (a) requires the expenditure of energy, (b) is diffusion of water from one region to another, (c) is diffusion of water across a differentially permeable membrane, (d) none of the above.

_____ **8.** Which of the following reagents does _not_ fit with the substrate being tested for? (a) biuret reagent–protein, (b) $BaCl_2$–starch, (c) $AgNO_3$–chloride ion, (d) albustix–protein.

_____ **9.** When the cytoplasm of a plant cell is pressed against the cell wall, the cell is said to be (a) turgid, (b) plasmolyzed, (c) hemolyzed, (d) crenate.

_____ **10.** If one solution contains 10% NaCl and another contains 30% NaCl, the 30% solution is _____ with respect to the 10% solution. (a) isotonic, (b) hypotonic, (c) hypertonic, (d) plasmolyzed.

ENZYMES: CATALYSTS OF LIFE

OBJECTIVES After completing this exercise you will be able to:

1. define catalyst, enzyme, activation energy, substrate, product, active site, denaturation, cofactor;

2. explain how an enzyme operates;

3. recognize benzoquinone as a brown substance formed in damaged plant tissue;

4. indicate the substrates for the enzyme catechol oxidase;

5. describe the effect of temperature on the rate of chemical reactions in general and on enzymatically controlled reactions in particular;

6. describe the effect that an atypical pH may have on enzyme action;

7. indicate how a cofactor might operate and identify a cofactor for catechol oxidase.

INTRODUCTION Life without enzymes is unimaginable. The energy required by your muscles simply to open your laboratory manual would take years to accumulate without enzymes. Due to the presence of enzymes, the myriad chemical reactions occurring in your cells at this very moment are being completed in a fraction of a second rather than the years or even decades that would be otherwise required.

Enzymes are proteins that function as biological catalysts. A **catalyst** is a substance that lowers the amount of energy necessary for a chemical reaction to proceed. You might think of this so-called **activation energy** as a hump to be negotiated. Enzymes decrease the size of the hump, in effect turning a mountain into a molehill (Fig. 5-1).

By lowering the activation energy, an enzyme affects the *rate* at which reaction occurs. Enzyme-boosted reactions may proceed from 100 thousand to 10 million times faster than they would without the enzyme.

In an enzyme-catalyzed reaction, the reactant (the substance being acted upon) is called the **substrate.** Substrate molecules combine with enzyme molecules

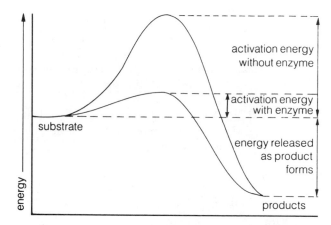

Figure 5-1 Enzymes and activation energy. (After Starr and Taggart, 1987.)

to form a temporary **enzyme-substrate complex. Products** are formed, and the enzyme molecule is released unchanged. Thus, the enzyme is not used up in the process and is capable of catalyzing the same reaction again and again. This can be summarized as follows:

$$\text{substrate} \xrightarrow{\text{enzyme}} \begin{array}{c}\text{enzyme-}\\\text{substrate}\\\text{complex}\end{array} \to \text{products} + \text{enzyme}$$

Although thousands of enzymes are present within cells, we will examine only one, catechol oxidase (also known as tyrosinase), to demonstrate the effects of several factors that influence enzyme action. These factors include:

1. temperature;

2. pH (hydrogen ion concentration of the environment);

3. specificity (how discriminating the enzyme is in catalyzing different potential substrates);

4. cofactor necessity (the need for a metallic ion for enzyme activity).

I. FORMATION AND DETECTION OF BENZOQUINONE

(Starr, p. 61)

Catechol oxidase is an enzyme that catalyzes the production of benzoquinone and water from catechol:

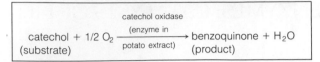

catechol $+\ 1/2\ O_2$ $\xrightarrow[\text{potato extract})]{\text{catechol oxidase}\\\text{(enzyme in}}$ benzoquinone $+\ H_2O$
(substrate)   (product)

This is an oxidation reaction, with catechol as the substrate. Hence, the enzyme gets its preferred name *catechol oxidase*.

Catechol and catechol oxidase are present in the cells of many plants, although in undamaged tissue they are separated in different compartments of the cells. Injury causes mixing of the substrate and enzyme, producing benzoquinone, a brown substance. You've probably noticed the brown coloration of a damaged apple or the blackening of an injured potato tuber. Benzoquinone inhibits the growth of certain microorganisms that cause rot.

In this experiment you will form the product, benzoquinone, and establish a color intensity scale to be used in subsequent experiments.

MATERIALS

Per student group (4):

- 3 test tubes
- test tube rack
- metric ruler
- ice bath with wash bottle of potato extract containing catechol oxidase
- wash bottle containing 1% catechol solution
- bottle of distilled water (dH_2O)
- china marker

Per lab room:

- 40°C waterbath
- vortex mixer (optional)

CAUTION   *Some of the chemicals (catechol, hydroquinone) used in these experiments may be hazardous to your health if they are ingested or taken in through your skin. Wear disposable plastic gloves for all experiments.*

PROCEDURE   Work in groups of four for all experiments in this exercise.

1. With a china marker, label three test tubes I_1, I_2, and I_3. Place your initials on each test tube for later identification.

2. Lay the test tubes against a metric ruler and mark lines on the tubes corresponding to 1 cm and 2 cm *from the bottom of each tube.*

3. Fill each tube as follows:

Tube I_1:
- 1 cm of potato extract containing catechol oxidase
- 1 cm of 1% catechol solution

Tube I_2:
- 1 cm of potato extract containing catechol oxidase
- 1 cm of dH_2O

Tube I_3:
- 1 cm of 1% catechol solution
- 1 cm of dH_2O

4. Shake all tubes (using a vortex mixer if your lab has one).

5. Record the color of the solution in each tube in the Time 0 spaces of Table 5-1.

6. Place the tubes in a 40°C waterbath.

7. At 5-minute intervals over the next 15 minutes, check the tubes and record the color of the solution of each in Table 5-1.

8. Remove the tubes from the waterbath and *save them for comparison of results of other experiments you will perform.*

Table 5-1 Formation and Detection of Benzoquinone			
Time (minutes)	Tube I_1 Potato Extract and Catechol	Tube I_2 Potato Extract and Water	Tube I_3 Catechol and Water
0			
5			
10			
15			

What is the brown-colored substance that appeared in test tube I_1?

_____

What was the substrate for the reaction that occurred in tube I_1?

_____

What was the product of the reaction in tube I_1?

_____

What substances lacking in tubes I_2 and I_3 account for the absence of the brown-colored substance?

I_2 _____ I_3 _____

What is the purpose of having tubes I_2 and I_3?

_____

After 15 minutes, the catechol should be completely oxidized. The color of the product in tube I_1 will be considered to be a "5" on a color intensity scale of 0 to 5, while the color of the substance in tubes I_2 and I_3 will be considered to be "0." You will use this scale to make comparisons in experiments II to V. Fill in Table 5-2.

Keep the contents in tubes I_1, I_2, and I_3, and refer to them in making comparisons in subsequent experiments.

Table 5-2 Color Intensity Scale		
Intensity	Tube No.	Color
0	I_2, I_3	
5	I_1	

II. ENZYME SPECIFICITY

(Starr, p. 62)

Generally, enzymes are substrate-specific, acting on one particular substrate or a small number of structurally similar substrates. This specificity is due to the three-dimensional structure of the enzyme. In order for the enzyme-substrate complex to form, the structure of the substrate must complement very closely that of the **active site** of the enzyme. The active site is a special region of the enzyme to which the substrate binds. Although the active site has a small amount of moldability, you can think of an enzyme as a key and the substrate as the particular lock into which it fits, as illustrated in Fig. 5-2.

This experiment demonstrates the ability of the enzyme catechol oxidase to catalyze the oxidation of two different but structurally similar substrates: catechol and hydroquinone.

Examine the chemical structure of each compound:

catechol   hydroquinone

You need not memorize these structural formulas, but do notice that both are ring structures with two hydroxyl (—OH) groups attached.

Keep this in mind as you do the next experiment, in which you will determine how specific (discriminating) catechol oxidase is for particular substrates.

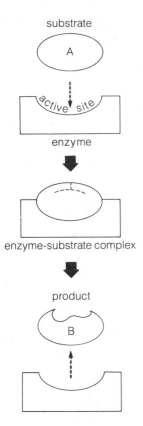

Figure 5-2 "Lock and key" fit of substrate (A) with enzyme. Substrate undergoes internal rearrangement to form product (B). (After Starr and Taggart, 1984.)

MATERIALS

Per student group (4):

- 2 test tubes
- test tube rack
- metric ruler
- china marker
- wash bottle containing 1% catechol
- wash bottle containing 1% hydroquinone
- ice bath with wash bottle of potato extract containing catechol oxidase

Per lab room:

- 40°C waterbath
- vortex mixer (optional)

PROCEDURE

1. With a china marker, label two clean test tubes II_1 and II_2. Include your initials for identification.

2. Lay the test tubes against a metric ruler and mark lines indicating 1 cm and 2 cm *from the bottom* of each test tube.

3. Fill each tube as follows:

Tube II₁:

- 1 cm of potato extract containing catechol oxidase
- 1 cm of 1% catechol

Tube II₂:

- 1 cm of potato extract containing catechol oxidase
- 1 cm of 1% hydroquinone

4. Gently shake the test tubes to mix the contents.

5. Compare the color intensity of the solution in each test tube *with the standards produced in experiment I* and record at Time 0 in Table 5-3.

6. Place the test tubes in a 40°C waterbath.

7. Examine the test tubes after 5 and 10 minutes, recording the color intensity (scale 0 to 5) of the contents of each in Table 5-3.

Table 5-3 Specificity of Catechol Oxidase		
	Relative Color Intensity on a Scale of 0 to 5	
Time (minutes)	Tube II₁: With Catechol	Tube II₂: With Hydroquinone
0		
5		
10		

Upon which substrate does catechol oxidase work best, forming the most benzoquinone in the shortest amount of time?

_____

_____

Based upon your knowledge of the structure of the two substrates, what apparently determines the specificity of catechol oxidase?

_____

_____

III. EFFECT OF TEMPERATURE ON ENZYME ACTIVITY   (Starr, p. 63)

The rate at which chemical reactions take place is largely determined by the temperature of the environment. *Generally, for every 10°C rise in temperature, the reaction rate doubles.* Within a rather narrow range, this is true for enzymatic reactions also. However, because enzymes are proteins, excessive temperature alters their structure, destroying their ability to function. When an enzyme's structure is changed sufficiently to destroy its function, the enzyme is said to be **denatured.** Most enzymatically controlled reactions have an **optimum** temperature and pH, that is, one temperature and pH where activity is maximized.

MATERIALS

Per student group (4):

- 6 test tubes
- test tube rack
- metric ruler
- china marker
- wash bottle containing 1% catechol
- ice bath with wash bottle of potato extract containing catechol oxidase
- 3 400-mL graduated beakers
- heat-resistant glove
- Centigrade thermometer

Per student group (table):

- 2 hotplates *or* burner, tripod support, wire gauze, and matches or striker
- boiling chips

Per lab room:

- source of room temperature water
- 40°C waterbath
- 60°C waterbath
- 80°C waterbath
- vortex mixer (optional)

PROCEDURE

1. Half fill one 400-mL beaker with tap water. Add a few boiling chips and turn on the hotplate to the highest temperature setting, *or*, if your lab is equipped with burners, light the burner. Bring the water to a boil, and then turn the heat down so that the water just continues to boil.

2. Put 150 mL of tap water into a second beaker, and then add ice to the water.

3. Half fill a third beaker with water from the source at room temperature.

4. With a china marker, label six test tubes III₁ through III₆. Include your initials for identification.

5. Lay the test tubes against a metric ruler and mark off lines indicating 1 cm and 2 cm *from the bottom* of each tube.

6. Fill each tube to the 1-cm mark with potato extract containing catechol oxidase.

Table 5-4   Effect of Temperature on Enzyme Action

Time (minutes)	Relative Color Intensity on a Scale of 0 to 5					
	Tube III$_1$ ___°C	Tube III$_2$ ___°C	Tube III$_3$ 40°C	Tube III$_4$ 60°C	Tube III$_5$ 80°C	Tube III$_6$ ___°C
0						
5						
10						
15						

7a. Place tube III$_1$ in the 400-mL beaker of ice water. Measure the water temperature and record here:

_____ °C.

b. Place tube III$_2$ in the 400 mL beaker containing room temperature water. Room temperature is

_____ °C.

c. Place tube III$_3$ in the 40°C waterbath.

d. Place tube III$_4$ in the 60°C waterbath.

e. Place tube III$_5$ in the 80°C waterbath.

f. Place tube III$_6$ in the 400-mL beaker containing boiling water. The boiling water is at

_____ °C.

8. Allow the test tubes to remain at the various temperatures for five minutes.

9. Remove the tubes and add catechol to the 2-cm line on each. Agitate the tubes (with a vortex mixer if available) to mix the contents.

CAUTION *Wear a Heat-Resistant Glove When Handling Heated Glassware.*

10. Record in Table 5-4 the relative color intensity (scale 0 to 5) of the solution in each tube, using the standard established in experiment I. Return each tube to its respective temperature bath immediately after recording.

11. Shake frequently (by hand) all tubes over the next 15 minutes, recording in Table 5-4 the relative color intensity at 5, 10, and 15 minutes after adding catechol.

12. Plot the data from Table 5-4 *for the 10-minute reading* on the graph in Fig. 5-3.

What is the temperature *range* over which catechol oxidase is active?

_____

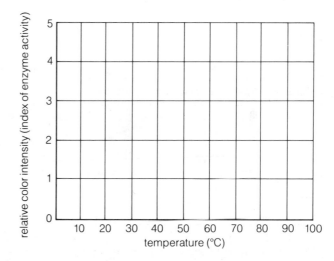

Figure 5-3   Effect of temperature on catechol oxidase activity

What is the *optimum* temperature for activity of this enzyme?

_____

What happens to enzyme activity at very high temperatures?

_____

IV. EFFECT OF pH ON ENZYME ACTIVITY   (Starr, p. 63)

Another factor influencing the rate of enzyme catalysis is the hydrogen ion concentration (pH) of the solution. pH, like temperature, affects the three-dimensional shape of enzymes, thus regulating their function. Most enzymes operate best when the pH of the solution is near neutrality (pH 7). Others, however, have pH optima in the acidic or basic range, corresponding to the environment in which they normally are found.

This experiment will allow you to determine the pH optimum of catechol oxidase.

Table 5-5 Effect of pH on Enzyme Activity

Time (minutes)	Relative Color Intensity on a Scale of 0 to 5						
	Tube IV$_1$ pH 2	Tube IV$_2$ pH 4	Tube IV$_3$ pH 6	Tube IV$_4$ pH 7	Tube IV$_5$ pH 8	Tube IV$_6$ pH 10	Tube IV$_7$ pH 12
0							
5							
10							
15							

MATERIALS

Per student group (4):

- 7 test tubes
- test tube rack
- metric ruler
- china marker
- wash bottle containing 1% catechol
- ice bath with wash bottle of potato extract containing catechol oxidase

Per lab room:

- 40°C waterbath
- phosphate buffer series, pH 2–12 (2, 4, 6, 7, 8, 10, 12)
- vortex mixer (optional)

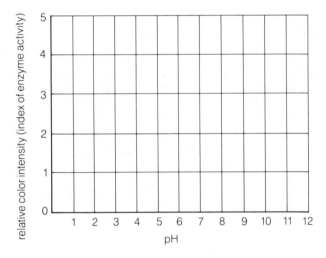

Figure 5-4 Effect of pH on catechol oxidase activity

PROCEDURE

1. With a china marker, label seven test tubes IV$_1$ through IV$_7$. Include your initials for identification.

2. Lay the test tubes against a metric ruler and mark lines indicating 4 cm, 5 cm, and 6 cm *from the bottom* of each tube.

3. Take your test tubes to the location of the phosphate buffer series and fill each tube according to the following directions:

Tube Number	Fill to the 4-cm Mark with Buffer of
1	pH 2
2	pH 4
3	pH 6
4	pH 7
5	pH 8
6	pH 10
7	pH 12

4. Return to your work area and add 1 cm of potato extract containing catechol oxidase to each of the seven tubes (thus bringing the total volume of each to the 5-cm mark). Agitate the tubes by hand.

5. Add 1% catechol to each of the seven tubes, bringing the total volume to the 6-cm mark. Agitate the contents of the tubes, using a vortex mixer if available.

6. Record in Table 5-5 at Time 0 the relative color intensity of each tube *immediately after adding the 1% catechol.*

7. Place the tubes in the 40°C waterbath.

8. Agitate the tubes frequently over the next 15 minutes. At 5-minute intervals, record in Table 5-5 the relative color intensity of each tube.

9. Plot the data from Table 5-5 *for your 10-minute reading* on the graph in Fig. 5-4.

What is the *range* of pH over which catechol oxidase catalyzes catechol to benzoquinone?

What is the *optimum* pH for activity of catechol oxidase?

V. NECESSITY OF COFACTORS FOR ENZYME ACTIVITY (Starr, p. 64)

Some enzymatic reactions occur only when the proper *cofactors* are present. **Cofactors** are metallic ions that are part of the structure of the active site, making possible the formation of the enzyme-substrate complex.

Catechol oxidase contains copper. In this experiment we will use phenylthiourea (PTU), which binds strongly to copper, to remove copper ions. Thus, we will be able to determine if copper is a cofactor necessary for producing benzoquinone from catechol.

MATERIALS

Per student group (4):

- 2 test tubes
- test tube rack
- metric ruler
- china marker
- ice bath with wash bottle of potato extract containing catechol oxidase
- wash bottle containing 1% catechol solution
- bottle of distilled water (dH$_2$O)
- china marker
- scoopula (small spoon)
- phenylthiourea crystals in small screw-cap bottle

Per lab room:

- 40°C waterbath
- vortex mixer (optional)
- bottle of 95% ethanol (at each sink)
- tissues (at each sink)

PROCEDURE

1. With a china marker, label two test tubes V$_1$ and V$_2$. Include your initials for identification.

2. Lay the test tube against a metric ruler and mark lines indicating 1 cm and 2 cm *from the bottom* of each tube.

3. Add potato extract containing catechol oxidase to the 1-cm mark of each test tube.

4. Using a scoopula, add five crystals of phenylthiourea (PTU) to tube V$_1$. Do not add anything to tube V$_2$.

CAUTION *PTU Is Poisonous.*

5. Agitate frequently by hand the contents of both test tubes during the next five minutes.

6. Add 1% catechol to the 2-cm mark of both test tubes, and agitate the contents of the tubes (with a vortex mixer if available). Record at Time 0 in Table 5-6 the relative color intensities (scale of 0 to 5).

7. Place the tubes in a 40°C waterbath. Agitate the tubes several times during the next 10 minutes.

8. Remove the tubes from the water bath and compare their relative color intensities. Record your observations in Table 5-6.

Table 5-6	Is Copper a Cofactor for Catechol Oxidase?	
Time (minutes)	Relative Color Intensity on a Scale of 0 to 5	
	Tube V$_1$: With PTU	Tube V$_2$: Without PTU
0		
10		

Did benzoquinone form in tube V$_1$?

In tube V$_2$?

From this experiment, what can you conclude about the necessity for copper for catechol oxidase activity?

After completing all experiments, take your dirty glassware to the sink and wash it following directions on page x. Use 95% ethanol to remove the china marker. Invert the test tubes in the test tube racks so they drain. Tidy up your work area, making certain all materials used in this exercise are there for the next class.

PRE-LAB QUESTIONS

_____ 1. Enzymes are (a) biological catalysts, (b) agents that speed up cellular reactions, (c) proteins, (d) all of the above.

2. Enzymes function by (a) being consumed (used up) in the reaction, (b) lowering the activation energy of a reaction, (c) combining with otherwise toxic substances in the cell, (d) adding heat to the cell to speed up the reaction.

3. The substance that an enzyme combines with is (a) another enzyme, (b) a cofactor, (c) a coenzyme, (d) the substrate.

4. Enzyme specificity refers to the (a) need for cofactors for the function of some enzymes, (b) fact that a single enzyme catalyzes only one particular reaction, (c) effect of temperature on enzyme activity, (d) effect of pH on enzyme activity.

5. For every 10°C rise in temperature, the rate of most chemical reactions will (a) double, (b) triple, (c) increase by 100 times, (d) stop.

6. When an enzyme becomes denatured, it (a) increases in effectiveness, (b) loses its requirement for a cofactor, (c) forms an enzyme-substrate complex, (d) loses its function.

7. An enzyme may lose its ability to function because of (a) excessively high temperatures, (b) a change in its three-dimensional structure, (c) a large change in the pH of the environment, (d) all of the above.

8. pH is a measure of (a) an enzyme's effectiveness, (b) enzyme concentration, (c) the hydrogen ion concentration, (d) none of the above.

9. Catechol oxidase (a) is an enzyme found in potatoes, (b) catalyzes the production of catechol, (c) has as its substrate benzoquinone, (d) is a substance that encourages the growth of microorganisms.

10. The relative color intensity used in the experiments of this exercise (a) is a consequence of production of benzoquinone; (b) is an index of enzyme activity; (c) may differ depending on the pH, temperature, or presence of cofactors, respectively; (d) all of the above.

EXERCISE 5
ENZYMES: CATALYSTS OF LIFE

POST-LAB QUESTIONS

1. Explain the effect of very high temperatures on the structure of an enzyme.

2. Explain what happens to catechol oxidase when the pH is on either side of the optimum.

3. As you demonstrated in this experiment, high temperatures inactivate catechol oxidase. How is it that some bacteria and algae live in the hot springs of Yellowstone Park at temperatures as high as 73°C?

4. What would you expect the pH optimum to be for an enzyme secreted into your stomach?

5. Why do you think high fevers alter cellular functions?

6. Some surgical procedures involve lowering a patient's body temperature during periods when blood flow must be restricted. What effect might this have on enzyme-controlled cellular metabolism?

7. At one time it was believed that individuals who had been submerged under water for longer than several minutes could not be resuscitated. Recently this has been shown to be false, especially if the person was in cold water. Explain why cold-water "drowning" victims might survive prolonged periods under water.

8. Is it necessary to have one enzyme molecule for every substrate molecule that needs to be catalyzed? Why or why not?

9. How does the induced fit model of enzyme-substrate interaction differ from the lock and key model?

10. Name the substrate for the following enzymes:

   a. urease _____

   b. sucrase _____

   c. DNA polymerase _____

   d. ATPase _____

PHOTOSYNTHESIS: CAPTURE OF LIGHT ENERGY

OBJECTIVES After completing this exercise you will be able to:

1. define photosynthesis, autotroph, heterotroph, chromatogram, chlorophyll, carotenoid;

2. describe the role of carbon dioxide in photosynthesis;

3. determine the effect of white light on the rate of photosynthesis;

4. determine the wavelengths absorbed by pigments;

5. identify the pigments in spinach chloroplast extract;

6. identify the carbohydrate produced in geranium leaves during photosynthesis;

7. identify the structures composing the chloroplast and indicate the function of each structure in photosynthesis.

INTRODUCTION **Photosynthesis,** the process by which light energy converts inorganic compounds to organic substances with the subsequent release of elemental oxygen, may very well be the most important biological event sustaining life. Without it, most living things would starve, and atmospheric oxygen would become depleted to a level incapable of supporting animal life. Ultimately, the source of light energy is the sun, although on a small scale we may substitute artificial light.

Nutritionally, two types of organisms exist in our world, autotrophs and heterotrophs. **Autotrophs** (*auto* means self, *troph* means feeding) synthesize organic molecules (carbohydrates) from inorganic carbon dioxide. The vast majority of autotrophs are photosynthetic organisms with which you are familiar—plants, as well as some protistans. These organisms use light energy to produce carbohydrates. (A few bacteria are able to produce their organic carbon compounds chemosynthetically, i.e., using chemical energy.)

By contrast, **heterotrophs** must rely directly or indirectly on autotrophs for their nutritional carbon

and metabolic energy. Heterotrophs include animals, fungi, many protistans, and most bacteria.

In both autotrophs and heterotrophs, carbohydrates originally produced by photosynthesis are broken down by *carbohydrate metabolism* (sometimes known as *cellular respiration;* Exercise 7), releasing the energy captured from the sun for metabolic needs.

The photosynthetic reaction can be conveniently summarized by the equation

$$2H_2O + CO_2 \xrightarrow[\text{enzymes chlorophyll}]{\text{light energy}} (CH_2O) + H_2O + O_2$$

(water) (carbon dioxide) (carbohydrate) (water) (oxygen)

If the carbohydrate produced is glucose ($C_6H_{12}O_6$), the reaction becomes

$$12H_2O + 6CO_2 \xrightarrow[\text{enzymes chlorophyll}]{\text{light energy}} C_6H_{12}O_6 + 6H_2O + 6O_2$$

Although glucose is often produced during photosynthesis, unless it is to be used immediately for carbohydrate metabolism, it is usually converted to another storage compound. In plants and many protistans, the most common storage carbohydrate is *starch*, a compound made up of numerous glucose units linked together. Starch is designated by the chemical formula $(C_6H_{12}O_6)_n$, where n indicates a large number.

The following experiments will acquaint you with the principles of photosynthesis.

I. ROLE OF LIGHT AND CARBON DIOXIDE IN PHOTOSYNTHESIS: LEAF DISK ASSAY

(Starr, pp. 68–78)

Work in groups of four.

During this experiment the overhead lights in the room will be turned off. Flood lamps will be used as light sources.

*Adapted with permission from Steucek, G. L., R. J. Hill, and Class/Summer 1982, 1985. *Am. Biol. Teacher* 471:96–99.

As photosynthesis proceeds, oxygen accumulates in the intercellular spaces of a plant. This floating leaf disk assay for photosynthesis utilizes the *rate* at which oxygen is produced or consumed as a measure of the rate at which photosynthesis is occurring as a whole. In this experiment, you will first *remove* that oxygen by vacuum infiltration, replacing it with a liquid. Thus, although leaf disks initially float, after infiltration they sink to the bottom of the test tube.

As photosynthesis takes place within the leaf disks, they rise once again.

MATERIALS

Per student group (4):

- 2 20-by-150-mm test tubes
- test tube rack
- china marker
- 5-mL pipet
- infiltration solution
- paper punch
- 15-cm wooden rod
- 2 25-mL graduated cylinders
- bottle of sodium bicarbonate ($NaHCO_3$) solution
- heat absorber (large beaker or chromatography chamber)
- floodlamp

Per lab room:

- several actively growing plants
- water aspirator set up at each sink
- source of distilled water (dH_2O)

PROCEDURE

1. Obtain two clean 20-by-150-mm test tubes, and number them using a china marker.

2. Into each test tube pipet 3.0 mL of "infiltration solution."

3. Using a paper punch, cut 20 leaf disks from recently expanded, young leaves of the plant provided, allowing them to fall onto a clean sheet of paper. Without delay, place 10 leaf disks in tube 1 and 10 disks into tube 2. The disks will float.

4. Now you will replace the intercellular air within the leaf disks with infiltration solution. Draw a vacuum on the contents of each test tube by attaching the tube to the water aspirator at the sink (Fig. 6-1). If the vacuum is very strong, the liquid in the tubes will "boil." (If the leaf disks are forced onto the side of the tube above the liquid level, break the vacuum and use a wooden rod to push the disks back into the liquid.)

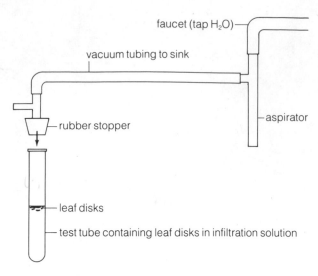

Figure 6-1 Method for vacuum infiltrating leaf disks

5. Evacuate the tubes three times for about 30 seconds each time, allowing a fresh charge of air to enter the tube after each evacuation. The disks should be sunken at this time. If not, repeat this step until they are.

6. Using two 25-mL graduated cylinders, measure out 15 mL of sodium bicarbonate ($NaHCO_3$) solution from the stock bottle into one cylinder, and 15 mL of distilled water (dH_2O) into the other. Sodium bicarbonate is a source of carbon dioxide.

7. Now carefully decant (pour off) the infiltration solution into the sink and *quickly* add the 15 mL of $NaHCO_3$ to test tube 1 and the 15 mL of dH_2O to test tube 2. Swirl the test tubes to mix the contents.

8. Place the test tubes in a test tube rack behind a fluid-filled heat absorber (large beaker or chromatography chamber) and turn on the floodlamp (Fig. 6-2).

CAUTION   *Separate the floodlamp bulb and heat absorber by 30 cm to avoid cracking the glass of the chamber.*

---

9. At 1-minute intervals for 20 minutes, count the number of leaf disks that are floating. *Swirl the contents of the test tubes at the end of each 1-minute interval* (after counting) so that all leaf disks are temporarily suspended in the vortex. The time required for a leaf disk to float is an index of the rate of photosynthesis in that leaf. Some disks will be "early floaters," others "late floaters." Record your data in Table 6-1.

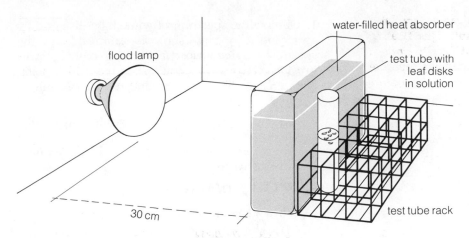

Figure 6-2 Arrangement of floodlamp, heat absorber, and test tube rack for photosynthesis assay

Table 6-1	Data from Floating Leaf Disk Assay Experiment						
	Tube 1 (NaHCO$_3$)			Tube 2 (dH$_2$O)		Tube 1 (NaHCO$_3$)			Tube 2 (dH$_2$O)	
ET*	NDF**	%***		NDF	%	ET*	NDF**	%***	NDF	%
1						11				
2						12				
3						13				
4						14				
5						15				
6						16				
7						17				
8						18				
9						19				
10						20				

  * = Elapsed time (minutes)
 ** = Number disks floating
*** = NDF/10 × 100%

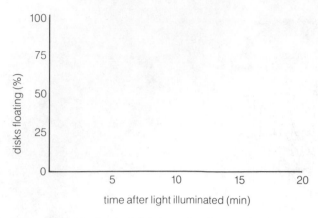

Figure 6-3 Floating leaf disk assay for photosynthesis

10. On the graph in Fig. 6-3, plot the data you accumulated in Table 6-1. Use a "+" for Tube 1 and an "O" for Tube 2.

What would you expect to happen if the floating leaf disks were placed in the dark? Explain.

II. ABSORPTION OF LIGHT BY CHLOROPLAST EXTRACT   (Starr, pp. 71–73)

We tend to think of sunlight as being white. However, as you will see in this experiment, white light consists of a continuum of wavelengths. If we see light of just one wavelength, that light will appear colored.

When light hits a pigmented surface, some of the wavelengths are absorbed and others are reflected or transmitted. This experiment demonstrates *which* wavelengths are absorbed, transmitted, or reflected by *particular* pigments, among them the plant pigment chlorophyll.

MATERIALS

Per lab room:

- several spectroscope setups (Fig. 6-4)
- sets of colored pencils (violet, blue, green, yellow, orange, red)
- colored filters (blue, green, red)
- small test tube containing pigment extract

PROCEDURE   Work alone for this experiment.

One means of separating light into its component parts is by viewing the light through a spectroscope. The spectroscope contains a prism that causes the formation of a spectrum of colors. A nanometer scale is imposed upon the spectrum to indicate the wavelength of each component of white light.

1. Observe the spectrum of white light given off by an incandescent bulb (showcase lamp) through the spectroscope (see Fig. 6-4). With the colored pencils provided, record the positions of the colors violet, blue, green, yellow, orange, and red on the scale of Fig. 6-5.

2. Slide the three colored filters into the clips of the spectroscope one at a time.

Which color(s) is (are) absorbed when a red filter is placed between the light and the prism?

Violet, orange

When a blue filter is used?

Red, orange

A green filter?

Violet, blue

Make a general statement concerning the color of a pigment (filter) and the absorption of light by that pigment.

3. Now obtain a small test tube containing spinach chloroplast pigment extract and place it in the clips of the spectroscope as shown in Fig. 6-4. By adjusting the height of the tube so that the upper portion of the light passes through the pigment extract and the lower portion is white light, you can compare the absorption spectrum of the pigment extract with the spectrum of white light. An **absorption spectrum** is a spectrum of light waves absorbed by a particular pigment.

4. Again using the colored pencils, record the absorption spectrum of the chloroplast extract on the scale in Fig. 6-6.

How does the absorption spectrum of the chloroplast extract compare with the absorption spectrum of the green filter?

How might you explain the difference in absorption by the green filter and by the chloroplast pigment extract? (You might want to do Part III before answering this question.)

III. SEPARATION OF PHOTOSYNTHETIC PIGMENTS BY PAPER CHROMATOGRAPHY   (Starr, pp. 71–73)

Paper chromatography allows substances to be separated from one another on the basis of their physical characteristics. A chloroplast pigment extract has been prepared for you by soaking spinach leaves in

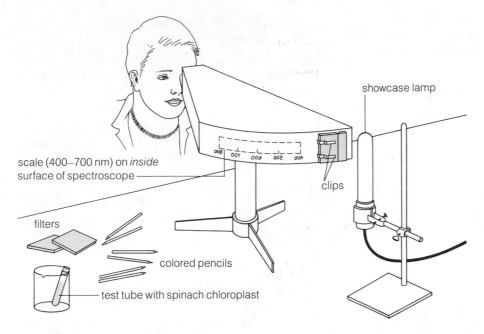

scale (400–700 nm) on *inside*
surface of spectroscope

showcase lamp

clips

filters

colored pencils

test tube with spinach chloroplast

Figure 6-4 Use of a spectroscope. (After Abramoff and Thomson, 1982.)

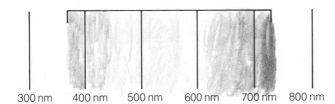

300 nm   400 nm   500 nm   600 nm   700 nm   800 nm

Figure 6-5 Spectrum of white light

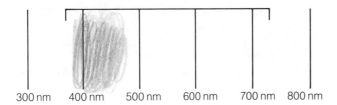

300 nm   400 nm   500 nm   600 nm   700 nm   800 nm

Figure 6-6 Absorption spectrum of chloroplast extract

cold acetone and ethanol. Although the extract appears green, other pigments present may be masked by the chlorophyll. In this experiment, you will use paper chromatography to separate any pigments present. Separation occurs due to the solubility of the pigment in the chromatography solvent and the affinity of the pigments for absorption to the paper surface. The finished product, showing separated pigments, is called a **chromatogram.**

MATERIALS

Per student:

- chromatography paper, 3-by-15-cm sheet
- metric ruler

Per student pair:

- chloroplast pigment extract in foil-wrapped dropping bottle
- chromatography chamber containing solvent
- colored pencils (green, blue-green, yellow, orange)

PROCEDURE

1. Obtain a 3-by-15-cm sheet of chromatography paper. *Touch only the edges of the paper,* because oil from your fingers may interfere with development of the chromatogram.

2. Using a ruler, make a *pencil* line (do *not* use ink) about 2 cm from the bottom of the paper.

3. Load the paper by applying a droplet of the chloroplast pigment extract near the center of the pencil line. Allow the pigment spot to dry for about 30 seconds. Several applications of extract on the same spot are necessary to get enough pigment for a good chromatogram. Be certain to allow the pigment to air-dry between applications.

4. Insert your "loaded" chromatography paper, spot side down, into a chromatography chamber, a bottle containing a solvent consisting of 10% acetone in petroleum ether. The level of the solvent should cover the bottom of the strip but no portion of the pigment spot. Seal the chromatography chamber and allow the solvent to rise on the paper.

CAUTION *Avoid inhaling the solvent vapors. Keep the chamber tightly capped whenever possible.*

5. Watch the separation take place over the next 10 minutes. When the solvent is within about 1 cm of the top of the paper, the separation is complete. Remove the strip, close the chromatography chamber, and allow the chromatogram to dry.

6. Using colored pencils, record the results as a sketch in Fig. 6-7, showing the relative position of the colors along the paper. Beginning nearest the original pigment spot, identify and label the yellow-green **chlorophyll b.** Moving upward, find the blue-green **chlorophyll a,** two yellow-orange **xanthophylls** in the middle, and an orange **carotene** at the top. Xanthophylls and carotenes belong to the class of pigments called **carotenoids.**

You may preserve your chromatogram for future reference by keeping it in a dark place (e.g., between the pages of your textbook). Light causes the chromatogram to fade.

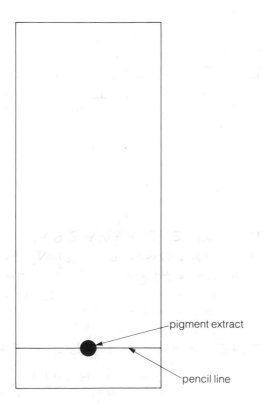

Labels: chlorophyll b, chlorophyll a, xanthophyll, carotene

Figure 6-7 Chloroplast pigment chromatogram

What pigments are contained within the chloroplasts of spinach leaves?

_____

What common "vegetable" is particularly high in carotenes?

_____

(Have you ever heard of a medical condition called "carotenosis"? If not, go to the library to look up this term in a medical dictionary.)

IV. RELATIONSHIP BETWEEN LIGHT AND PHOTOSYNTHETIC PRODUCTS   (Starr, pp. 69–70)

As indicated by the overall formula of photosynthesis, one end product is a carbohydrate, CH_2O. But a number of different carbohydrates have the empirical formula CH_2O. In this experiment, you will perform a test to determine the specific carbohydrate stored in photosynthesizing geranium leaves.

MATERIALS

Per student group (4):

- china marker
- 4 400-mL graduated beakers
- hotplate
- heat-resistant glove
- 2 20-by-150-mm test tubes
- 25-mL graduated cylinder
- iodine (I_2KI) solution in dropping bottle
- bottle of starch solution
- bottle of 95% ethanol
- forceps
- petri dish
- I_2KI solution in foil-wrapped stock bottle

Per lab room:

- source of distilled water (dH_2O)
- light-grown geranium plant or leaves of geranium plant
- dark-grown geranium plant or leaves of geranium plant (kept in dark place)

PROCEDURE Work in groups of four.

1. With a china marker, label two 400-mL beakers, one "L" (for leaf kept in light), the other "D" (for leaf placed in dark). Add 150 mL of tap water to each beaker, set them on a hotplate, and turn on the hotplate to the highest setting. Allow the water to come to a boil. (You will use the boiling water a bit later.)

2. Obtain two clean 20-by-150-mm test tubes. With a graduated cylinder, measure out 5 mL of distilled water, add it to one test tube, and then add three drops of iodine (I_2KI) solution.

3. Measure out 5 mL of starch solution, add it to the second test tube, and then add three drops of I_2KI.

4. Record the results of this test in Table 6-2.

Table 6-2 Reaction of I_2KI with Water and Starch	
Solution	Color after Addition of I_2KI
Water	
Starch	

What does the blue-black color indicate?

_____

_____

5. Select a leaf from each of two geranium plants. One plant has been growing in bright light for several hours; the other has been kept in the dark for a day or more. Both leaves should have had an area of the lamina (blade) masked by an opaque design.

6. Remove the opaque cover. Place each leaf for about 30 seconds in the appropriately labeled beaker of boiling water. Using a heat-resistant glove, remove the beaker from the hotplate and pour the water down the drain.

7. Pour about 100 mL of 95% ethanol into each of the two beakers, place on the hotplate, and bring the alcohol to a gentle boil.

_____

CAUTION *Ethanol is highly flammable. Use only electric hotplates, never open flame.*

_____

8. When the pigments have been extracted (one to two minutes), remove the leaves from the alcohol with forceps and place them in two appropriately labeled petri dishes containing water.

9. After 60 seconds, gently pour the water off into one of the beakers and flood the leaves with I_2KI solution *from the larger stock bottle.*

10. After several minutes, pour the I_2KI into the sink, rinse the leaves in cold water, and observe the pattern of staining. Show the distribution of the stain by shading in and labeling Fig. 6-8. In the blank provided in the legend for Fig. 6-8, record the *substance* that I_2KI stains.

What does the blue-black coloration of the leaf indicate?

_____

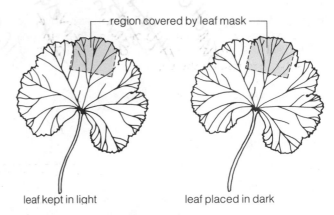

region covered by leaf mask

leaf kept in light   leaf placed in dark

Figure 6-8 Distribution of the photosynthetic product _____

_____

_____

Why did the masked area fail to stain?

_____

_____

V. STRUCTURE OF THE CHLOROPLAST
(Starr, pp. 70–71)

Work individually.

The chloroplast is the organelle concerned with photosynthesis. Study Fig. 6-9, an artist's conception of the three-dimensional structure of a chloroplast.

Like the mitochondrion and the nucleus, the chloroplast is surrounded by two membranes, comprising the chloroplast **envelope.** Within the **stroma** (semifluid matrix), identify the **thylakoid disks** stacked into **grana** (a single stack is a **granum**). The chloroplast pigment molecules are located on the surface of the thylakoid disks. It is within the interior of the disks where sunlight energy is trapped and ATP is

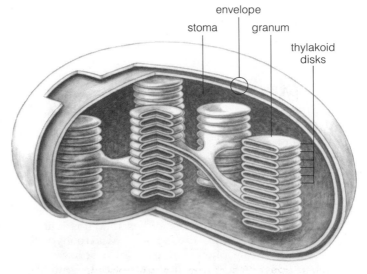

envelope
stoma   granum
thylakoid disks

Figure 6-9 The arrangement of membranes and compartments inside a chloroplast. (After Wolfe, 1985.)

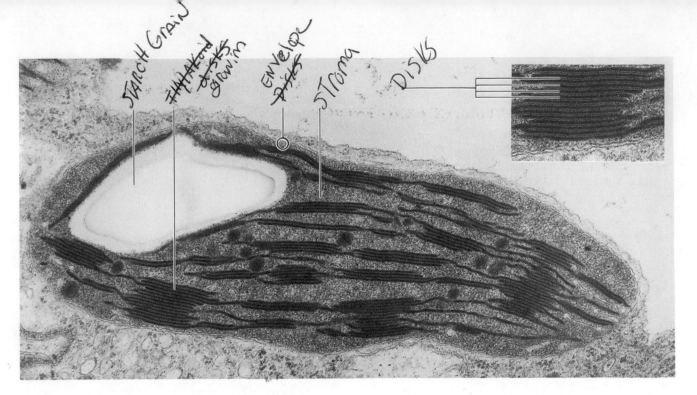

(handwritten labels on micrograph: Starch Grain, thylakoid disks granum, envelope, stroma, disks)

Labels: envelope, thylakoid disks, granum, stroma, starch grain

Figure 6-10 Electron micrograph of chloroplast. Inset: a single granum. (Photo courtesy R. R. Dute.)

formed. The ATP is transported to the stroma, where it is used to generate organic compounds. These compounds may be converted to carbohydrates, lipids, and amino acids from carbon dioxide, water, and other raw materials.

Now examine Fig. 6-10, a high-magnification electron micrograph of a chloroplast. With the aid of Fig. 6-9, label the electron micrograph.

If the plant is killed and fixed for electron microscopy after being exposed to strong light, the chloroplasts will contain **starch grains.** Note the large starch grain present in this chloroplast. (Starch grains appear as ellipsoidal white structures in electron micrographs.)

PRE-LAB QUESTIONS

_____ **1.** The raw materials used for photosynthesis include (a) O_2, (b) $C_6H_{12}O_6$, (c) $CO_2 + H_2O$, (d) CH_2O.

_____ **2.** A device useful for viewing the spectrum of light is a (a) spectroscope, (b) volumeter, (c) chromatogram, (d) chloroplast.

_____ **3.** Which of the following is *not* true of the floating leaf disk assay for photosynthesis? (a) it is a direct means for measuring the amount of carbohydrate produced during photosynthesis; (b) it utilizes oxygen production as an indication of photosynthesis; (c) the number of floating leaves is an indication of the rate of photosynthesis; (d) infiltration solution is

used to replace the intercellular oxygen before photosynthesis is measured.

_____ **4.** A paper chromatogram is useful for (a) measuring the amount of photosynthesis, (b) determining the amount of gas evolved during photosynthesis, (c) separating pigments based upon their physical characteristics, (d) determining the distribution of chlorophyll in a leaf.

_____ **5.** Which of the following pigments would you find in a geranium leaf? (a) chlorophyll, xanthophyll, phycobilins; (b) chlorophyll a, chlorophyll b, carotenoids; (c) phycocyanin, xanthophyll, fucoxanthin; (d) carotenoids, chlorophylls, phycoerythrin.

_____ **6.** Which reagent would you use to determine the distribution of the carbohydrate stored in leaves? (a) starch, (b) Benedict's solution, (c) chlorophyll, (d) I_2KI.

_____ **7.** An example of a heterotrophic organism is (a) a plant, (b) a geranium, (c) a human, (d) none of the above.

_____ **8.** Organisms capable of producing their own food are known as (a) autotrophs, (b) heterotrophs, (c) omnivores, (d) herbivores.

_____ **9.** Grana are (a) the same as starch grains, (b) the site of ATP production within chloroplasts, (c) part of the chloroplast envelope, (d) contained within mitochondria and nuclei.

_____ **10.** The ultimate source of energy trapped during photosynthesis is (a) CO_2, (b) H_2O, (c) O_2, (d) sunlight.

RESPIRATION: ENERGY CONVERSION

OBJECTIVES After completing this exercise you will be able to:

1. define metabolism, reaction, metabolic pathway, respiration, ATP, phosphorylation, exergonic reaction, endergonic reaction, glycolysis, Krebs cycle;

2. give the overall balanced equations for aerobic respiration and alcoholic fermentation;

3. distinguish between the products and efficiency of aerobic respiration and fermentation;

4. describe the structure and list the functions of the mitochondrion;

5. explain the relationship between photosynthesis and respiration.

INTRODUCTION The first law of thermodynamics states that energy can neither be created nor destroyed, only converted from one form to another. Because all living organisms have a constant energy requirement, they have mechanisms to gather, store, and use energy. Collectively, these mechanisms are called **metabolism.** A specific metabolic step is a **reaction,** and a sequence of such reactions a **metabolic pathway.**

During Exercise 6, we investigated the metabolic pathways by which green plants capture light energy and use it to make carbohydrates such as glucose. Carbohydrates are temporary energy stores. The process by which energy stored in carbohydrates is released to the cell is **respiration.**

The energy needed for living processes is stored in the chemical bonds holding carbohydrate atoms together. However, the cell cannot *directly* use the chemical bond energy of carbohydrates. Rather, the energy must be converted by a metabolic pathway to form **adenosine triphosphate (ATP),** the so-called *universal energy currency* of the cell. The bond energy of carbohydrates is transferred to ATP during **phosphorylation,** the addition of a phosphate group to adenosine diphosphate (ADP). When the bond holding this new phosphate group is broken during respiration,

the energy released is available for a great variety of cellular reactions. Thus, the needs of a cell are linked by the energy (ATP)releasing **exergonic reactions** of respiration to the energy (ATP)-requiring **endergonic reactions.** This last group of reactions is important for maintaining or synthesizing cellular structures, or doing cellular work.

Both autotrophs and heterotrophs undergo respiration. Autotrophs utilize the carbohydrates they have produced to build new cells and maintain cellular machinery. Heterotrophic organisms may obtain materials for respiration two ways: by digesting plant material or by digesting the tissues of animals that have previously digested plants.

There are four categories of respiration: (1) **aerobic respiration,** an oxygen-dependent pathway common in most organisms; (2) **anaerobic electron transport,** a pathway utilized by some bacteria; (3) **alcoholic fermentation,** an ethanol-producing process occurring in some yeasts; and (4) **lactate fermentation,** a pathway taken by some animal cells that normally rely on aerobic respiration, but which are subjected to oxygen-deficient conditions. (During strenuous activity, your skeletal muscles may switch to lactate fermentation for energy.)

Perhaps the most important aspect to remember about these four processes is that aerobic respiration is by far the most energy-efficient. **Efficiency** refers to the amount of energy captured in the form of ATP relative to the amount available within the bonds of the carbohydrate.

For aerobic respiration, the general equation is:

$$C_6H_{12}O_6 + 6O_2 \xrightarrow{\text{enzymes}} 6CO_2 + 6H_2O + 36ATP*$$

glucose   oxygen   carbon   water   chemical
                             dioxide   energy

---

*Depending upon the tissue, as many as 38 ATP may be found.

If glucose is broken down completely to CO_2 and H_2O, about 686,000 calories of energy are released. Each ATP molecule represents about 7,500 calories of usable energy. The 36 ATP represent 270,000 calories of energy (36 × 7,500 calories). Thus, aerobic respiration is about 39% efficient [(270,000/686,000) × 100%].

By contrast, fermentation and anaerobic electron transport yield only 2 ATP. Thus, these processes are only about 2% efficient [(2 × 7,500/686,000) × 100%]. Obviously, breaking down carbohydrates by aerobic respiration gives a bigger payback than the other means.

Regardless of the specific respiratory pathway an organism uses, one series of metabolic steps occurs: **glycolysis.** The word *glycolysis* should be a tip-off concerning what happens during this process. The Greek word *glykos* means "sweet," referring to sugar, while *lysis* means "loosening." During glycolysis the 6-carbon sugar glucose, $C_6H_{12}O_6$, is split into two 3-carbon pyruvate molecules. This universal event occurs within the cytoplasm of all living cells, whether bacteria, protistans, fungi, plants, or animals, including humans. The net energy yield from glycolysis is 2 ATP per molecule of glucose.

I. AEROBIC RESPIRATION (Starr, pp. 80–86)

The fate of pyruvate molecules produced by glycolysis depends upon the organism and environmental conditions. If oxygen is abundant and the organism normally undergoes aerobic respiration, pyruvate is further metabolized by a cyclic (circular) pathway known as the **Krebs cycle,** which generates a small amount of ATP and releases CO_2. For the most part, the Krebs cycle functions to reduce (donate electrons to) special electron carriers. These electron carriers eventually become oxidized (lose electrons) during **electron transport phosphorylation,** where large amounts of ATP are produced. The Krebs cycle and electron transport phosphorylation occur in the mitochondrion.

Why must oxygen be present? Whenever one substance is oxidized (loses electrons), another must be reduced (accept, or gain, those electrons). The final electron acceptor of electron transport phosphorylation is oxygen. Tagging along with the electrons as they pass through the electron transport process are protons (H^+). When the electrons and protons are captured by oxygen, water (H_2O) is formed:

$$2H^+ + 2e^- + 1/2O_2 \rightarrow H_2O$$

In the following experiments we examine aerobic respiration in two sets of seeds.

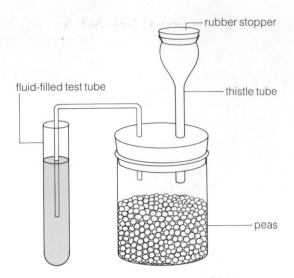

Figure 7-1 Respiration bottle apparatus

A. Carbon Dioxide Production

Seeds contain stored food material in the form of some carbohydrate. When a seed germinates, the carbohydrate is broken down, liberating energy (ATP) needed for growth of the enclosed embryo into a seedling.

Yesterday one set of dry pea seeds was soaked in water to start the germination process. Another set was not soaked. This experiment will compare carbon dioxide production between germinating pea seeds, germinating pea seeds that have been boiled, and ungerminated (dry) pea seeds.

MATERIALS

Per student group (4):

- 600-mL beaker
- hot plate *or* burner, wire gauze, tripod, and matches
- heat-resistant glove
- respiration bottle apparatus (Fig. 7-1)
- china marker
- phenol red solution

Per lab room:

- germinating pea seeds
- ungerminated (dry) pea seeds

PROCEDURE Work in groups of four.

1. Place about 250 mL of tap water in a 600-mL beaker, put the beaker on a heat source, and bring the water to a boil.

2. Obtain three respiration bottle setups (Fig. 7-1). With a china marker, label one "Germ" for germinating pea seeds, the second "Germ-Boil" for those you will boil, and the third "Ungerm" for ungerminated seeds.

3. From the class supply, obtain and put enough germinating pea seeds in the two appropriately labeled respiration bottles to fill them approximately halfway. Fill the third bottle half full with ungerminated (dry) pea seeds.

4. Dump the germinating peas from the "Germ-Boil" bottle into the boiling water bath; continue to boil for 5 minutes. After 5 minutes, turn off the heat source, put on a heat-resistant glove, and remove the water bath. Pour the water off into the sink and cool the boiled peas by pouring cold water into the beaker. Allow 5 minutes for the peas to cool, then pour off the water. Now replace the peas into the "Germ-Boil" respiration bottle.

5. Fit the rubber stopper with attached glass tubes into the respiration bottles. Add enough water to the test tube to cover the end of the glass tubing that comes out of the respiration bottle. (This keeps gases from escaping from the respiration bottle.)

6. Insert rubber stoppers into the thistle tubes.

7. Set the three bottles aside for the next hour and a quarter and do the other experiments in this exercise.

Now start the next series of experiments while you allow this one to proceed.

8. After 1¼ hours, pour the water in each test tube into the sink and replace it with an equal volume of dilute phenol red solution. Phenol red solution, which should appear pinkish in the stock bottle, will be used to test for the presence of carbon dioxide (CO_2) within the respiration bottles. If CO_2 is bubbled through water, carbonic acid (H_2CO_3) forms, as shown by the following equation:

$$CO_2 + H_2O \rightarrow H_2CO_3$$

Phenol red solution is mostly water. When the phenol red solution is basic (pH > 7), it is pink; when it is acidic (pH < 7), the solution is yellow. The phenol red solution in the stock bottle is

_____ (color);

therefore, the stock solution is

_____ (acidic/basic).

9. Put several hundred mL of tap water in the 600-mL beaker.

10. Remove the stopper plugging the top of the thistle tube and *slowly* pour water from the beaker into each thistle tube. The water will force out gases present within the bottles. If CO_2 is present, the phenol red will become yellow.

11. Record your observations in Table 7-1.

Table 7-1 CO_2 Evolution by Pea Seeds		
Pea Seeds	Indicator color (phenol red)	Conclusion (CO_2 present or absent)
Germinating-unboiled	_____	_____
Germinating-boiled	_____	_____
Ungerminated	_____	_____

Which set of seeds was undergoing respiration?

_____

What happened during boiling that caused the results you found? (Hint: enzymes.)

_____

B. Oxygen Consumption

One set of pea seeds has been soaked in water for the past 24 hours to initiate germination. In this experiment you will measure the respiratory rate of germinating and ungerminated seeds as determined by oxygen consumption.

MATERIALS

Per student group (4):

- volumeter (Fig. 7-2)
- china marker
- 80 germinating pea seeds
- 80 ungerminated (dry) pea seeds
- glass beads
- nonabsorbent cotton
- metric ruler
- bottle of potassium hydroxide (KOH) pellets
- 1/4 teaspoon measure
- marker fluid in dropping bottle

PROCEDURE Work in groups of four.

1. Obtain a volumeter set up as in Fig. 7-2. Skip to step 6 if your instructor has already assembled the volumeters as described by steps 2 through 5.

2. Remove the test tubes from the volumeter. With a china marker, number the tubes and then fill as follows:

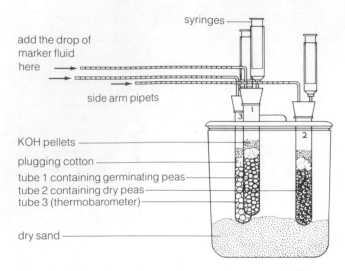

syringes

add the drop of
marker fluid
here →

side arm pipets

KOH pellets

plugging cotton

tube 1 containing germinating peas
tube 2 containing dry peas
tube 3 (thermobarometer)

dry sand

Figure 7-2 Volumeter

Tube 1: 80 germinating (soaked) pea seeds

Tube 2: 80 ungerminated (dry) pea seeds plus enough glass beads to bring the total volume equal to that of Tube 1

Tube 3: Enough glass beads to equal volume of Tube 1. This tube serves as a thermobarometer and is used to correct experimental reading to account for changes in temperature and barometric pressure taking place during the experiment.

3. Pack cotton *loosely* into each tube to a thickness of about 1.5 cm above the peas/beads.

4. Measure out 1 cubic centimeter (cm³) (about 1/4 teaspoon) of potassium hydroxide (KOH) pellets and pour them atop the cotton.

_____

_____

CAUTION *Potassium hydroxide can cause burns. Do not get any on your skin or clothing. If you do, wash immediately with copious amounts of water.*

_____

KOH absorbs carbon dioxide (CO_2) given off during aerobic respiration. Since the volumeter measures change in gas volume, any gas *given off* during respiration must be removed from the tube so an accurate measure of O_2 consumption can be made.

5. Insert the stopper-syringe assembly in place.

6. Add a small drop of marker fluid to each side arm pipet by touching the dropper to the end of each. The drop should be taken into the side arm by capillary action. Gently withdraw the plunger of each syringe and adjust the position of the drop so it is between 0.80 and 0.90 cm³ on the scale of the graduated pipet.

7. Adjust each side arm pipet so it is parallel to the table top. Wait 5 minutes before starting data collection.

8. At time 0, record in Table 7-2 the position of the marker droplet within each pipet. Record readings for each tube at 5-minute intervals for the next 60 minutes. (If respiration is rapid and the marker moves too near the end of the scale, *carefully* depress the syringe to readjust its position. The new readings are then added to the old readings as the data are being recorded.)

9. To determine total change in volume of gas within each tube, subtract each subsequent reading from the first (time 0).

10. At the end of the experiment, correct for any volume changes caused by changes in temperature or barometric pressure by using the reading obtained from the thermobarometer. If the thermobarometric marker moves *toward* the test tube (decrease in volume), *subtract* the volume change from the last total oxygen consumption measurement of the pea-containing test tubes. If the marker droplet moves *away* from the test tube (increase in volume), *add* the volume change to the last total oxygen consumption measurement for each pea-containing test tube.

How do the respiratory rates for germinating and nongerminating seeds compare?

_____

_____

How do you account for this difference?

_____

_____

It takes 820 cm³ of oxygen to completely oxidize 1 gram of glucose. How much glucose are the germinating peas consuming per hour?

_____

II. FERMENTATION (Starr, pp. 87–88)

Despite relatively low energy yield, fermentation provides sufficient energy for certain organisms to survive. Alcoholic fermentation by yeast is the basis for the brewing industry. It's been said that yeast and alcoholic fermentation have made Milwaukee famous.

The chemical equation for this process is:

$$C_6H_{12}O_6 \xrightarrow{\text{enzymes}} 2\ CH_3CH_2OH + 2\ CO_2 + 2\ ATP$$

glucose   ethanol   carbon   energy
                                     dioxide

Table 7-2   Respiratory Rate as Measured by Oxygen Consumption

	Thermobarometer			Tube 1: Germinating Peas			Tube 2: Dry Peas		
Time (minutes)	Reading	Total Change In Volume		Reading	Total Change In Volume	Total Oxygen Consumption	Reading	Total Change In Volume	Total Oxygen Consumption
0									
5									
10									
15									
20									
25									
30									
35									
40									
45									
50									
55									
60									

Starch (amylose), a common storage carbohydrate in plants, is a polymer consisting of a chain of repeating glucose ($C_6H_{12}O_6$) units. The polymer has the chemical formula $(C_6H_{12}O_6)_n$,* where n represents a large number. Starch is broken down by the enzyme amylase into individual glucose units. To summarize:

$$(C_6H_{12}O_6)_n \xrightarrow{\text{amylase}} C_6H_{12}O_6 + C_6H_{12}O_6 + C_6H_{12}O_6 + \ldots$$
$$\text{starch} \qquad\qquad\qquad \text{glucose}$$

This experiment demonstrates the action of yeast cells on carbohydrates.

MATERIALS

Per student group (4):

- china marker
- 3 50-mL beakers
- 25-mL graduated cylinder
- bottle of 10% glucose
- bottle of 1% starch

- 0.5% amylase in bottle fitted with graduated pipet
- 3 glass stirring rods
- 1/4 teaspoon measure (optional)
- 3 fermentation tubes
- 15-cm metric ruler

Per lab room:

- 0.5-gram pieces of fresh yeast cake *or* a bottle of dry yeast
- scale and weighing paper (optional)
- 37°C incubator

PROCEDURE   Work in groups of four.

1. Using a china marker, number three 50-mL beakers.

2. With a *clean* 25-mL graduated cylinder, measure out and pour 15 mL† of the following solutions into each beaker:

---

*A number of carbohydrates share this same chemical formula but differ slightly in the arrangement of their atoms. These carbohydrates are called structural isomers.

†The amount of fluid needed to fill the fermentation tube depends upon its size. Your instructor may indicate the required volume.

Beaker 1: 15 mL of 10% glucose
Beaker 2: 15 mL of 1% starch
Beaker 3: 15 mL of 1% starch; then, using the graduated pipet to measure, add 5 mL of 0.5% *amylase.*

3. To each beaker add a 0.5-gram piece of fresh cake yeast *or* 1/4 teaspoon dry yeast. Stir with *separate* glass stirring rods.

4. When each is thoroughly mixed, pour the contents into three correspondingly numbered fermentation tubes (Fig. 7-3). Cover the opening of the fermenta-

tion tube with your thumb and invert each fermentation tube so that the "tail" portion is filled with the solution.

5. Place the tubes in a 37°C incubator.

6. At intervals of 20, 40, and 60 minutes after the start of the experiment, remove the tubes and, using a metric ruler, measure the distance from the tip of the tail to the fluid level. Record your results in Table 7-3. Calculate the volume of gas evolved using the formula at the bottom of Table 7-3. (If time is short, do your calculations later.)

What gas accumulates in the tail portion of the fermentation tube?

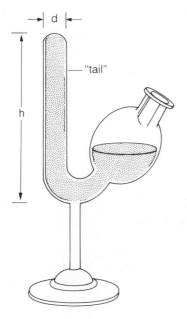

Figure 7-3 Fermentation tube

III. ULTRASTRUCTURE OF THE MITOCHONDRION (Starr, p. 84)

Study Fig. 7-4, an artist's interpretation of the three-dimensional structure of a mitochondrion.

Now observe Fig. 7-5, a high magnification electron micrograph of the mitochondrion. Identify and label the **outer membrane, outer compartment, inner membrane** (which is folded into **cristae;** singular is **crista**), and **inner compartment** (matrix).

After completing all experiments, take your dirty glassware to the sink and wash it following the directions given in "Instructions for Washing Laboratory Glassware," p. ix. Invert the test tubes in the test tube racks so that they drain. Tidy up your work area, making certain all equipment used in this exercise is there for the next class.

Table 7-3 Evolution of Gas by Yeast Cells					
Tube	Solution	Distance from tip of tube to fluid level (mm)			Volume* of gas evolved (mm³)
		20 min	40 min	60 min	
1	10% glucose + yeast				
2	1% starch + yeast				
3	1% starch + yeast + amylase				

*To calculate the volume of gas evolved, use the following equation: $V = \pi r^2 h$, where $\pi = 3.14$, r = radius of tail of fermentation tube ($r = \frac{1}{2}d$), h = distance from top of tail to level of solution.

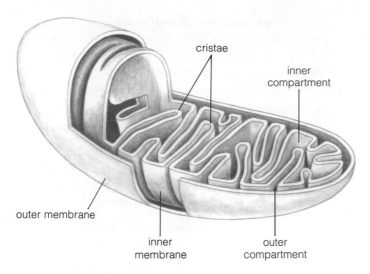

Figure 7-4 The membranes and compartments of a mitochondrion. (After Wolfe, 1985.)

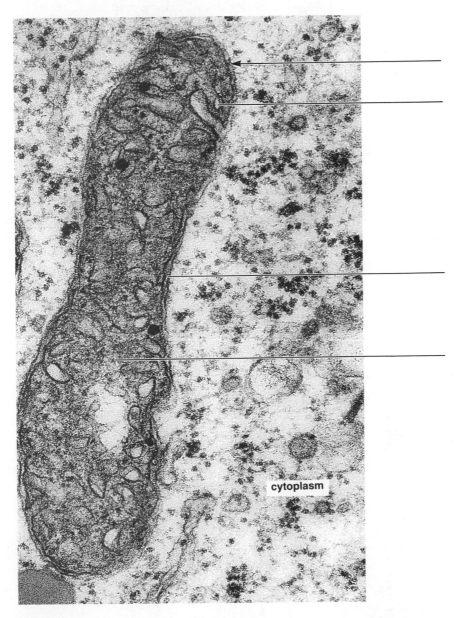

cytoplasm

Labels: outer membrane, outer compartment, crista, inner compartment

Figure 7-5 Transmission electron micrograph of a mitochondrion (23,250×). (Photo courtesy S. E. Eichhorn.)

PRE-LAB QUESTIONS

___ 1.   A metabolic pathway is a (a) single, specific reaction that starts with one compound and ends up with another, (b) sequence of chemical reactions that are part of the metabolic process, (c) series of events that occurs only in autotrophs, (d) all of the above.

___ 2.   The "universal energy currency" of the cell is (a) glucose, (b) $C_6H_{12}O_6$, (c) ATP, (d) H_2O.

___ 3.   The most efficient type of respiration is (a) fermentation, (b) anaerobic electron transport, (c) aerobic respiration, (d) lactate production.

___ 4.   When muscle cells are subjected to periods of strenuous activity, during which oxygen is not replaced within the cells as fast as it's used, the muscle cells switch from (a) lactate fermentation to aerobic recpiration, (b) lactate fermentation to alcoholic fermentation, (c) alcoholic fermentation to lactate fermentation, (d) aerobic respiration to lactate fermentation.

___ 5.   The purpose of the thermobarometer in a volumeter is to (a) judge the amount of O_2 evolved during respiration, (b) determine the volume changes as a result of respiration, (c) indicate oxygen consumption by germinating pea seeds, (d) indicate volume changes resulting from changes in temperature or barometric pressure.

___ 6.   Phenol red is used in the experiments as (a) an O_2 indicator, (b) a CO_2 indicator, (c) a sugar indicator, (d) an enzyme.

___ 7.   If the pH of a phenol red solution is 2, (a) phenol red will be pink, (b) the solution is acidic, (c) phenol red will be yellow, (d) b and c above.

___ 8.   Which of the following enzymes breaks down starch into glucose? (a) kinase, (b) maltase, (c) fructase, (d) amylase.

___ 9.   Oxygen is necessary for life because (a) photosynthesis depends upon it, (b) it serves as the final electron acceptor during aerobic respiration, (c) it is necessary for glycolysis, (d) all of the above.

___ 10.   Yeast cells produce (a) ATP, (b) ethanol, (c) CO_2, (d) all of the above.O

EXERCISE 7
RESPIRATION: ENERGY CONVERSION

POST-LAB QUESTIONS

1. During aerobic respiration, glucose ($C_6H_{12}O_6$) is broken down to form several end products. Which end products contain:

    a. the carbon atoms from glucose? _____

    b. the hydrogen atoms from glucose? _____

    c. the oxygen atoms from glucose? _____

    d. the energy stored in the glucose molecules? _____

2. Compare aerobic respiration, anaerobic electron transport, and fermentation in terms of:

    a. efficiency of obtaining energy from glucose

    b. end products

3. Besides the product formed in the alcoholic fermentation experiment, what other products might be formed by other types of fermentation?

4. How much ATP is derived when one molecule of glucose goes through glycolysis? _____
   What does this tell you about the source of ATP obtained during fermentation and anaerobic electron transport pathways?

5. Re-examine Fig. 7-5, the electron micrograph of the mitochondrion.

   a. Where does the Krebs cycle occur? _____

   b. Where does electron transport phosphorylation occur? _____

   c. Does glycolysis occur within this organelle? _____ If not, where does it occur? _____

6. How would you explain this statement: "The ultimate source of our energy is the sun"?

7. Oxygen is used during aerobic respiration. How is that oxygen returned to the atmosphere?

8. Carbohydrates that we eat in plant and animal tissue must be broken down before they are used for respiration. Where within our bodies are the enzymes located that accomplish that breakdown?

9. The first law of thermodynamics seems to conflict with what we know about ourselves. For example, after strenuous exercise we run out of "energy." We must eat to replenish our energy stores. Where has that energy gone? What form has it taken?

10. As a plant grows, not all of the carbohydrates produced by photosynthesis are stored as starch, nor are they all respired to produce ATP. What cellular constituents would be synthesized by some of the carbohydrate produced?

MITOSIS AND CYTOKINESIS: NUCLEAR AND CYTOPLASMIC DIVISION

OBJECTIVES   After completing this exercise you will be able to:

1. define fertilization, zygote, fission, DNA, chromosome, mitosis, cytokinesis, chromatid, nucleoprotein, centromere, meristem, derivative, aster, astral spindle, centriole;

2. identify the stages of the cell cycle;

3. distinguish between mitosis and cytokinesis as they take place in animal and plant cells;

4. identify the structures involved in nuclear and cell division and describe the role each plays.

INTRODUCTION   "All cells arise from preexisting cells." This is one tenet of the cell theory. It is easy to understand this concept when thinking of a single-celled *Amoeba* or bacterium. Each cell divides to give rise to two entirely new individuals. But it is enormously fascinating that each of us began life as **one** single cell and developed into an astonishingly complex animal.

In higher plants and animals, **fertilization,** the fusion of egg and sperm nuclei, produces a single-celled **zygote.** The zygote divides into two cells, these two into four, and so on to produce a multicellular organism. During cell division each new cell receives a complete set of hereditary information *and* an assortment of cytoplasmic components.

Recall from Exercise 3 that there are two basic cell types, prokaryotic and eukaryotic. Because of their simple genetic material, prokaryotes reproduce primarily as a result of **fission,** the splitting of a preexisting cell into two, with each new cell receiving a full complement of the genetic material.

In eukaryotes, the process of cell division is more complex, primarily because of the much more complex nature of the hereditary material, **DNA (deoxyribonucleic acid)** and the proteins complexed to it. In these cells, the genetic material is organized into **chromosomes.** Cell division usually involves two processes: **mitosis (nuclear division)** and **cytokinesis** (cytoplasmic division). Whereas mitosis results in the production of two nuclei, both containing identical chromosomes, cytokinesis ensures that each new cell contains all the metabolic machinery necessary for sustenance of life.

Dividing cells pass through a regular sequence of events called the cell cycle (Fig. 8-1). Notice that the majority of the time is spent in interphase and that actual nuclear division—mitosis—is but a brief portion of the cycle.

Interphase is broken into three parts (Fig. 8-1): the G_1 period, during which cytoplasmic growth takes place; the S period, when the DNA is duplicated; and the G_2 period, when structures directly involved in mitosis are synthesized.

Unfortunately, because of the apparent relative inactivity that early microscopists observed, interphase was given the misnomer "resting stage." In fact, we know now that interphase is anything but a resting period. The cell is producing new DNA, assembling proteins from amino acids, synthesizing or breaking down carbohydrates. In short, interphase is a very busy time in the life of a cell.

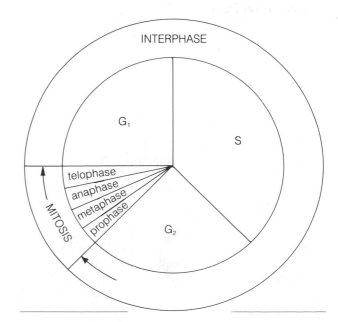

Figure 8-1   The eukaryotic cell cycle

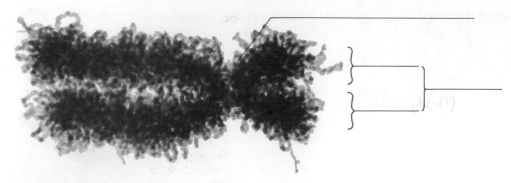

Labels: chromatid (2), centromere, duplicated chromosome

Figure 8-2 Electron micrograph of a chromosome. (Photo by E. J. DuPraw.)

I. CHROMOSOMAL ULTRASTRUCTURE

(Starr, pp. 94–95)

During much of a cell's life, each DNA-protein complex, the **nucleoprotein,** is extended as a thin strand within the nucleus. In this form it is called **chromatin.** Prior to the onset of nuclear division, the genetic material duplicates itself, and the chromatin condenses. These two identical condensed nucleoproteins are called **sister chromatids** and are attached at the **centromere.** The centromere gives the appearance of dividing each chromatid into two "arms." Collectively, the two attached sister chromatids are referred to as a duplicated **chromosome.**

Examine Fig. 8-2, an electron micrograph of a human chromosome. Label the two chromatids, centromere, and the duplicated chromosome.

NOTE *Use illustrations in your textbook to aid you in the following study.*

II. THE CELL CYCLE IN PLANT CELLS: ONION ROOTS

(Starr, pp. 97–101)

Nuclear and cell divisions in plants are, for the most part, localized in specialized regions called meristems. **Meristems** are regions of active growth.

A meristem contains cells that have the capability to divide repeatedly. Each division results in two cells. One of these, the **derivative,** eventually differentiates (becomes specialized for a particular function), generally losing its ability to divide again.* The other cell, however, remains meristematic and eventually divides again. This process, summarized by Fig. 8-3, accounts for the unlimited or prolonged growth of vegetative (nonreproductive) plant meristems.

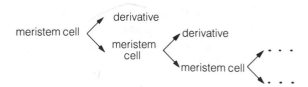

Figure 8-3 Cell division in plant meristems

Plants have two types of meristems: apical and lateral. Apical meristems are found at the tips of plant organs (shoots and roots) and increase length. Lateral meristems, located beneath the bark of woody plants, increase girth.

MATERIALS

Per student:

- prepared slide of onion (*Allium*) root tip mitosis
- compound microscope

PROCEDURE Obtain a prepared slide of a longitudinal section of an *Allium* (onion) root tip. This slide has been prepared from the terminal several millimeters of an actively growing root. It was "fixed" (killed) by chemicals to preserve the cellular structure and stained with dyes that have an affinity for the structures involved in nuclear division.

Focus first with the low power objective of your compound microscope to get an overall impression of the root's morphology.

Concentrate your study in the region about 1 mm behind the actual tip. This region is the apical meristem of the root (Fig. 8-4).

A. Interphase and Mitosis

1. Interphase. Use the medium power objective to scan the apical meristem. Note that most of the nuclei are in interphase.

---

*In some instances, recently formed derivatives can also divide, increasing the number of derivatives.

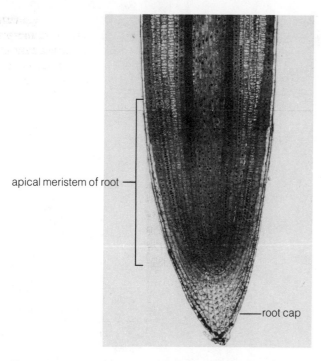

apical meristem of root

root cap

Figure 8-4 Root tip, l.s. (70×). (Photo courtesy J. W. Perry and R. F. Evert.)

Switch to the high-dry objective, focusing on a single interphase cell. Note the distinct **nucleus,** with one or more **nucleoli,** and the **chromatin** dispersed within the bounds of the **nuclear envelope.** Label these features in cell 1 of Fig. 8-5.

2. Mitosis.

a. *Prophase.* During **prophase** the chromatin condenses, rendering the chromosomes visible as thread-like structures. At the same time, microtubules outside the nucleus are beginning to assemble into **spindle fibers.** Collectively, the spindle fibers make up the **spindle,** a three-dimensional structure widest in the middle and tapering to a point at the two **poles** (opposite ends of the cell). *You will not see the spindle during prophase.*

Find a nucleus in prophase. Draw and label a prophase nucleus in cell 2 of Fig. 8-5.

The transition from prophase to metaphase is marked by the fragmentation and disappearance of the nuclear envelope. At about the same time the nucleoli disappear.

b. *Metaphase.* When the nuclear envelope is no longer distinct, the cell is in **metaphase.** Identify a metaphase cell by locating a cell with the duplicated chromosomes, each consisting of two **chromatids,** lined up midway between the two poles. This imaginary midline is called the **spindle equator.** (You will not be able to distinguish the chromatids.) The spindle has moved into the space the nucleus once occupied. The microtubules have become attached to the chromosomes within the centromere region. Find a cell in metaphase. Label cell number 3 of Fig. 8-5.

c. *Anaphase.* During **anaphase** the sister chromatids separate, each moving toward an opposite pole.

Find an early anaphase cell, recognizable by the slightly separated chromatids. Notice that the chromatids begin separating at the centromere. The last point of contact before separation is complete is at the ends of the "arms" of each chromatid. Although incompletely understood, the mechanism of chromatid separation is based upon action of the spindle-fiber microtubules. Once separated, each chromatid is referred to as an individual chromosome. Note that now the chromosome consists of a *single* chromatid.

Find a later anaphase cell and draw it in cell 4 of Fig. 8-5.

d. *Telophase.* When the chromosomes (formerly sister chromatids) arrive at opposite poles, the cell is in **telophase.** The spindle disorganizes. The chromosomes expand again, and a nuclear envelope reforms around each newly formed daughter nucleus.

Find a telophase cell and label individual chromosomes, nuclei, and nuclear envelopes on cell 5 of Fig. 8-5.

B. Cytokinesis in Onion Cells

Cytokinesis, division of the cytoplasm, usually follows mitosis. In fact, it often overlaps with telophase. Find a cell undergoing cytokinesis in the onion root tip. In plants, cytokinesis takes place by **cell plate formation** (Fig. 8-6). During this process Golgi body-derived vesicles migrate to the spindle equator, where they fuse. Their contents contribute to the formation of a new cell wall, and their membranes make up the new plasma membrane. In most plants, cell plate formation starts in the *middle* of the cell.

Find a cell undergoing cytokinesis. With your light microscope, the developing **cell plate** appears as a line running horizontally between the two newly formed nuclei. Return to cell 5 of Fig. 8-5 and label the developing cell plate.

Recently divided cells are often easy to distinguish by their square, boxy appearance. Find two recently divided **daughter cells;** then draw and label their contents in cell 6 of Fig. 8-5. Include cytoplasm, nuclei, nucleoli, nuclear envelopes, and chromatin. What is the difference between chromatin and chromosomes?

_____

_____

_____

Following cytokinesis, the cell undergoes a period of growth and enlargement, during which time the nucleus is in interphase. Interphase may be followed by another mitosis and cytokinesis, or in some cells interphase may persist for the rest of a cell's life.

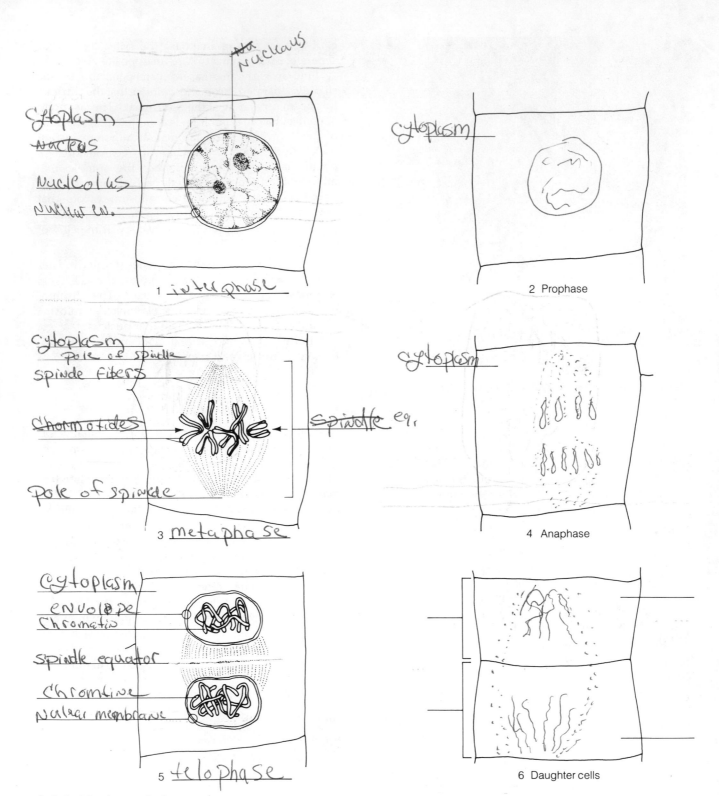

1 interphase

Cytoplasm
Nucleus
Nucleolus
Nuclear en.
nucleus

2 Prophase

cytoplasm

3 metaphase

Cytoplasm
Pole of spindle
spindle fibers
Chromatides
spindle eq.
Pole of spindle

4 Anaphase

cytoplasm

5 telophase

Cytoplasm
envolope
Chromatin
spindle equator
Chromline
Nuclear membrane

6 Daughter cells

Labels: interphase, cytoplasm, nucleus, nucleolus, chromatin, nuclear envelope, metaphase, spindle fibers, spindle, pole, spindle equator (between arrows), chromatids, telophase and cell plate formation, chromosome, cell plate, daughter cell (Note: some terms are used more than once.)

Figure 8-5 Interphase, mitosis, and cytokinesis in onion root tip cells. (After H. Clark, 1937.)

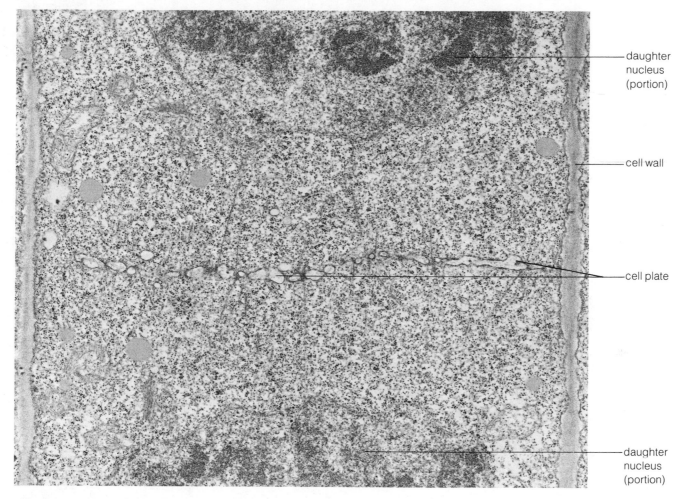

Figure labels:
- daughter nucleus (portion)
- cell wall
- cell plate
- daughter nucleus (portion)

Figure 8-6 Transmission electron micrograph of cytokinesis by cell plate formation in a plant cell. (Photo by W. P. Wergin, courtesy E. H. Newcomb.)

III. THE CELL CYCLE IN ANIMAL CELLS: WHITEFISH BLASTULA (Starr, pp. 97–101)

Fertilization of an ovum by a sperm produces a zygote. In animal cells, the zygote undergoes a special type of cell division (*cleavage*) in which no increase in cytoplasm occurs between divisions. A ball of cells called a blastula is produced by cleavage. Within the blastula, repeated nuclear and cytoplasmic divisions take place; consequently, the whitefish blastula is an excellent example in which to observe the cell cycle of an animal.

Note a difference between plants and animals: Whereas plants have meristems where divisions continually take place, animals do not have specialized regions to which mitosis and cytokinesis are limited. Indeed, divisions occur continually throughout an animal's body, replacing worn-out cells.

With several important exceptions, mitosis in animals is remarkably like that in plants. These exceptions will be pointed out as we go through the cell cycle.

MATERIALS

Per student:

- prepared slide of whitefish blastula mitosis
- compound microscope

Per student pair:

- scissors
- tape *or* glue

PROCEDURE Obtain a slide labeled "whitefish blastula." Scan it with the low power objective. This slide has numerous sections of a blastula. Select one section and then switch to the high-dry objective for detailed observation.

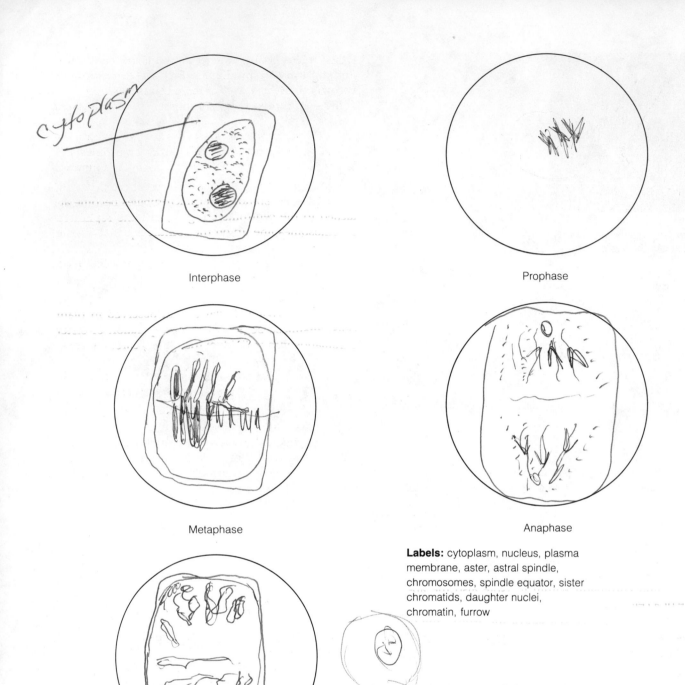

cytoplasm

Interphase

Prophase

Metaphase

Anaphase

Labels: cytoplasm, nucleus, plasma membrane, aster, astral spindle, chromosomes, spindle equator, sister chromatids, daughter nuclei, chromatin, furrow

Telophase and cytokinesis

Figure 8-7 Drawings of cell cycle stages in whitefish blastula

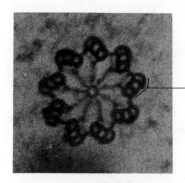

Figure 8-8 Transmission electron micrograph of centriole in cross section. (Photo courtesy I. R. Gibbons.)

— triplet of microtubules

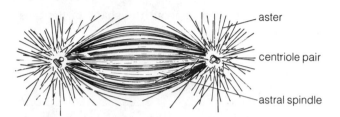

Figure 8-9 Aster and astral spindle. (From Starr and Taggart, 1984.)

aster

centriole pair

astral spindle

As you examine the slides, draw the cells to show the correct sequence of events in the cell cycle of whitefish blastula.

A. Interphase and Mitosis

1. Interphase. Locate a cell in **interphase.** As you observed in the onion root tip, note the presence of the nucleus and chromatin within it. Note also the absence of a cell wall.

Draw an interphase cell above the word "Interphase" in Fig. 8-7 and label cytoplasm, nucleus, and plasma membrane.

2. Mitosis.

a. *Prophase.* The first obvious difference between mitosis in plants and animals is found in **prophase.** Unlike the onion cells, those of whitefish contain **centrioles** (Fig. 8-8). As seen with the electron microscope, centrioles are barrel-shaped structures consisting of nine radially arranged triplets of microtubules.

One pair of centrioles was present in the cytoplasm in the G_1 stage of interphase. These centrioles duplicated during the S stage of interphase. Subsequently, one new and one old centriole migrated to each pole.

Although the centrioles are too small to be resolved with your light microscope, you can see a starburst pattern of spindle fibers that appear to radiate from the centrioles. These fibers are called the **aster.**

Other microtubules extend between the centrioles, forming an **astral spindle** (Fig. 8-9). The chromosomes become visible as the chromatin condenses.

Find a prophase cell, identifying the aster and astral spindle.

Draw the prophase cell in the proper location on Fig. 8-7. Label aster, astral spindle, chromosomes, cytoplasm, and the position of the plasma membrane.

b. *Metaphase.* As was the case in plant cells, during **metaphase** the spindle fiber microtubules become attached to the centromere region, and the replicated chromosomes line up on the spindle equator. Locate a metaphase cell.

Draw the metaphase cell in the proper location on Fig. 8-7. Label the chromosomes on spindle equator, aster, astral spindle, and plasma membrane.

c. *Anaphase.* Again similar to that observed in plant cells, **anaphase** begins with the separation of sister chromatids into individual (daughter) chromosomes. Observe a blastula cell in anaphase.

Draw the anaphase cell in the proper location on Fig. 8-7. Label the separating sister chromatids, astral spindle, cytoplasm, and plasma membrane.

d. *Telophase.* Telophase is characterized by the arrival of the individual (daughter) chromosomes at the poles. A nuclear envelope forms around each daughter nucleus. Find a telophase cell. Are the astral spindles still visible?

_____

Is there any evidence of a nuclear envelope forming around the chromosomes?

_____

Draw the telophase cell. Label daughter nuclei, chromatin, cytoplasm, and plasma membrane.

B. Cytokinesis in Animal Cells

A second major distinction between cell division in plants and animals occurs during cytoplasmic division. Cell plates are absent in animal cells. Instead, cytokinesis takes place by **furrowing.**

To visualize how furrowing takes place, imagine wrapping a string around a balloon and slowly tightening the string until the balloon has been pinched in two. In life, the animal cell is pinched in two, forming two discrete cytoplasmic entities, each with a single nucleus. Figure 8-10 illustrates furrowing in a frog zygote.

Find a cell in the blastula undergoing cytokinesis. The telophase cell that you drew in Fig. 8-7 may also show an early stage of cytokinesis. Label the furrow if it does.

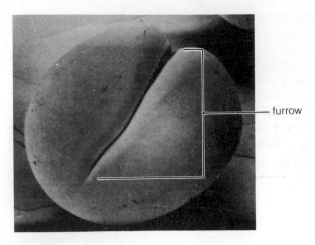

— furrow

Figure 8-10 Scanning electron micrograph of cytokinesis by furrowing in frog zygote. (Photo from R. Kessel and C. Shih, *Scanning Electron Microscopy in Biology*, Springer-Verlag, 1974.)

IV. CHROMOSOME SQUASHES

(Starr, pp. 97–101)

You can make your own chromosome squash preparation quite simply. Cytologists and taxonomists do this routinely to count chromosomes. Observing whole sets of chromosomes is useful for studying chromosomal abnormalities and for determining if two organisms are different species.

MATERIALS

Per student:

- onion or daffodil root tips
- sharp razor blade
- 2 dissecting needles
- microscope slide and coverslip
- compound microscope

Per student pair:

- acetocarmine stain in dropping bottle
- iron alum in dropping bottle
- burner and matches

PROCEDURE We will use onion or daffodil root tips that have been fixed, preserved, and softened to make squashes.

1. Obtain a single root and place it on a clean microscope slide. Notice that the terminal 2 mm or so is opaque white. This is the apical meristem.

2. With a sharp razor blade, separate the apical meristem from the rest of the root. Discard all *but* this meristem region.

3. Add a drop of acetocarmine stain and tease the tissue apart with dissecting needles.

4. Add a drop of iron alum. The iron intensifies the staining of the chromosomes.

5. Place a coverslip over the root tip. Spread the cells out by gently pressing down on the coverslip with your finger or a pencil eraser. Gently heat the slide over a flame.

BE CAREFUL *You Don't Want to Cause the Fluid to Boil Away!*

Examine the preparation with your light microscope. Identify all stages of the cell cycle that have been described above.

What do you notice about the shape of the cells after this preparation?

_____

_____

_____

PRE-LAB QUESTIONS

____ 1. Reproduction in prokaryotes occurs primarily through the process known as (a) mitosis, (b) cytokinesis, (c) furrowing, (d) fission.

____ 2. The genetic material (DNA) of eukaryotes is organized into (a) centrioles, (b) spindles, (c) chromosomes, (d) microtubules.

____ 3. The process of cytoplasmic division is known as (a) meiosis, (b) cytokinesis, (c) mitosis, (d) fission.

____ 4. The product of nucleoprotein duplication is (a) two chromatids, (b) two nuclei, (c) two daughter cells, (d) two spindles.

____ 5. The correct sequence of stages in *mitosis* is (a) interphase, prophase, metaphase, anaphase, telophase; (b) prophase, metaphase, anaphase, telophase; (c) metaphase, anaphase, prophase, telophase; (d) prophase, telophase, anaphase, interphase.

____ 6. During prophase, duplicated chromosomes (a) consist of chromatids, (b) contain centromeres, (c) consist of nucleoproteins, (d) all of the above.

____ 7. During the S period of interphase (a) cell growth takes place, (b) nothing occurs because this is a resting period, (c) chromosomes divide, (d) synthesis (or replication) of the nucleoproteins takes place.

____ 8. Chromatids separate during (a) prophase, (b) telophase, (c) cytokinesis, (d) anaphase.

____ 9. Cell plate formation (a) occurs in plant cells but not in animal cells, (b) begins during telophase, (c) is a result of fusion of Golgi vesicles, (d) all of the above.

____ 10. An aster and astral spindle would be found in (a) both plant and animal cells, (b) only plant cells, (c) only animal cells, (d) none of the above.

MEIOSIS:
BASIS OF SEXUAL REPRODUCTION

OBJECTIVES After completing this exercise you will be able to:

1. define meiosis, homologue (homologous chromosome), diploid, haploid, gene, gene pair, allele, gamete, ovum, sperm, gametic meiosis, fertilization, sporic meiosis, locus, synapsis, zygote;

2. indicate the differences and similarities between meiosis and mitosis;

3. describe the basic differences between the life cycles of higher plants and higher animals;

4. describe the process of meiosis, recognizing the events that occur during each stage;

5. describe the significance of crossing over, independent assortment, and segregation;

6. identify the meiotic products in male and female animals.

INTRODUCTION Like mitosis, meiosis is a process of nuclear division. During mitosis, the number of chromosomes within the daughter nuclei remains the same as was present in the parental nucleus. In meiosis, however, the genetic complement is halved, resulting in daughter nuclei containing only one-half the number of chromosomes as the parental nucleus. Thus, where mitosis is sometimes referred to as an *equational division*, meiosis is often called *reduction division*. Moreover, while mitosis is completed after a single nuclear division, two divisions, called meiosis I and meiosis II, occur during meiosis. Table 9-1 summarizes the differences between mitosis and meiosis.

In the body cells of most eukaryotes, chromosomes exist in pairs called **homologues** (homologous chromosomes), i.e., there are two chromosomes that are physically similar and contain genetic information for the same traits. When both homologues are *in the same nucleus*, the nucleus is **diploid** (2N); when only one of the homologues is present, the nucleus is **haploid** (N). If the parental nucleus normally contains the diploid (2N) chromosome number before meiosis, all four daughter nuclei contain the haploid (N) number at the completion of meiosis.

Table 9-1 Comparison of Mitosis and Meiosis	
Mitosis	Meiosis
Equational division: amount of genetic material remains constant	Reduction division: amount of genetic material is halved
Completed in one division	Requires two divisions for completion
Produces two genetically identical nuclei	Produces two to four genetically different nuclei
Generally produces cells not directly involved in sexual reproduction	Produces cells for sexual reproduction

Each chromosome bears **genes,** which are units of inheritance. Genes may exist in two or more alternative forms called **alleles.** Thus each homologue bears *genes* for the same traits; these are called **gene pairs.** However, the homologues may or may not have the same *alleles*. An example will help here.

Suppose the trait in question is flower color and that a flower has only two possible colors, red or white. The gene is coding (providing the information) for flower color. Now there are two homologues in the same nucleus, so each bears the gene for flower color. *But,* on one homologue, the *allele* might code for red flowers, while the allele on the other homologue might code for white flowers. There are two other possibilities. The alleles on *both* homologues might be coding for red flowers, or they *both* might be coding for white flowers. (These three possibilities are mutually exclusive.)

The reduction in chromosome number is the basis for sexual reproduction. In animals, the cells containing the daughter nuclei produced by meiosis are called **gametes**—**ova** (singular is ovum) if the parent is female, **sperm** cells if male. As you probably know, gametes are produced in the gonads—ovaries and testes, respectively. In fact, this is the *only* place where meiosis occurs in higher animals. The simple diagram in Fig. 9-1 illustrates the life cycle of a higher animal.

Note where meiosis has occurred—during gamete production. Consequently, this is called **gametic meiosis.** During **fertilization** (the fusion of a sperm nucleus with an ovum nucleus), the diploid chromosome number is restored as the two haploid gamete nuclei fuse.

What about plants? Do plants have sex? Indeed they do. However, the plant life cycle is a bit more complex than that of animals. Plants of a single species have two completely different body forms. The primary function of one is production of gametes. This plant is called a *gametophyte* ("gamete-producing plant") and is haploid. Because the entire plant is haploid, gametes are produced in specialized organs (gametangia) by mitosis. The other body form is diploid and is called a *sporophyte*. This diploid sporophyte has specialized organs (sporangia) where meiosis occurs, producing haploid meiospores (hence the name *sporophyte*, "spore-producing plant"). When spores germinate, they grow into gametophytes.

Examine Fig. 9-3, which illustrates the gametophyte and sporophyte of a fern plant. Remember, the gametophyte and sporophyte are different, free-living stages of the *same* species of fern. Look at Fig. 9-2, a diagram of a typical plant life cycle. Again, note the consequence of meiosis. In plants it results in the production of meiospores. Hence, this type of meiosis is called **sporic meiosis.**

You should understand an important concept from these diagrams: *Meiosis always reduces the chromosome number. The diploid chromosome number is eventually restored during fertilization.*

Understanding meiosis is an absolute necessity for understanding the patterns of inheritance in Mendelian genetics. Gregor Mendel, an Austrian monk, spent years deciphering the complexity of simple genetics. Although he knew nothing of genes and chromosomes, he noted certain patterns of inheritance and formulated three principles, now known as Mendel's principles of recombination, segregation, and independent assortment. The following activities will demonstrate the events of meiosis and the genetic basis for Mendel's principles.

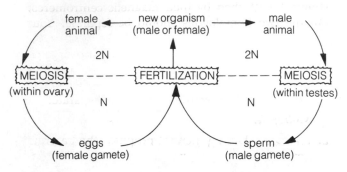

Figure 9-1 Life cycle of a higher animal

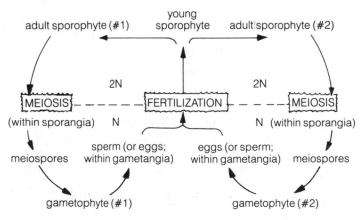

Figure 9-2 Life cycle of a plant

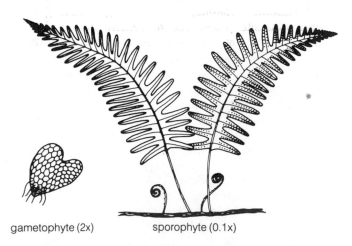

Figure 9-3 Gametophyte and sporophyte of the same fern plant

gametophyte (2x)   sporophyte (0.1x)

I. DEMONSTRATION OF MEIOSIS USING POP BEADS

(Starr, pp. 104–113)

MATERIALS

Per student pair:

- 8 chains of simulated chromosomes consisting of pop beads with magnetic centromeres
- marking pens ("Sharpie")
- 8 pieces of string, each 40 cm long
- meiotic diagram cards similar to those illustrated within this exercise
- colored pencils

Per student group (table):

- bottle of 95% ethanol to remove marking ink
- tissues

PROCEDURE   Work in pairs.

A. Meiosis without Crossing Over

Obtain four chains of pop beads (see Fig. 9-4). The beads on two chains should be one color while those of the other two chains should be another color. All beads *within* a chain should be the *same* color.

The chains of beads represent chromatids of duplicated homologous chromosomes (homologues), each bead a gene, and the magnet the centromere. We start by assuming that these chromosomes represent the diploid condition. The two colors represent the origin of the chromosomes: One homologue (color) came from the male parent, and the other homologue (color) came from the female parent.

You have four chains of beads because chromosome replication occurred during the S stage of interphase, prior to the onset of meiosis.

How many sister chromatids are there in a duplicated chromosome?

_____

How many chromosomes are represented by four sister chromatids?

_____

What is the diploid number of the starting (parental) nucleus? (Hint: Count the number of homologues to obtain the diploid number.)

_____

As mentioned previously, genes may exist in two or more alternative forms called alleles. The location of an allele on a chromosome is its **locus** (plural: loci). Using the marking pen, mark two loci on each chromatid with letters to indicate alleles for a common trait. For example, suppose the homologous chromosomes code for 2 traits, skin pigmentation and the presence of attached earlobes in humans. Let the capital letter *A* represent the allele for normal pigmentation, lower case *a* the allele for albinism (the absence of skin pigmentation); let *F* represent free earlobes and *f* attached earlobes.

A suggested sequence is illustrated in Fig. 9-4.

Using the meiotic diagram in Fig. 9-5, manipulate your model chromosomes through the stages of meiosis described below.

1. Interphase. During interphase the nuclear envelope is intact, and the chromosomes are randomly distributed throughout the nucleoplasm (semifluid substance within the nucleus). Both duplicated chromosomes (four chromatids) should be in the parental nucleus, indicating that DNA duplication has taken place. The sister chromatids of each homologue should be attached by their magnetic centromeres, but the two homologues should be separate. Your model nucleus contains a diploid number (2N) = 2.

The pop bead chromosomes should appear during interphase in the parental nucleus as shown in Fig. 9-5. Be sure to mark the location of the alleles. Use different colors to keep the homologues separate.

2. Meiosis I.

a. *Prophase I.* During the first prophase the parental nucleus contains two duplicated homologous chromosomes, each made up of two sister chromatids joined at their centromeres. The homologues pair with each other. This pairing is called **synapsis.** Slide the two homologues together.

Twist the chromatids about one another to simulate synapsis.

b. *Metaphase I.* Homologous chromosomes now move toward the spindle equator, the centromeres of each homologue coming to lie *on either side of the equator.* Spindle fibers, consisting of aggregations of microtubules, attach to the centromeres. One homologue becomes attached to microtubules extending from one pole, and the other homologue becomes attached to microtubules extending from the opposite spindle pole.

To simulate the spindle fibers, attach one piece of string to each centromere. Then lay the free ends of one pair of strings toward one spindle pole and the ends of the other pair toward the opposite pole.

c. *Anaphase I.* During anaphase I, the homologous chromosomes separate, each homologue moving toward opposite poles. The movement of the chromosomes is apparently the result of shortening of the

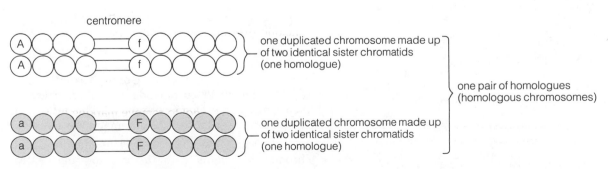

Figure 9-4   One pair of homologous pop-bead chromosomes

parental nucleus

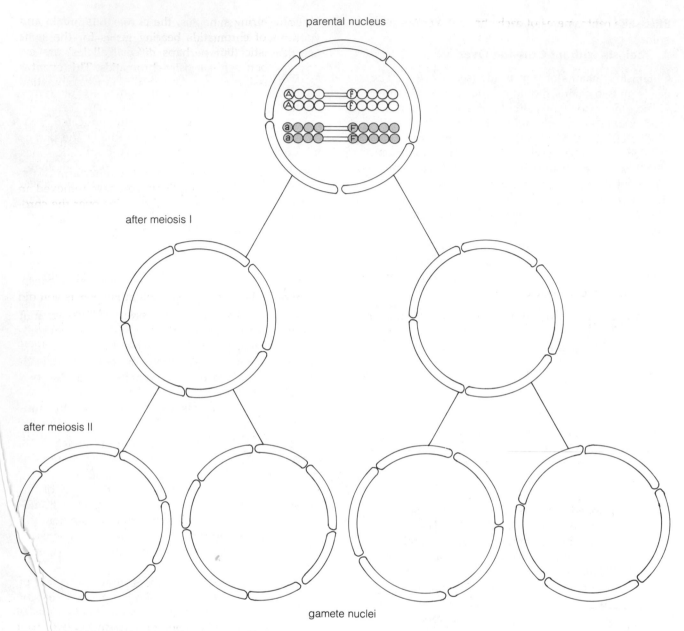

after meiosis I

after meiosis II

gamete nuclei

Figure 9-5 Meiosis without crossing over

spindle fibers. Each homologue is still in the duplicated form, consisting of two sister chromatids.

Pull the two strings of one homologous pair toward its spindle pole and the other toward the opposite spindle pole, separating the homologues from one another.

d. *Telophase I.* Continue pulling the string spindle fibers until each homologue is now at its respective pole. The first meiotic division is now complete. You should have two nuclei, each containing a single chromosome consisting of two sister chromatids.

Draw your pop bead chromosomes as they appear after meiosis I on the two nuclei labeled "after meiosis I" of Fig. 9-5. Depending on the organism involved, an interphase (interkinesis) and cytokinesis may pre-

cede the second meiotic division, or each nucleus may proceed directly into meiosis II.

It is important to note here that DNA synthesis *does not* occur following telophase I (between meiosis I and meiosis II).

Before meiosis II the spindle is rearranged into two spindles, one for each nucleus.

3. Meiosis II.

a. *Prophase II.* At the beginning of the second meiotic division, the sister chromatids are still attached by their centromeres. During prophase II, the nuclear envelope disorganizes and the chromatin recondenses.

b. *Metaphase II.* Within each nucleus, the duplicated chromosome aligns with the equator, the centromeres lying *on the equator.* Spindle fiber microtubules attach the centromeres of each chromatid to opposite spindle poles.

Your string spindle fibers should be positioned just as they were during prophase I. Note that each nucleus contains only *one* duplicated chromosome consisting of *two* sister chromatids.

c. *Anaphase II.* The sister chromatids separate, moving to opposite poles. Pull on the string until the two sister chromatids separate. After the sister chromatids separate, each is an individual (not duplicated) daughter chromosome.

d. *Telophase II.* Continue pulling on the string spindle fibers until the two daughter chromosomes are at opposite poles. The nuclear envelope re-forms around each chromosome. Four daughter nuclei now exist. Note that each nucleus contains one individual chromosome (formerly a chromatid) originally present within the parental nucleus.

Draw your pop-bead chromosomes as they appear after meiosis II in the "gamete nuclei" of Fig. 9-5. Your diagram should indicate the genetic (chromatid) complement *before* meiosis and *after* each meiotic division, *not* the stages of each division.

Remember that meiosis takes place in both male and female organisms. If the parental nucleus was from a male, what is the gamete called?

_____

If female?

_____

Is the parental nucleus diploid or haploid?

_____

Are the nuclei produced after the *first* meiotic division diploid or haploid?

_____

Are the nuclei of the gametes diploid or haploid?

_____

If you answered the above questions correctly, you might logically ask, "If the chromosome number of the gametes is the same as that produced after the first meiotic division, why bother to have two separate divisions? After all, the genes present are the same in both gametes and first division nuclei." The answer to this apparent paradox is that although you have simulated meiosis, you have done so without showing what happens in real life. That's the next step . . .

B. Meiosis with Crossing Over

A very important event that results in a reshuffling of alleles on the chromatids occurs during prophase I. Recall that synapsis results in pairing of the homologues. During synapsis, the chromatids break, and portions of chromatids bearing genes for the same characteristic (but perhaps *different* alleles) are exchanged between *non-sister* chromatids. This event is called **crossing over,** and it results in recombination (shuffling) of alleles. Look again at Fig. 9-4. Distinguish between sister and non-sister chromatids.

To simulate crossing over, break the beads of the arms of two non-sister chromatids at the centromere and then re-attach them so that two of the four chains have one arm of one color and the other arm of the opposite color. (For simplicity, you have removed an entire arm, but during actual crossing over the chromosomes may break anywhere within the arms.)

Crossing over is virtually a universal event in meiosis.

Manipulate your model chromosomes through meiosis I and II again, this time watching what happens to the distribution of the alleles as a consequence of the crossing over. Fill in Fig. 9-6 as you did before, but this time, show the effects of crossing over. Again, use different colors in making your sketches.

Is the distribution of alleles present in the gamete nuclei after crossing over the same as that which was present without crossing over?

_____

Is the distribution of alleles present in the gamete nuclei after crossing over the same as that in the nuclei after the first meiotic division?

_____

Crossing over provides for genetic recombination. What is genetic recombination?

Recall that the parental nucleus contained a pair of homologues, each homologue consisting of two sister chromatids. Because sister chromatids are identical in all respects, they have the same alleles of a gene (see Fig. 9-4). As your models showed, the alleles on non-sister chromatids may not (or may) be identical; they bear the same gene but may have different alleles.

What is the difference between a gene and an allele?

Let's look at a single set of alleles that are on your model chromosomes, say, the alleles for pigmentation, *A* and *a*. Both alleles were present in the parental nucleus. How many are present in the gametes?

_____

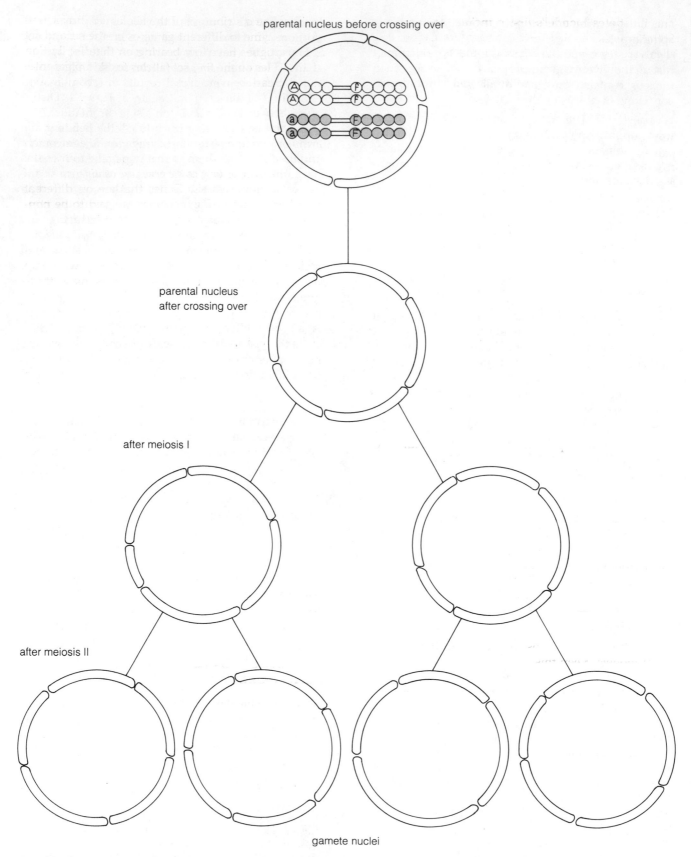

Figure 9-6 Meiosis after crossing over

This illustrates Mendel's first principle, segregation. *Segregation* means that during gamete formation, the alleles are separated (segregated) from each other and end up in different gametes.

C. Demonstrating Independent Assortment

You have just demonstrated meiosis where only one pair of homologues was present (N = 1, 2N = 2). Now obtain another set of model chromosomes (four more chains with magnetic centromeres). These two chromosomes should be distinct from the original set. The easiest way to accomplish this is to make the chains different colors, different lengths, and/or with different numbers of beads on the arms on either side of the centromeres.

Let's assign a gene to our second set of homologues. Suppose this gene codes for the production of an enzyme necessary for metabolism. On one homologue (consisting of two chromatids), mark the letter *P*, representing the allele causing production of the enzyme. On the other homologue, let *p* represent the allele that interferes with normal enzyme production. (For now, it is not important to remember these traits; they're real situations used simply as examples.)

We now have a parental nucleus where there are two sets of homologous chromosomes (four homologues). Here the diploid number (2N) is 4. Count the number of duplicated chromosomes. How many are there?

This is the 2N number.

You know that meiosis is reduction division, so you can predict the number of individual chromosomes (the haploid number) each gamete will have after meiosis II.

Manipulate your model chromosomes through meiosis I and II. Simulate crossing over with the original set of models. Keep this in mind: Crossing over and recombination occur between non-sister chromatids of homologous chromosomes, but *not* between *non-homologous* chromosomes.

Fill in Fig. 9-7, showing the outcome of meiosis in a nucleus with two sets of homologues.

How many individual chromosomes does each gamete contain?

Are the gametes the same genetically or different from each other?

Go through meiosis again, searching for different possibilities in chromosome distribution that would make the gametes different.

Does the distribution of the alleles for production of the enzyme to different gametes on the second set of homologues have any bearing on the distribution of the alleles on the first set (alleles for skin pigmentation and earlobe condition)?

This demonstrates the principle of independent assortment, which states that segregation of alleles into gametes is independent of the segregation of alleles for other traits, *as long as the genes are on different sets of homologous chromosomes.* Genes that are on different (non-homologous) chromosomes are said to be **non-linked.** By contrast, genes for different traits that are on the same chromosome are **linked.**

Because the genes for enzyme production and those for skin pigmentation and earlobe attachment are on different homologous chromosomes, these genes are

_____ ,

while the genes for skin pigmentation and earlobe attachment are

_____ ,

because they are on the same chromosome.

In reality, most organisms have many more than two sets of chromosomes. Humans have 23 pairs (2N = 46), while some plants literally have hundreds!

A thorough understanding of meiosis is necessary to understand genetics. With this basis you will find doing problems involving Mendelian genetics easy and fun.

Remove Marking Ink from Pop Beads with 95% Ethanol and Tissues.

II. MEIOSIS IN ANIMAL AND PLANT CELLS
(Starr, pp. 104–113)

Now that you have a conceptual understanding of meiosis, let's see the actual divisions as they occur in living organisms.

In animals, as mentioned previously, meiosis results in the production of gametes, ova in females and sperm in males.

MATERIALS

Per lab room:

- set of demonstration slides of meiosis in grasshopper testes and lily anther
- set of models illustrating meiosis and fertilization in roundworm

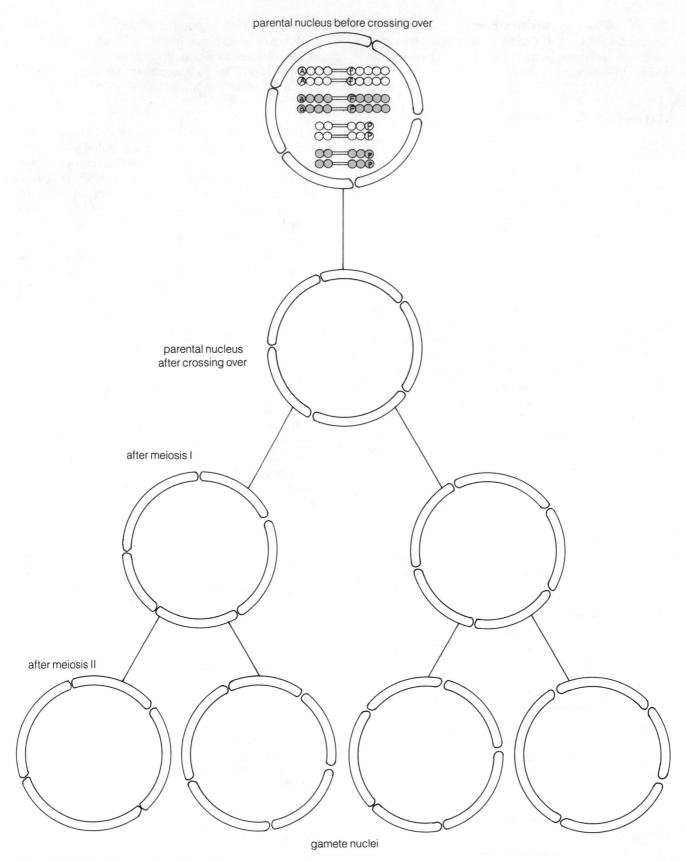

Figure 9-7 Meiosis in a nucleus where 2N = 4

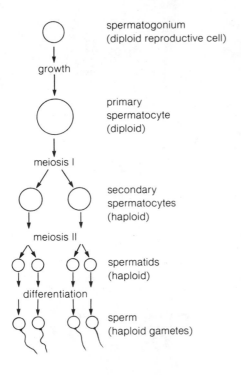

a spermatogenesis

Figure 9-8 Gametogenesis in animals. (From Starr and Taggart, 1987.)

b oogenesis

Per student pair:

- scissors
- tape or glue

A. Meiosis in Male Animals

In male animals meiosis occurs in the testes.

Examine Fig. 9-8a. A diploid reproductive cell, the *spermatogonium*, first enlarges into a *primary spermatocyte*. The primary spermatocyte undergoes meiosis I to form two haploid *secondary spermatocytes*. After meiosis II, four haploid *spermatids* are produced, which develop flagella during differentiation into four *sperm* cells. This process is called *spermatogenesis*.

Examine the demonstration slide of spermatogenesis in grasshopper testes.

B. Meiosis in Female Animals

In the ovaries of female animals, *ova* (eggs) are produced by meiosis during the process called oogenesis (Fig. 9-8b). Unlike spermatogenesis, only one of the meiotic products becomes a gamete.

Examine Fig. 9-8b. The diploid reproductive cell, called an *oogonium*, grows into a *primary oocyte*. The primary oocyte undergoes meiosis I, one product being the *secondary oocyte*, the other a *polar body*. Notice the difference in size of the secondary oocyte and the polar body. This is because the secondary oocyte

ends up with nearly all of the cytoplasm after meiosis I. Following meiosis II, only the secondary oocyte becomes a mature, haploid *ovum;* depending on the species, the polar body may or may not undergo meiosis II. In any case, the polar bodies are extremely small and do not function as gametes.

Observe the models illustrating oogenesis in the demonstration series. These models represent the events as they occur in the roundworm, an organism that has only two pairs of homologous chromosomes (2N = 4).

As you study the models and read the description for each stage, cut out the photographs of the models (page 111) and tape or glue them in the proper sequence in Fig. 9-9. Label each stage in your correctly sequenced photographs.

1. Meiosis in the primary oocyte does not begin until a *sperm* penetrates the cytoplasm. In this model, note that the oogonium's nucleus is intact. A photograph of this stage has been inserted in Fig. 9-9 to get you started.

2. *Prophase I.* The second model represents prophase I. Note that the nuclear envelope has disorganized. Each dark dot represents a chromatid. How many chromatids are there?

_____

How many chromosomes does this represent?

_____

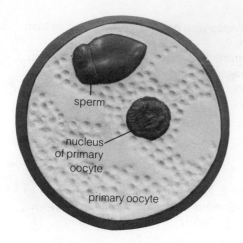

sperm

nucleus of primary oocyte

primary oocyte

Sperm entrance

Labels: primary oocyte, sperm, homologous chromosomes, sister chromatids

Prophase I

Labels: primary oocyte, homologous chromosomes, centrioles, spindle, sister chromatids, sperm nucleus

Late Metaphase (Early Anaphase I)

Labels: primary oocyte, homologous chromosomes, sister chromatids, spindle

Later Anaphase I (Telophase I)

Labels: 1st polar body, 4 chromatids, secondary oocyte, sperm nucleus

1st polar body formation

Labels: 1st polar body, centrioles, 4 chromatids, spindle, sperm nucleus

Later Metaphase II (Early Anaphase II)

Figure 9-9 Animal meiosis, ovum formation, and fertilization

Labels: 1st polar body, unduplicated chromosomes, mature ovum, sperm nucleus

Telophase II and cytokinesis

Labels: sperm nucleus, ovum nucleus, zygote

Fertilization

3. *Late Metaphase I (or Early Anaphase I).* (The third model represents a transition between metaphase I and anaphase I.) During metaphase I, the *homologous chromosomes* become located on either side of the spindle equator. The *spindle* is distinct, the component fibers seemingly attached to the *centrioles*, here represented by two small dots. The homologous chromosomes are beginning to separate. Note the *sperm nucleus* within the cytoplasm of the primary oocyte.

4. *Later Anaphase I (or Early Telophase I).* The homologous chromosomes move toward opposite spindle poles. Remember, each homologous chromosome consists of two sister chromatids. The sperm nucleus remains "lying in wait."

5. *Formation of the first polar body.* Cytokinesis takes place, separating the homologous chromosomes. One set of homologues resides in a small cell with relatively little cytoplasm. This is the first polar body.

Two non-homologous chromosomes (four chromatids) remain in the larger cell, which is now called the *secondary oocyte.* A nuclear envelope does not form about these chromosomes, so essentially the secondary oocyte is in prophase II.

6. *Late Metaphase II (or Early Anaphase II).* Now the sister chromatids of the chromosomes within the secondary oocyte line up on the spindle equator. (The models show them on opposite sides of the equator.) A new spindle with centrioles is present as the sperm nucleus remains in wait. In the roundworm, the polar body does not undergo meiosis II.

7. *Telophase II and cytokinesis.* A thin line represents cytokinesis occurring to form the second polar body. How many unduplicated chromosomes (formerly sister chromatids) does the mature haploid *ovum* contain?

(The models do not show formation of the second polar body.)

8. *Fertilization.* The final model represents fertilization, the fusion of the ovum nucleus with the sperm nucleus. With fertilization, the large ovum becomes the first diploid cell, the zygote. How many chromosomes does the zygote contain?

C. Meiosis in Plants

For the sake of brevity, we will examine meiosis in the male reproductive structure of flowering plants only. Recall from our earlier discussion that meiosis in plants results in meiospore production, not directly into gametes. The details of the life cycle of flowering plants will be considered in Exercise 26.

Examine the demonstration series of meiosis beginning with the diploid *microsporocytes.* Microsporocytes are the cells within a flower that undergo meiosis to produce haploid *microspores.* Eventually these microspores develop into pollen grains, which in turn produce sperm.

As you examine the slides, cut out the photomicrographs on pages 113 and 115 and arrange them on Fig. 9-10 to depict the meiotic events leading to microspore formation.

1. *Interphase.* During interphase the *nucleus* of each diploid *microsporocyte* is distinct, containing granular-appearing chromatin. The cells are compactly arranged.

2. *Early Prophase I.* Now the chromatin has begun to condense into discrete *chromosomes,* which have the appearance of fine threads within the nucleus.

3. *Mid-Prophase I.* Additional condensation of the *chromosomes* has taken place. Pairing of homologous chromosomes is taking place.

4. *Late Prophase I.* The chromosomes have condensed into short, rather fat structures. Synapsis and crossing over are taking place. Note that the nuclear envelope has disorganized.

5. *Metaphase I.* The homologous chromosomes lie in the region of the *spindle equator.* The *spindle,* composed of *spindle fibers,* can be discerned as fine lines running toward the *poles.* (Note the absence of centrioles in plant cells.)

6. *Early Anaphase I.* Separation of homologous chromosomes is beginning to take place.

7. *Later Anaphase I.* Homologous chromosomes have nearly reached the opposite poles. Reduction division has occurred.

8. *Telophase I.* The homologous chromosomes have aggregated at opposite poles. The spindle remains visible.

9. *Cytokinesis I.* The *cell plate* is forming in the midplane of the cell. Spindle fibers, which are aggregations of microtubules, are visible running perpendicularly through the cell plate. The microtubules are directing the movement of Golgi vesicles, which contain the materials that form the cell plate.

A nuclear envelope has re-formed about the chromosomes, resulting in a well-defined nucleus in each *daughter cell.*

10. *Interkinesis.* In these plant cells, a short stage exists between meiosis I and II. Distinct nuclei are apparent in the two daughter cells. A cell wall has formed across the entirety of the midplane.

11. *Prophase II.* The chromosomes in each nucleus of the two daughter cells condense again into distinct, threadlike bodies. As was the case at the end of prophase I, the nuclear envelope disorganizes.

12. *Metaphase II.* Chromosomes consisting of sister chromatids line up on the spindle equator in both cells. (The photomicrograph shows the very early stages of separation of the chromatids.)

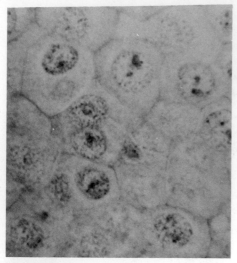

Interphase

Early Prophase I

Mid-Prophase I

Late Prophase I

Metaphase I

Early Anaphase I

Later Anaphase I

Telophase I

Cytokinesis I

Interkinesis   Prophase II   Metaphase II

Anaphase II   Telophase II & cytokinesis

Labels: diploid microsporocyte, nucleus, chromosomes, spindle equator, spindle, spindle fibers, pole, cell plate, daughter cells, sister chromatids (unduplicated chromosomes), haploid microspores

Figure 9-10   Meiosis and microsporogenesis in the anther

13. *Anaphase II.* The sister chromatids (now more appropriately considered *unduplicated chromosomes*) are being drawn to their respective poles in each cell.

Before anaphase II begins, sister chromatids are attached to each other along their length. Shortening of the spindle fibers, which are attached to the chromatids at their centromeres, causes separation of the chromatids, beginning in the region of the centromere. This causes a V-shaped configuration of the chromosomes.

14. *Telophase II and cytokinesis.* Nuclear envelopes are now re-forming around each of the four sets of chromosomes. Cell plate formation is occurring perpendicular to the cell wall that was formed after telophase I.

After cell wall formation is complete, the four haploid cells (microspores) will separate. Subsequently, each will develop into a pollen grain inside which sperm cells will be formed.

PRE-LAB QUESTIONS

_____ **1.** In meiosis the number of chromosomes _____, while in mitosis, it _____. (a) is halved, is doubled, (b) is halved, remains the same, (c) is doubled, is halved, (d) remains the same, is halved.

_____ **2.** The term "2N" means that (a) the diploid chromosome number is present, (b) the haploid chromosome number is present, (c) within a single nucleus chromosomes exist in homologous pairs, (d) a and c.

_____ **3.** In higher animals, meiosis results in the production of (a) egg cells (ova), (b) gametes, (c) sperm cells, (d) all of the above.

_____ **4.** Recombination of alleles on non-sister chromatids occurs during (a) anaphase I, (b) tetrad formation, (c) telophase II, (d) crossing over.

_____ **5.** Alternative forms of genes are called (a) homologues, (b) locus, (c) loci, (d) alleles.

_____ **6.** If both homologous chromosomes of each pair exist in the same nucleus, that nucleus is (a) diploid, (b) unable to undergo meiosis, (c) haploid, (d) none of the above.

_____ **7.** DNA duplication occurs during (a) interphase, (b) prophase I, (c) prophase II, (d) interkinesis.

_____ **8.** A daughter chromosome (a) is formed during anaphase II, (b) is the same as a homologous chromosome, (c) is the result of separation of chromatids, (d) a and c.

_____ **9.** Humans (a) don't undergo meiosis, (b) have 46 chromosomes, (c) produce gametes by mitosis, (d) all of the above.

_____ **10.** Gametogenesis in female animals results in (a) four sperm, (b) one gamete and three polar bodies, (c) four functional ova, (d) a haploid ovum and three diploid polar bodies.

NAME _____   SECTION NUMBER _____

EXERCISE 9
MEIOSIS: BASIS OF SEXUAL REPRODUCTION

POST-LAB QUESTIONS

1. If a cell of an organism had 46 chromosomes before meiosis, how many chromosomes would exist in each nucleus after meiosis?

2. Suppose one sister chromatid of a chromosome has the allele *H*. What allele will the other sister chromatid have? (Assume crossing over has not taken place.)

3. Suppose that two alleles on one homologous chromosome are *A* and *B*, and the other homologous chromosome's alleles are *a* and *b*. How many different genetic types of gametes would be produced *without* crossing over?

_____

What are those types?

If crossing over were to occur, how many different genetic types of gametes could occur?

_____

List them.

4. List two differences between mitosis and meiosis.
   1. _____
   2. _____

5. Describe the similarities between mitosis and meiosis.

6. From a genetic viewpoint, of what significance is fertilization?

7. In animals, meiosis results directly in gamete production, while in plants meiospores are produced. Where do the gametes come from in the life cycle of a plant?

8. What basic difference exists between the life cycles of higher plants and higher animals?

9. How would you argue that sporic meiosis is the basis for sexual reproduction in plants, even though the *direct* result is a meiospore rather than a gamete?

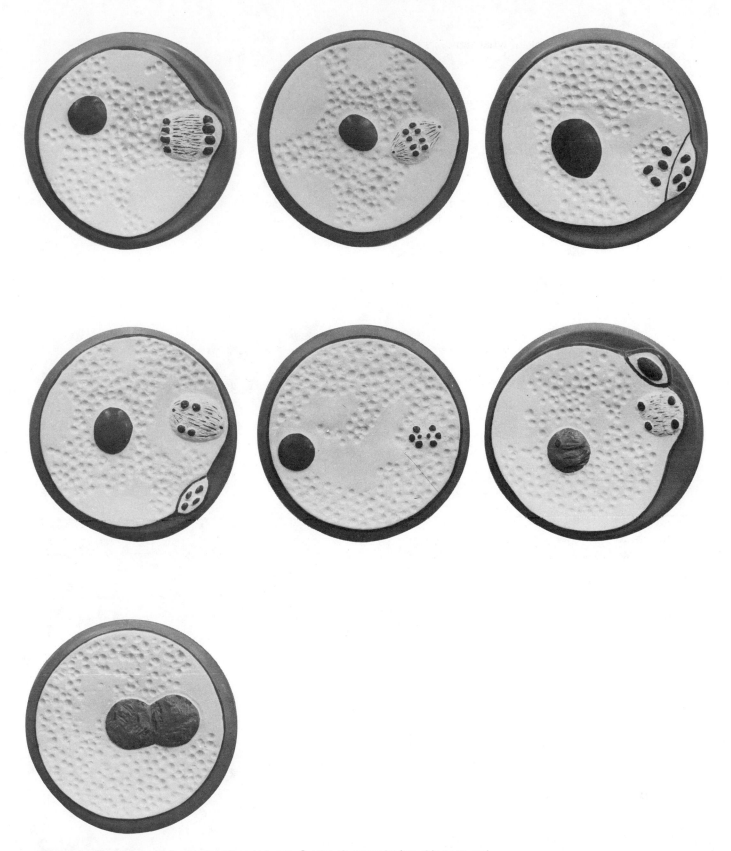

Photos for Figure 9-9 Meiosis in female roundworm. Cut the photographs from this page and arrange them in the proper sequence in Fig. 9-9. (Photos by J. W. Perry.)

Photos for Figure 9-10 Meiosis in flowering plants. Cut from
this page and arrange in proper sequence in Fig. 9-10.
(Photos by J. W. Perry.) *Continues.*

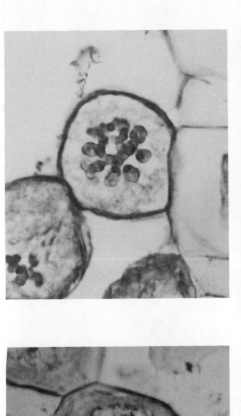

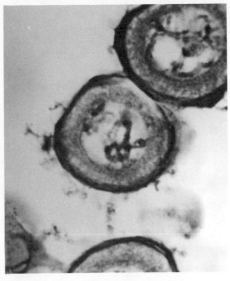

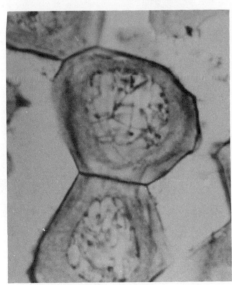

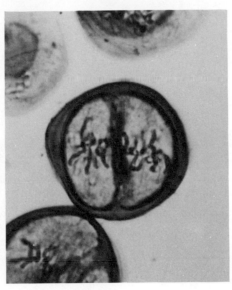

Photos for Figure 9-10   *Continued*

MENDELIAN GENETICS

OBJECTIVES After completing this exercise you will be able to:

1. define true-breeding, hybrid, monohybrid cross, diploid, haploid, genotype, phenotype, dominant, recessive, complete dominance, homozygous, heterozygous, incomplete dominance, codominance, sex-linked, dihybrid cross, probability, multiple alleles;

2. solve problems illustrating monohybrid and dihybrid crosses, including those with incomplete dominance, codominance, sex-linkage, and problems involving multiple alleles;

3. determine your phenotype for traits used in this exercise and give your probable genotype for these traits.

INTRODUCTION In 1866 an Austrian monk, Gregor Mendel, presented the results of painstaking experiments on the inheritance of the garden pea. Those results were heard, but probably not understood, by Mendel's audience. Now, more than 100 years later, biologists take the importance of Mendel's work for granted. In this era of genetic engineering—the incorporation of foreign DNA into chromosomes of unrelated species—it is easy to lose sight of the basics of the process that makes it all possible.

Recent advances in molecular genetics have resulted in production of insulin and human growth hormone by genetic engineering techniques. This newfound technology has not been without controversy. In 1983 researchers at the University of California–Berkeley proposed to release into the environment the first genetically engineered organism, a bacterium that was designed to prevent ice damage to certain crop plants. The proposal was met with strong opposition. In the future, *you* will be called upon to help make decisions about issues like this. To make an educated judgment, you must understand the basics, just as Mendel did. The genetics problems in this exercise should take you to that point.

MATERIALS

Per student group (table):

- genetic corn ears illustrating monohybrid and dihybrid crosses
- pop beads used in Exercise 9 (optional)
- cotton balls
- bottle of 70% ethanol

Per student:

- blood typing card
- bottles of antiserum (A, B, D)
- sterile blood lancet
- PTC taste-testing paper

I. MONOHYBRID PROBLEMS WITH COMPLETE DOMINANCE (Starr, pp. 117–122)

Garden peas have both male and female parts in the same flower and are able to self-fertilize. For his experiments, Mendel chose parental plants that were **true-breeding,** meaning that all self-fertilized offspring displayed the same form of a trait as their parent. For example, if a true-breeding purple-flowered plant is allowed to self-fertilize, all of the offspring will have purple flowers.

When parents that are true-breeding for *different* forms of a trait are crossed—for example, purple flowers and white flowers—the offspring are called **hybrids.** When only one trait is being studied, the cross is called a **monohybrid cross.**

We'll look first at monohybrid problems.

1. Recall that most organisms are diploid; i.e., they contain homologous chromosomes with genes for the same traits. The location of a gene on a chromosome is its *locus* (plural: *loci*). Two genes at homologous loci

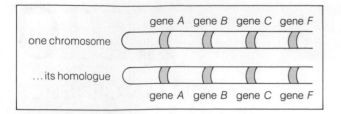

Figure 10-1 Arrangement of genes on homologous chromsomes

Locus location of where the gene is (handwritten margin note)

are called a *gene pair*. Chromosomes have numerous genes, as illustrated in Fig. 10-1. Recall also that genes may exist in different forms, called *alleles*. Let's consider one gene pair at the F locus. There are three possibilities for the allelic makeup at the F locus:

Both alleles are *FF*.

Allele/as a form it the gene (handwritten margin note)

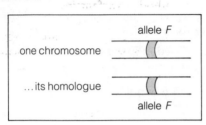

Both alleles are *ff*.

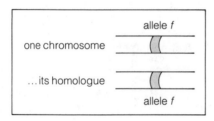

One allele is *F* and the other is *f*.

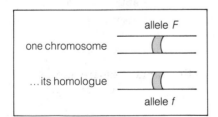

sea cull (handwritten margin note)

Gametes, on the other hand, are haploid, containing only one of the two homologues, and thus only one of the two alleles for a specific trait.

The **genotype** of an organism is its genetic constitution, i.e., the alleles present.

For each of the following diploid genotypes, indicate the possible genotypes of the gametes.

two genes two chromosomes respectively (handwritten margin note)

Diploid genotype	Gamete genotype
FF	homozygous
ff	_____
Ff	_____ , _____

Homozygous & heterozygous (handwritten margin notes)

If you don't understand the process that gives rise to the gamete genotypes, return to the pop bead models that you used in the meiosis exercise. Attach labels to one bead of each chromosome and go through the meiotic divisions that give rise to the gametes.

It is imperative that you understand meiosis before attempting to do genetics problems.

2. During fertilization, two gamete nuclei fuse, and the diploid condition is restored. Give the diploid genotype produced by fusion of the following gamete genotypes.

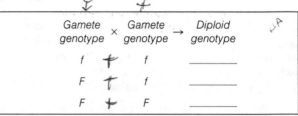

3. Now let's attach some meaning to genotypes. As you see from the previous problems, the genotype is an expression of the actual genetic makeup of the organism. The **phenotype** is the observable result of the genotype, i.e., what the organism looks like because of its genotype. (Although phenotype is determined primarily by genotype, in many instances environmental factors can modify phenotype.)

Human earlobes are either attached or free (Fig. 10-2). This trait is determined by a single gene consisting of two alleles, *F* and *f*. An individual whose genotype is *FF* or *Ff* has free earlobes. This is the **dominant** condition. Note that the presence of one *or* two *F* alleles results in the dominant phenotype, free earlobes. The allele *F* is said to be dominant over its allelic partner, *f*. The **recessive** phenotype, attached earlobes, occurs only when the genotype is *ff*. In the case of **complete dominance,** the dominant allele completely masks the expression or effect of the recessive allele.

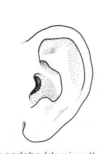

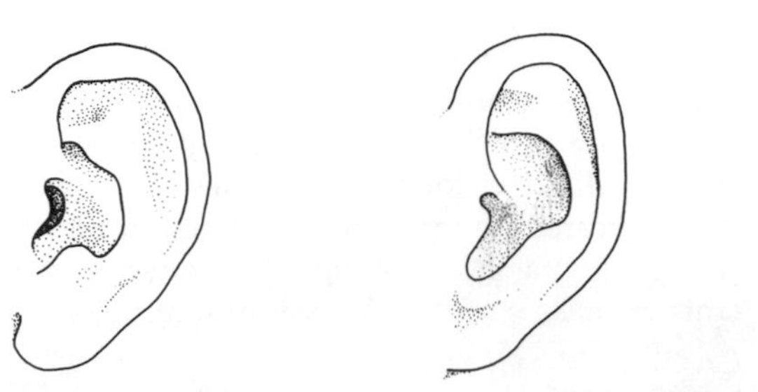

free earlobe (dominant)   attached earlobe (recessive)

Figure 10-2 Free and attached earlobes in humans

Suppose a man has the genotype *FF*. What is the genotype of his gamete (sperm) nuclei?

_~~FF~~ F_

When both alleles in the nucleus are identical, the condition is **homozygous**. Those having both dominant alleles are homozygous dominant.

Suppose a woman has attached earlobes. What is her genotype?

_ff_

Her gametes (ova) carry what allele(s)?

_f_

When both recessives are present in the same nucleus, the individual is said to be *homozygous recessive* for the trait.

Suppose these two individuals produce a child. Show the genotype of the child by doing the cross:

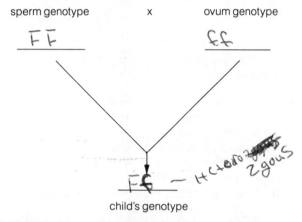

sperm genotype ___ x ___ ovum genotype

_FF_   _ff_

_Ff_ — Heterozygous

child's genotype

When both the dominant and recessive alleles are present within a single nucleus, the individual is **heterozygous** for that trait.

What is the phenotype of the child? (i.e., does this child have attached or free earlobes?)

Ff free earlobes

4. In garden peas, purple flowers are dominant over white flowers. Let *P* represent the allele for purple flowers, *p* the allele for white flowers.

a. What is the phenotype (color) of the flowers with the following genotypes:

Genotype	Phenotype
PP	_____
pp	_____
Pp	_____

CAUTION *Always Be Sure to Distinguish Clearly between Upper- and Lower-case Letters.*

A white-flowered garden pea is crossed with a homozygous dominant purple-flowered plant.

b. What is the genotype of the gametes of the white-flowered plant?

_w_

c. What is the genotype of the gametes of the purple-flowered plant?

_~~P~~ w_

d. What is the genotype of the plant produced by the cross?

_W~~P~~w_

e. What is the phenotype of the plant produced by the cross?

_____

A convenient method of performing the mechanics of a cross is to use a Punnett square. The circles along the top and side of the Punnett square represent the gamete nuclei. Insert the proper letters indicating the genotypes of the gamete nuclei for the above cross in the circles and then fill in the Punnett square.

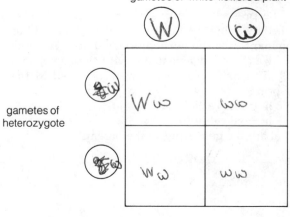

gametes of white-flowered plant

A heterozygous plant is crossed with a white-flowered plant. Fill in the Punnett square and give the genotypes and phenotypes of the offspring.

gametes of white-flowered plant

Genotypes: _____

Phenotypes: _____

(Draw a line from the genotype to its respective phenotype.)

For the remaining problems, you may wish to draw your own Punnett squares on a separate sheet of paper.

5. In mice, black fur (B) is dominant over brown fur (b). Breeding a brown mouse and a homozygous black mouse produces all black offspring.

a. What is the genotype of *the gametes* produced by the brown-furred parent?

_____ *b* _____

b. What is the genotype of the brown-furred parent?

_____ *bb* _____

c. What is the genotype of the black-furred parent?

_____ *B B* _____

d. What is the genotype of the black-furred offspring?

_____ *Bb* _____

By convention, P stands for the parental generation. The offspring are called the "first filial generation," abbreviated F_1. If these F_1 offspring are crossed, their offspring are called the "second filial generation," designated F_2. Note the following diagram.

P x P   and F_1 x F_1

Bb Bb

↓   ↓

F_1 F_2

e. If two of the F_1 mice are bred with one another, what will be the phenotype of the F_2 and in what proportion? *genotypes BB Bb bb,*

Phenotype *Black & brown*

Proportion *3 : 1*

6. The presence of horns on Hereford cattle is controlled by a single gene. The hornless (H) condition is dominant over the horned (h) condition. A hornless cow was crossed repeatedly with the same horned bull. The following results were obtained in the F_1 offspring:

   8 hornless cattle
   7 horned cattle

What are the genotypes of the parents?

cow _____ *Hh* _____

bull _____ *hh* _____

7. In fruit flies, red eyes (R) are dominant over purple eyes (r). Two red-eyed fruit flies were crossed, producing the following offspring:

   76 red-eyed flies
   24 purple-eyed flies

a. What is the approximate ratio of red-eyed to purple-eyed flies? *3 : 1*

b. Based upon your experience with previous problems, what two genotypes give rise to this ratio?

2 heterozygous

c. What are the genotypes of the parents?

Rr   Rr

d. What is the genotypic ratio of the F_1?

1 RR : 2 Rr : 1 rr

e. What is the phenotypic ratio of the F_1?

3 : 1
reds   purple

II. MONOHYBRID PROBLEMS WITH INCOMPLETE DOMINANCE   (Starr, pp. 122–123)

8. Petunia flower color is governed by two alleles, but neither allele is truly dominant over the other. Petunias with the genotype R^1R^1 are red-flowered, those that are heterozygous (R^1R^2) are pink, while those with the R^2R^2 genotype have white flowers. This is an example of **incomplete dominance.** (Note that superscripts are used rather than upper- and lower-case letters to describe the alleles.)

a. If a white-flowered plant is crossed with a red-flowered petunia, what is the genotypic ratio of the F_1?

2 : 1

b. What is the phenotypic ratio of the F_1?

_____

c. If two of the F_1 offspring were crossed, what phenotypes would appear in the F_2?

_____

d. What would be the genotypic ratio in the F_2 generation?

_____

III. MONOHYBRID PROBLEMS ILLUSTRATING CODOMINANCE

(Starr, pp. 122–123)

9. Another type of monohybrid inheritance involves the expression of *both* phenotypes in the heterozygous situation. This is called **codominance.**

One of the best known examples of codominance occurs in the blue Andalusian chicken. "Blue" birds are heterozygous (B^1B^2) and result from the mating between a black bird (B^1B^1) and a "splashed white" bird (B^2B^2). Splashed whites are predominantly white but have some feathers with black margins. Blue birds do not really have blue feathers, instead having a mixture of black and white feathers that reflects light to appear blue. Thus, the "blue" coloration is not a consequence of blending of the pigments (after all, the mix is not gray) but rather the result of *both* colors existing on the same bird. That is, both phenotypes occur on the same individual.

a. If a blue Andalusian hen is mated with a splashed white rooster, what will be the genotypic and phenotypic ratios in the F_i generation?

genotypic ratio _____

phenotypic ratio _____

b. List the parental genotypes of crosses that could produce at least some:

splashed white offspring $B^1\ B^2\quad B''B^1$

black offspring _____

IV. MONOHYBRID, SEX-LINKED PROBLEMS
(Starr, pp. 128–131)

10. In humans, as well as in other primates, sex is determined by special sex chromosomes. An individual containing two X chromosomes is a female, while an individual possessing an X and a Y chromosome is a male. (Rare exceptions of XY females and XX males have recently been discovered.)

a. What sex chromosomes do you have?

_____

b. In terms of sex chromosomes, what type of gametes (ova) does a female produce?

_____

c. What are the possible sex chromosomes in a male's sperm cells?

_____

d. The gametes of which parent will determine the sex of the offspring?

_____

11. The sex chromosomes bear alleles for traits, just like the other chromosomes in our bodies. Genes that occur on the sex chromosomes are said to be **sex-linked**. More specifically, the genes present on the X chromosome are said to be X-linked. There are many more genes present on the X chromosome than is found on the Y chromosome. Nonetheless, those genes found on the Y chromosome are said to be Y-linked.

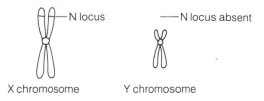

Figure 10-3 Diagrammatic representation of a sex-linked trait

The Y chromosome is smaller than its homologue, the X chromosome. Consequently, some of the loci present on the X chromosome are absent on the Y chromosome.

In humans, color vision is X-linked; the gene for color vision is located on the X chromosome but is absent from the Y chromosome. Figure 10-3 illustrates the appearance of duplicated sex chromosomes, each consisting of two sister chromatids.

In Fig. 10-4 sketch the appearance and distribution of the sex chromosomes as they would appear in gametes after meiosis.

Normal color vision (X^N) is dominant over color blindness (X^n). Suppose a color-blind man fathers children of a woman with the genotype X^NX^N.

a. What is the genotype of the father?

_____

b. What proportion of daughters would be color blind?

_____

c. What proportion of sons would be color blind?

_____

12. One of the daughters from the above problem marries a color-blind man.

a. What proportion of their sons will be color blind? (Another way to think of this is to ask what are the *chances* that their sons will be color blind.)

_____

b. Explain how a color-blind daughter might result from this marriage.

_____

_____

Examine the genetic corn demonstration. This illustrates a monohybrid cross between plants producing purple kernels and plants producing yellow kernels. Note that all the first-generation kernels (F_1) are purple while the second-generation ear (F_2) has both purple kernels and yellow kernels. Count the purple kernels and then the yellow kernels. _____ *purple:* _____*yellow.* When reduced to the lowest common denominator, is this ratio closest to 1:1, 2:1, 3:1, or 4:1? _____. This is called the *phenotypic ratio.*

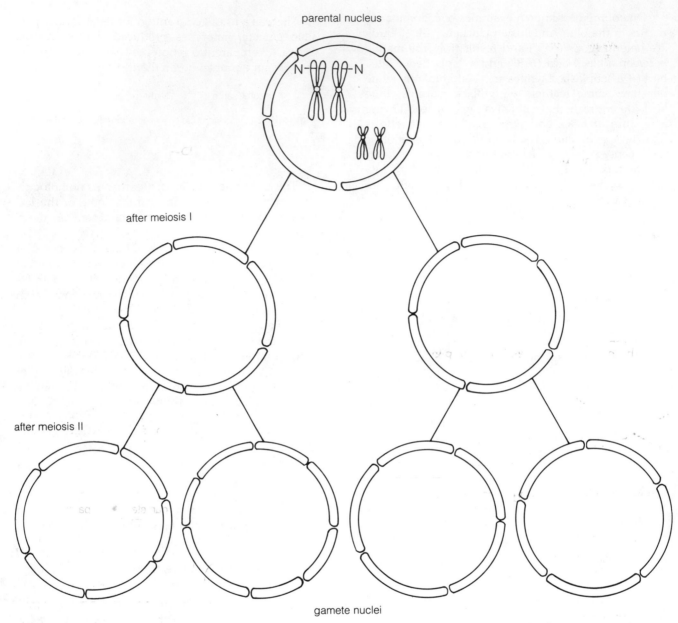

parental nucleus

after meiosis I

after meiosis II

gamete nuclei

Figure 10-4 Distribution of sex chromosomes after meiosis

13. A corncob represents the products of multiple instances of sexual reproduction. Each kernel represents a single instance; fertilization of one egg by one sperm produced *each* kernel. Thus each kernel represents a different cross.

a. What genotypes produce a purple phenotype?

_____

b. Which allele is dominant?

_____

c. What is the genotype of the yellow kernels on the F_2 ear?

_____

d. Suppose you were given an ear with purple kernels. How could you determine its genotype with a single cross?

_____

_____

V. DIHYBRID PROBLEMS   (Starr, pp. 122–123)

All the problems so far have involved the inheritance of only one trait, i.e., they were monohybrid problems. We will now examine cases in which two traits are involved: **dihybrid problems.**

NOTE *We will assume that the genes for these traits are carried on **different** chromosomes.*

Examine the demonstration of dihybrid inheritance in corn. Notice that not only are the kernels two different colors (one trait), they are also differently shaped (second trait). Kernels with starchy endosperm (the carbohydrate-storing tissue) are smooth, while those with sweet endosperm are shriveled. Notice that all *four* possible phenotypic combinations of color and shape are present in the F_2 generation.

14. In humans, a pigment in the front part of the eye masks a blue layer at the back of the iris. The dominant allele P causes production of this pigment. Those who are homozygous recessive (pp) lack the pigment, and the back of the iris shows through, resulting in blue eyes. (Other genes determine the color of the pigment, but in this problem we'll consider only the presence or absence of *any* pigment at the front of the eye.)

Dimpled chins (D = allele for dimpling) are dominant over undimpled chins (d = allele for lack of dimple).

a. List all possible genotypes for an individual with pigmented iris and dimpled chin.

b. List the possible genotypes for an individual with pigmented iris but lacking a dimpled chin.

c. List the possible genotypes of a blue-eyed, dimple-chinned individual.

d. List the possible genotypes of a blue-eyed individual lacking a dimpled chin.

15. Suppose an individual is heterozygous for both traits (eye pigmentation and chin form).

a. What is the genotype of such an individual?

b. What are the possible genotypes of that individual's gametes?

If determining the answer for the last question was difficult, recall from Exercise 9 that the principle of independent assortment states that genes on different chromosomes are separated out independently of one another during meiosis. That is, the occurrence of an allele for eye pigmentation in a gamete has no bearing on which allele for chin form will occur in that same gamete.

There is a useful convention for determining possible gamete genotypes produced during meiosis from a given parental genotype. Using the genotype *PpDd* as an example, here's the method:

Follow the four arrows to determine the four gamete genotypes.

c. Suppose two individuals heterozygous for both eye pigmentation and chin form have children. What are the possible genotypes of their children?

You can set up a Punnett square to do dihybrid problems just as you did with monohybrid problems. However, depending upon the parental genotypes, the square may have as many as 16 boxes rather than just 4. Insert the possible genotypes of the gametes from one parent in the top circles and the gamete genotypes of the other parent in the circles to the left of the box.

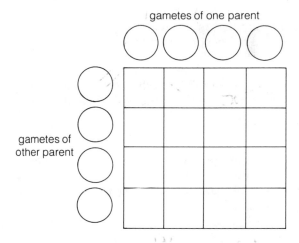

gametes of one parent

gametes of other parent

Possible genotypes of children produced by two parents heterozygous for both eye pigmentation and chin form are

d. What is the ratio of the genotypes?

e. What is the phenotypic ratio?

16. You would probably agree that it is unlikely that a family will have 16 children. In fact, one of the most useful facets of problems such as these is that they allow you to *predict* what the chances are for a phenotype occurring. Genetics is really a matter of **probability,** the likelihood of the occurrence of any particular outcome.

To take a simple example, consider that the probability of coming up with heads in a single toss of a coin is one chance in two, or 1/2.

Now apply this example to the question of the probability of having a certain genotype. Look at your Punnett square in problem 15. The probability of having a genotype is the sum of all occurrences of that genotype. For example, the genotype *PPDD* occurs in 1 of the 16 boxes. The probability of having the genotype *PPDD* is 1/16.

f. What is the probability of an individual from the above problem having the genotype:

ppDD _____

PpDd _____

PPDd _____

To extend this idea, let's consider the probability of flipping heads twice in a row with our coin. The chance of flipping heads the first time is 1/2. The same is true for the second flip. The chance (probability) that we will flip heads twice in a row is $1/2 \times 1/2 = 1/4$. The probability that we could flip heads three times in a row is $1/2 \times 1/2 \times 1/2 = 1/8$.

g. Returning to eye color and chin form, state the probability that three children born to these parents will have the genotype *ppdd*.

_____

h. What is the probability that three children born to these parents will have dimpled chins and pigmented eyes?

_____

i. What is the genotype of the F_1 generation when the father is homozygous for both pigmented eyes and dimpled chin, and the mother has blue eyes and no dimple?

_____

j. What is the phenotype of the individual(s) you determined in letter i. above?

_____

17. A pigment-eyed, dimple-chinned man marries a blue-eyed woman without a dimpled chin. Their first-born child is blue-eyed and has a dimpled chin.

a. What are the possible genotypes of the father?

_____

b. What is the genotype of the mother?

_____

c. What alleles may have been carried by the father's sperm?

_____

18. Suppose a dimple-chinned, blue-eyed man whose father lacked a dimple marries a woman who is homozygous recessive for both traits.

a. What would be the expected genotypic ratio of children produced in this marriage?

_____

b. What would be the expected phenotypic ratio?

_____

19. In his original work on the genetics of garden peas, Mendel found that yellow seed color (*YY, Yy*) was dominant over green seeds (*yy*) and that round seed shape (*RR, Rr*) was dominant over shrunken seeds (*rr*). Mendel crossed pure breeding (homozygous) yellow, round-seeded plants with green, shrunken-seeded plants.

a. What would be the genotype and phenotype of the F_1 produced from such a cross?

_____

b. If the F_1 plants are crossed, what would be the expected phenotypic ratio of the F_2?

_____

VI. MULTIPLE ALLELES (Starr, pp. 386–387)

20. The major blood groups in humans are determined by **multiple alleles,** that is, there are *more than* two possible alleles, any one of which can occupy a locus.

In this ABO blood group system, a single gene can exist in any of three allelic forms: I^A, I^B, or i. The alleles A and B code for production of antigen A and antigen B (two proteins) on the surface of red blood cells. Alleles A and B are codominant, while allele i is recessive.

Four blood groups (phenotypes) are possible from combinations of these alleles (Table 10-1).

Table 10-1	The ABO Blood Groups		
Blood Type	Antigens Present	Antibody Present	Genotype
O	neither A nor B	A and B	*ii*
A	A	B	$I^A I^A$ or $I^A i$
B	B	A	$I^B I^B$ or $I^B i$
AB	AB	neither A nor B	$I^A I^B$

a. Is it possible for a child with blood type O to be produced by two AB parents?

_____

Explain.

_____

_____

b. In a case of disputed paternity, the child is type O, the mother type A. Could an individual of the following blood types be the father?

O _____

A _____

B _____

AB _____

VII. DETERMINING YOUR BLOOD TYPE: THE ABO AND Rh BLOOD GROUPS
(Starr, pp. 386–387)

_____

CAUTION _Some people are very sensitive to the sight of blood. Determining your blood type is optional. Be seated while you do this exercise and have a lab partner at your side to support you should you become faint._

_____

_____

CAUTION _Blood is a potential source of disease-causing organisms that can be transmitted with a puncture. If you have been exposed to serum hepatitis or AIDS, or have hemophilia, do not do this procedure. Do not touch any material contaminated by another person's blood. Never reuse a lancet or use a contaminated lancet._

_____

PROCEDURE

1. On a blood typing card, place one drop of anti-A serum in the correspondingly labeled circle, and one drop of anti-B and anti-D serums in their respective circles (Fig. 10-5a). Antigen D determines Rh type; the abbreviation "Rh" comes from the origin of its discovery, the _Rh_esus monkey.

2. Clean the finger from which you are going to draw blood by swabbing it with a cotton ball soaked in 70% ethanol.

3. Shake downward the hand with the finger from which you will draw blood.

4. Place your hand on the desk and use a sterile blood lancet to puncture your cleaned finger (Fig. 10-5b).

You may wish to have an assistant puncture your finger since it is often difficult to draw blood from oneself.

_____

CAUTION _Do not touch the pointed end of the lancet with any object before pricking the finger._

_____

5. Place one drop of blood in each circle on the test card labeled "Blood" (Fig. 10-5c).

6. Use a clean toothpick _for each drop of blood_ to mix the blood and serum drops (Fig. 10-5d).

7. Gently rock the test card back and forth for a thorough mixing of blood and antiserum (Fig. 10-5e).

8. Tilt the test card toward one corner to allow the mixtures to drain, but don't tilt so steeply that they drain out of the circles (Fig. 10-5f).

9. Examine the blood films for signs of agglutination. Compare your results with Fig. 10-5g to determine your blood type.

Agglutination (clumping together of the blood cells) appears when the antibody combines with its respective antigen present on the surface of the red blood cell. If you have a particular antigen, your blood plasma _lacks_ that respective antibody, but it has the opposite antibody. Table 10-1 summarizes the relationship between the ABO blood groups, antigens, and antibodies.

Knowing blood type is critical in blood transfusions. If the blood from a group A donor is mixed (transfused) with that of a group B recipient, the red blood cells will agglutinate and become trapped in the recipient's capillaries, where, after several days, they will rupture. Breakdown products may clog vital organs and may cause death in extreme cases.

Inheritance of the Rh blood group is complex, with different theories advancing different explanations. This blood group is determined by the presence or absence of antigen D. Persons having the antigen are considered Rh +; those without it, Rh −.

Fill in Table 10-2.

Table 10-2	Summary of My Blood Data			
My Blood Type	ABO Antigens Present in My Blood	ABO Antibodies Present in My Blood	My Possible Genotype for ABO Blood Group	Rh Antigens Present in My Blood

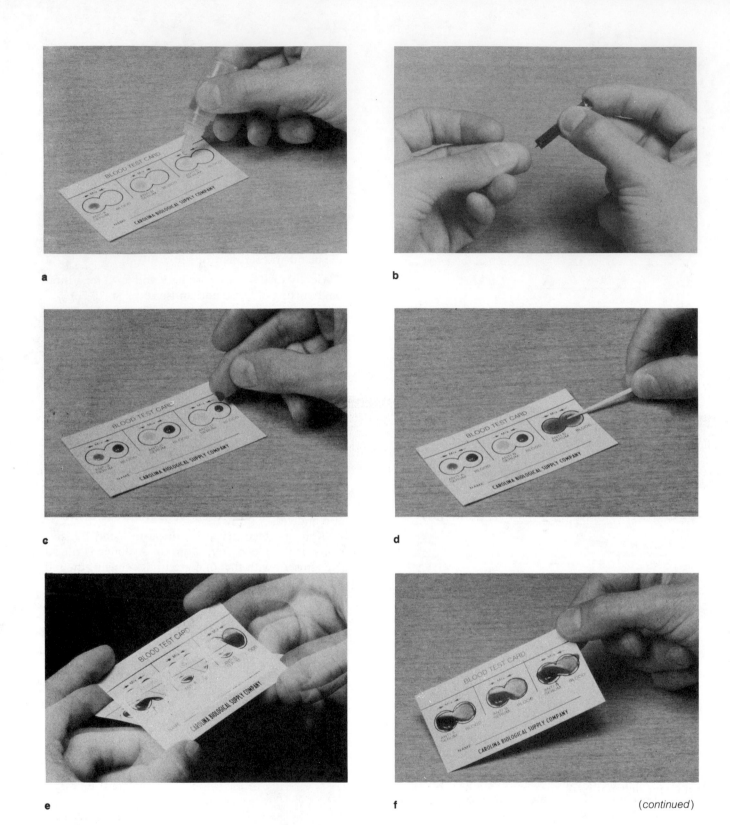

a

b

c

d

e

f

(continued)

Figure 10-5 Method for determining blood type. (Photos from *Carolina Tips*, courtesy Carolina Biological Supply Company.)

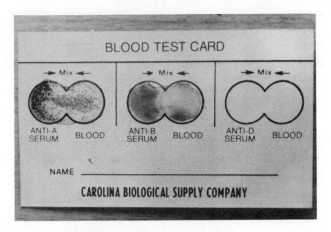

g (Blood type A)

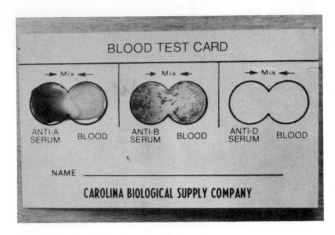

h (Blood type B)

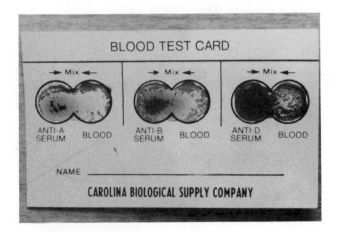

i (Blood type AB⁺)

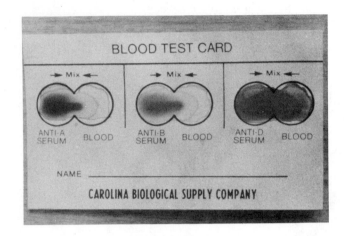

j (Blood type O⁻)

Figure 10-5 *continued*

21. Confine your attention to the ABO blood groups in this problem. Pick a lab partner of the opposite sex and write his/her blood group here:

_____

a. Would you be able to receive a blood transfusion from your lab partner without agglutination of your red blood cells occurring?

_____

b. Would your lab partner be able to receive a blood transfusion from you?

_____

c. Suppose you and your lab partner were to have children. List the possible ABO phenotypes your children might have.

_____

d. List the respective genotypes for your children's phenotypes.

_____

VIII. SOME READILY OBSERVABLE HUMAN TRAITS

In the preceding problems, we examined several human traits that are fairly simple and that follow the Mendelian pattern of inheritance. Most of our traits are much more complex, involving many genes or interactions between genes. As an example, hair color is determined by at least four genes, each one coding for the production of melanin, a brown pigment. Because the effect of these genes is cumulative, hair color can range from blond (little melanin) to very dark brown (much melanin).

Clearly, human traits are most interesting to humans. A number of traits listed below exhibit Mendelian inheritance. For each, examine your phenotype and fill in Table 10-3. List your possible genotype(s) for each trait. When convenient, examine your parents' phenotypes and attempt to determine your actual genotype.

Table 10-3 Summary of My Mendelian Traits

Trait	My Phenotype	My Possible Genotype(s)	Mom's Phenotype	Mom's Possible Genotype	Dad's Phenotype	Dad's Possible Genotype	My Possible or Probable Genotype
Mid-digital hair							
Tongue rolling							
Widow's peak							
Earlobe attachment							
Hitchhiker's thumb							
Relative finger length							

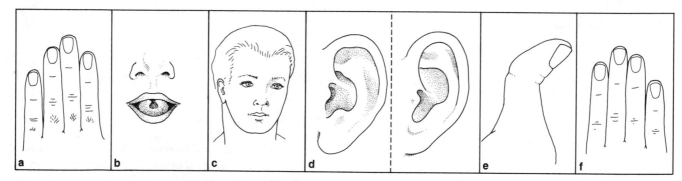

Figure 10-6 Some readily observable human Mendelian traits

1. *Mid-digital hair (Fig. 10-6a).* Examine the middle joint of your fingers for the presence of hair, the dominant condition (*MM, Mm*). Complete absence of hair is due to the homozygous-recessive condition (*mm*). You may need a hand lens to determine your phenotype. Even the slightest amount of hair indicates the dominant condition.

2. *Tongue rolling (Fig. 10-6b).* The ability to roll one's tongue is due to a dominant allele, *T*. The homozygous-recessive condition results in inability to roll one's tongue.

3. *Widow's peak (Fig. 10-6c).* Widow's peak describes a distinct downward point in the frontal hairline and is due to the dominant allele, *W*. The recessive allele, *w*, results in a continuous hairline. (Omit study of this trait if baldness is affecting the hairline.)

4. *Earlobe attachment (Fig. 10-6d).* Most individuals have free earlobes (*FF, Ff*). Homozygous recessives (*ff*) have earlobes attached directly to the head.

5. *Hitchhiker's thumb (Fig. 10-6e).* Although considerable variation exists in this trait, we will consider those individuals who *cannot* extend their thumbs backward to approximately 45° to be carrying the dominant allele, *H*. Homozygous-recessive persons (*hh*) can bend their thumbs at least 45°, if not farther.

6. *Relative finger length (Fig. 10-6f).* An interesting sex-influenced (*not sex-linked*) trait relates to the relative lengths of the index and ring finger. In males, the allele for a short index finger (*S*) is dominant. In females, it is recessive. In rare cases each hand may be different. If one or both index fingers are greater than or equal to the length of the ring finger, the recessive genotype is present in males, and the dominant present in females.

PRE-LAB QUESTIONS

____ **1.** In a monohybrid cross (a) only one trait is being considered, (b) the parents are always homozygous, (c) the parents are always heterozygous, (d) no hybrid is produced.

____ **2.** The genetic makeup of an organism is its (a) phenotype, (b) genotype, (c) locus, (d) gamete.

____ **3.** An allele whose expression is completely masked by the expression or effect of its allelic partner is (a) incompletely dominant, (b) homozygous, (c) dominant, (d) recessive.

____ **4.** The physical appearance of an organism, resulting from interactions of its genetic makeup and its environment, is (a) phenotype, (b) hybrid vigor, (c) dominance, (d) genotype.

____ **5.** An organism that is heterozygous for a trait is (a) haploid, (b) homozygous, (c) diploid, (d) all of the above.

____ **6.** Codominance occurs when (a) the phenotype for both (or all) alleles is expressed, (b) the individual is heterozygous, (c) the organism is homozygous recessive, (d) a and b above.

____ **7.** A gene located only on the female (X) chromosome having no allelic partner on the Y chromosome would be (a) incompletely dominant, (b) codominant, (c) sex-linked, (d) heterozygous.

____ **8.** The sex chromosome determining maleness is (a) the Y chromosome, (b) the X chromosome, (c) sex-linked, (d) heterozygous.

____ **9.** A nucleus containing only one of the two homologues is (a) sex-linked, (b) an improbable event, (c) diploid, (d) haploid.

____ **10.** An example of a trait controlled by multiple alleles is (a) baldness in males, (b) color blindness, (c) the ABO blood groups, (d) blue Andalusian chickens.

EXERCISE 10
MENDELIAN GENETICS

POST-LAB QUESTIONS

1. Explain the implications of the principle of independent assortment as it applies to distribution of alleles in gametes.

2. What does it mean to say that certain traits are sex-linked?

3. Distinguish between incomplete dominance and codominance.

4. Define the term *multiple alleles*.

5. Suppose you have two traits controlled by genes on separate chromosomes. If sexual reproduction occurs between two heterozygous parents, what will be the genotypic ratio of all possible gametes?

6. What is the probability that five sons and no daughters will be born to parents?

7. How does probability differ from actuality?

8. Studies have suggested (although not proved) that whether you are right- or left-handed may be hereditary. Homozygous-dominant (*RR*) people are strongly right-handed and are not easily influenced to change preferences. Homozygous-recessive individuals are strongly left-handed. Heterozygous individuals are more variable. They are potentially ambidextrous but are easily influenced by environment or training.

  a. Would you characterize handedness as an example of complete dominance, incomplete dominance, or codominance?

  b. Would it be possible for a left-handed person to be heterozygous?

  c. Would it be possible for two left-handed parents to have a right-handed child? Explain.

9. For this problem, assume that one allele is completely dominant over the other.

  Suppose two individuals heterozygous for a *single* trait have children. What is the expected phenotypic ratio of the offspring?

  If two individuals heterozygous for *two* traits have children, what would be the expected phenotypic ratio of the offspring?

  Remember that the gene for each trait is located at a locus, a physical region on the chromosome. Suppose that crossing two individuals heterozygous for two traits resulted in the same phenotypic ratio as for a single trait. Are the genes for these two traits on separate chromosomes or on the same chromosome? Explain your answer.

NUCLEIC ACIDS: BLUEPRINTS FOR LIFE

OBJECTIVES After completing this exercise you will be able to:

1. define DNA, RNA, purine, pyrimidine, principle of base pairing, replication, transcription, translation, codon, anticodon, peptide bond, gene, genetic engineering, recombinant DNA, bacterial conjugation, plasmid;

2. identify the components of deoxyribonucleotides and ribonucleotides;

3. distinguish between DNA and RNA according to their structure and function;

4. describe DNA replication, transcription, and translation;

5. give the base sequence of DNA or RNA when presented with the complementary strand;

6. identify a codon and anticodon on RNA models and describe the location and function of each;

7. give the base sequence of an anticodon when presented with that of a codon, and vice versa;

8. describe what is meant by the "one gene, one polypeptide hypothesis";

9. describe the process of DNA recombination by bacterial conjugation;

10. explain the difference between DNA recombination by bacterial conjugation and the technique by which eukaryotic gene products are produced by bacteria.

INTRODUCTION By 1900 the patterns of inheritance had been demonstrated by Gregor Mendel, based solely on careful experimentation and observation. Mendel had no idea how the traits he observed were passed from generation to generation, although the seeds of that knowledge had been sown as early as 1869, when the physician-chemist Friedrich Miescher isolated the chemical substance of the nucleus. Miescher found the substance to be an acid with a large phosphorus content and named it "nuclein." Subsequently, nuclein was identified as **DNA**, short for **deoxyribonucleic acid.** Some 75 years would pass before the significance of DNA would be revealed.

Few would argue that the demonstration of DNA as the genetic material and subsequent determination of its molecular structure are among the most significant discoveries of the twentieth century. Since the early 1950s, when James Watson and Francis Crick built their first model of DNA, tremendous advances in molecular biology have occurred, many of them based upon the structure of DNA. Today we speak of gene therapy and genetic engineering in household conversations. In the minds of some, these topics raise hopes for curing or preventing many of the diseases plaguing humanity. For others, thoughts turn to "playing with nature, undoing the deeds of God, or creating monstrosities that will wipe humanity off the face of the earth."

This exercise will familiarize you with the basic structure of nucleic acids and their role in the cell. Understanding the function of nucleic acids—both DNA and **RNA** (**ribonucleic acid**)—is central to understanding life itself. We hope you will gain an understanding that will allow you to form educated opinions concerning what science should do with its new-found technology.

In this exercise, we are concerned with three processes: *replication, transcription,* and *translation.* But before we study these three per se, let's formulate an idea of the structure of DNA itself.

I. MODELING THE STRUCTURE AND FUNCTION OF NUCLEIC ACIDS AND THEIR PRODUCTS (Starr, pp. 145–156)

MATERIALS

Per student pair:

- DNA puzzle kit

Per lab room:

- DNA model

PROCEDURE

A. Nucleic Acid Structure

Work in pairs.

NOTE *Clear Your Work Surface of Everything Except Your Lab Manual and the DNA Puzzle Kit.*

---

1. Obtain a nucleic acid construction kit. It should contain the following parts:

- 18 deoxyribose sugars
- 9 ribose sugars
- 18 phosphate groups
- 4 adenine bases
- 6 guanine bases
- 6 cytosine bases
- 4 thymine bases
- 2 uracil bases

- 3 transfer RNA (tRNA)
- 3 amino acids
- 3 activating units
- ribosome template sheet

2. Group the components into separate stacks. Select a single deoxyribose sugar, an adenine base (labeled "A"), and a phosphate, fitting them together as shown in Fig. 11-1. This is a single nucleotide (specifically a *deoxy*ribonucleotide), a unit consisting of a sugar (deoxyribose), a phosphate group, and a nitrogen-containing base (adenine).

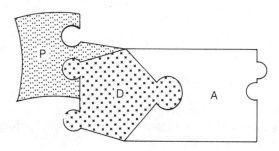

Figure 11-1 One deoxyribonucleotide

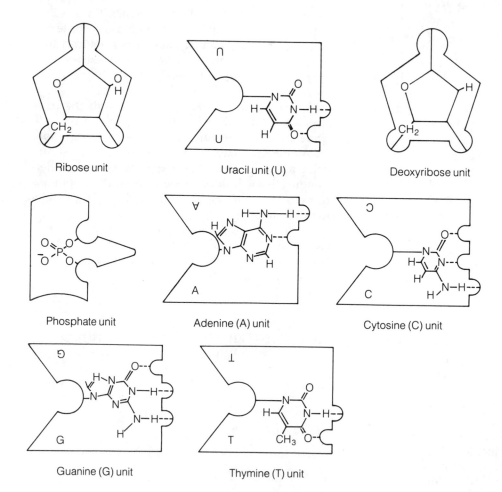

Let's examine each component of the nucleotide. Deoxyribose (Fig. 11-2) is a sugar compound containing five carbon atoms. Four of the five are joined by covalent bonds into a ring. Each carbon is given a number, indicating its position in the ring. (These numbers are read "1-prime, 2-prime," etc. "Prime" is used to distinguish the carbon atoms from the position of atoms in the nitrogen-containing bases.)

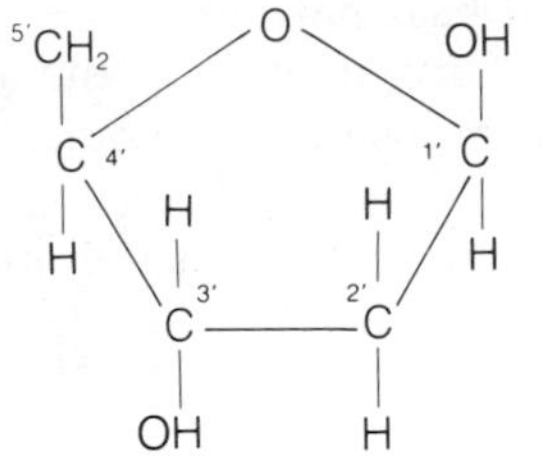

Figure 11-2 Deoxyribose

This structure is usually illustrated in a simplified manner, without actually showing the carbon atoms within the ring.

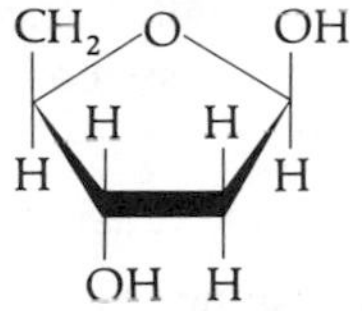

Figure 11-3 Simplified representation of deoxyribose

There are four kinds of nitrogen-containing bases in DNA. Two are **purines** and are double-ring structures. Specifically, the two purines are *adenine* and *guanine* (abbreviated A and G, respectively).

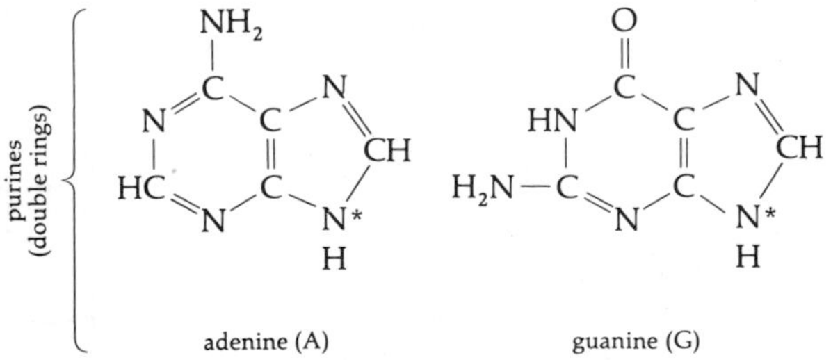

purines (double rings)

adenine (A)   guanine (G)

Figure 11-4 Double-ringed purines found in DNA

The other two nitrogen-containing bases are **pyrimidines,** specifically *cytosine* and *thymine* (abbreviated C and T, respectively). Pyrimidines are single-ring compounds.

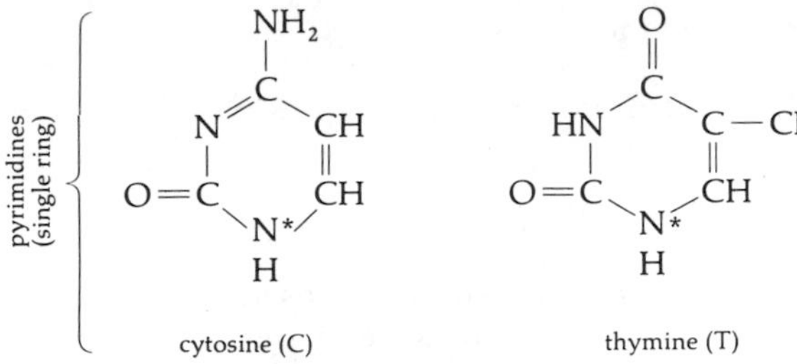

pyrimidines (single ring)

cytosine (C)   thymine (T)

Figure 11-5 Pyrimidines found in DNA

The symbol * indicates where a bond forms between each nitrogen-containing base and the 1' carbon atom of the sugar ring structure. Although deoxyribose and the nitrogen-containing bases are organic compounds (they contain carbon), the phosphate group is an inorganic compound, with the structural formula shown in Fig. 11-6:

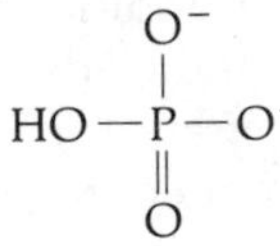

Figure 11-6 Phosphate group found in nucleic acids

The phosphate end of the deoxyribonucleotide is referred to as the 5' end, because the phosphate group bonds to the 5' carbon atom.

There are four kinds of deoxyribonucleotides, each differing only in the type of base it possesses. Construct the other three kinds of deoxyribonucleotides. Draw each in Fig. 11-7b through d. Rather than drawing the somewhat complex shape of the model, in this and other drawings just give the correct position and letters. Use D for deoxyribose, P for a phosphate group, and A, C, G, and T for the different bases (as shown in Fig. 11-7a).

Deoxyribonucleotide containing adenine	Deoxyribonucleotide containing guanine	Deoxyribonucleotide containing cytosine	Deoxyribonucleotide containing thymine
a	b	c	d

Figure 11-7 Drawings of deoxyribonucleotides containing guanine, cytosine, and thymine

Note the small notches and projections in the nitrogen-containing bases. Will the notches of adenine and thymine fit together?

_____

Of guanine and cytosine?

_____

Of adenine and cytosine?

_____

Of thymine and guanine?

_____

The notches and projections represent bonding sites. Make a conclusion about which bases will bond with one another.

_____

_____

_____

Will a purine base bond with another purine?

_____

Will a purine base bond with both types of pyrimidines?

_____

3. Assemble the three additional deoxyribonucleotides, linking them with the adenine-containing unit, to form a nucleotide strand of DNA. Note that the sugar backbone is bonded together by phosphate groups. Your strand should appear as shown in Fig. 11-8.

4. Now assemble a second four-nucleotide strand, similar to that of Fig. 11-8. However, this time make the base sequence T — A — C — G, from top to bottom.

DNA molecules consist of *two* strands of nucleotides, each strand the *complement* of the other.

5. Assemble the two strands by attaching (bonding) the nitrogen bases of complementary strands. Note that the adenine of one nucleotide always pairs with the thymine of its complement; similarly, guanine always pairs with cytosine. This phenomenon is called the **principle of base pairing.**

On Fig. 11-9, attach letters to the model pieces indicating the composition of your double-stranded DNA model.

What do you notice about the *direction* in which each strand is running?

_____

_____

_____

(Does the second strand of your drawing show this? It should.)

In life, the purines and pyrimidines are joined together by hydrogen bonds. Note again that the sugar backbone is linked by phosphate groups. Your model illustrates only a very small portion of a DNA molecule. The entire molecule may be tens of thousands of nucleotides in length!

6. Slide your DNA segment aside for the moment.

7. Examine the three-dimensional model of DNA on display in the laboratory. Notice that the two strands of DNA are twisted into a spiral staircase-like pattern.

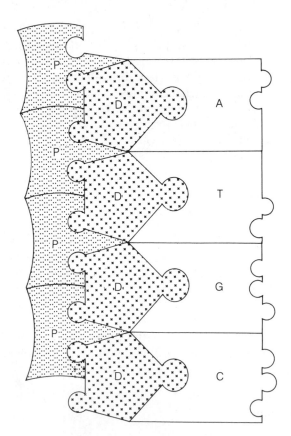

Figure 11-8 Four-nucleotide strand of DNA

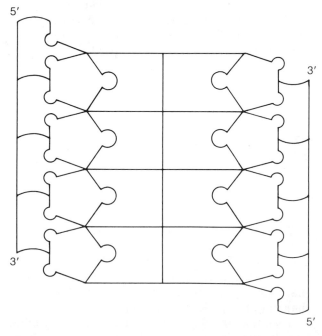

Labels: A, T, G, C, D, P (all used more than once)

Figure 11-9 Drawing of a double strand of DNA

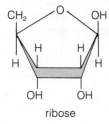

ribose

Figure 11-10 Ribose

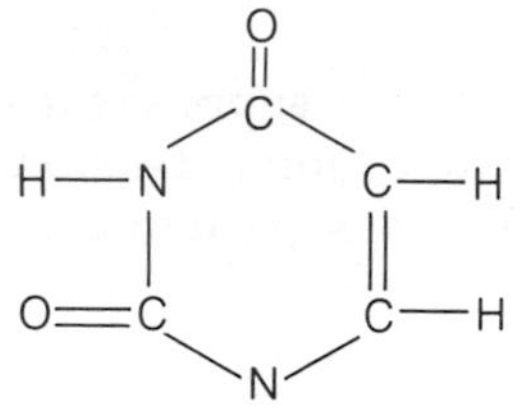

Figure 11-11 The pyrimidine uracil

This is why DNA is known as a **"double helix."** Identify the deoxyribose sugar, nitrogen-containing bases, hydrogen bonds linking the bases, and the phosphate groups.

The second type of nucleic acid is RNA, short for ribonucleic acid. There are three important differences between DNA and RNA:

a. RNA is a *single strand* of nucleotides.

b. The sugar of RNA is **ribose.**

c. RNA lacks the nucleotide that contains thymine. Instead, it has one containing the pyrimidine uracil (U) (see Fig. 11-11).

Compare the structural formulas of ribose (Fig. 11-10) and deoxyribose (Fig. 11-3). How do they differ?

_____

_____

_____

Why is the sugar of DNA called *deoxy*ribose?

_____

_____

_____

8. From the remaining pieces of your model kit, select four ribose sugars—an adenine, uracil, guanine, and cytosine—and four phosphate groups. Assemble the four **ribonucleotides** and draw each in Fig. 11-12. (Use the convention illustrated in Fig. 11-7 rather than drawing the actual shapes.)

Disassemble the RNA models after completing your drawing.

Figure 11-12 Drawings of four possible ribonucleotides

B. Modeling DNA Replication

DNA **replication** takes place during the S stage of interphase of the cell cycle (see Exercise 8, page 85). Recall that the DNA is aggregated into chromosomes. Before mitosis, the chromosomes duplicate themselves so that the daughter nuclei formed by mitosis will have the same number of chromosomes (and hence the same amount of DNA) as did the parent cell.

Replication begins when hydrogen bonds between nitrogen bases break and the two DNA strands "unzip." Free nucleotides within the nucleus bond to the exposed bases, thus creating *two* new strands of DNA (as described below). The process of replication is controlled by enzymes called **DNA polymerases.**

1. Construct eight more deoxyribonucleotides (two of each kind), but don't link them into strands.

2. Now return to the double-stranded DNA segment you constructed earlier. Separate the two strands, imagining the zipper-like fashion in which this occurs within the nucleus.

3. Link the free deoxyribonucleotides to each of the "old" strands. When you are finished, you should have two double-stranded segments.

Note that one strand of each is the parental ("old") strand and that the other is newly synthesized from free nucleotides. This illustrates the *semiconservative* nature of DNA replication. Each of the parent strands remains intact—it is "conserved"—and a new complementary strand is formed on it. Two "half-old, half-new" DNA molecules result.

4. Draw the two replicated DNA molecules in Fig. 11-13, labeling the old and new strands. (Once again, use the convention illustrated in Fig. 11-7.)

What is the sequence from left to right of nitrogen bases on the mRNA strand?

left _____ right

After the mRNA is synthesized within the nucleus, the hydrogen bonds between the nitrogen bases of the deoxyribonucleotides and ribonucleotides break.

4. Separate your mRNA strand from the DNA strand. (You can disassemble the deoxyribonucleotides now.) The mRNA now moves out of the nucleus and into the cytoplasm.

By what avenue do you suppose the mRNA exits the nucleus? (Hint: Re-examine the structure of the nuclear membrane, as described in Exercise 3.)

_____

_____

_____

"To transcribe" means to make a copy of. Is transcription of RNA from DNA the formation of an *exact* copy?

_____

_____

Explain.

_____

_____

_____

You will use this strand of mRNA in the next section. Keep it close at hand.

D. Translation: RNA to Polypeptides

Once in the cytoplasm, mRNA strands attach to *ribosomes*, on which translation occurs. "To translate" means to change from one language to another. In the biological sense, **translation** is the conversion of the linear message encoded on mRNA to a linear strand of amino acids to form a polypeptide. (A *peptide* is two or more amino acids linked by a peptide bond.)

Translation is accomplished by the interaction of mRNA, ribosomes, and **transfer RNA (tRNA)**, another type of RNA. The tRNA molecule is formed into a four-cornered loop. You can think of tRNA as a baggage-carrying molecule. Within the cytoplasm, tRNA attaches to specific free amino acids. This occurs with the aid of activating enzymes, represented in your model kit by the pieces labeled "glycine activating" or "alanine activating." The amino acid-carrying tRNA then positions itself on ribosomes where the amino acids become linked together to form polypeptides.

1. Obtain three tRNA pieces, three amino acid units, and three activating units.

Figure 11-13 Drawing of two replicated DNA segments, illustrating their semiconservative nature

C. Transcription: DNA to RNA

DNA is an "information molecule" residing *within* the nucleus. The information it provides is for assembling proteins *outside* the nucleus, within the cytoplasm. The information does not go directly from the DNA to the cytoplasm. Instead, RNA serves as an intermediary, carrying the information from DNA to the cytoplasm.

Synthesis of RNA takes place within the nucleus by **transcription.** During transcription, the DNA double helix unwinds and unzips, and a single strand of RNA, designated **messenger RNA (mRNA)**, is assembled using the nucleotide sequence of *one* of the DNA strands as a pattern (template). Let's see how this happens.

1. Disassemble the replicated DNA strands into their component deoxyribonucleotides.

2. Construct a new DNA strand consisting of nine deoxyribonucleotides. With the purines and pyrimidines pointing away from you, lay the strand out horizontally in the following base sequence:
T — G — C — A — C — C — T — G — C.

3. Now assemble RNA ribonucleotides on the exposed nitrogen bases of the DNA strand. Don't forget to substitute the pyrimidine uracil for thymine.

2. Join the amino acids first to the activating units and then to the tRNA. Will any tRNA bond with any amino acid, or is each tRNA specific?

_____

3. Now let's do some translating. In the space below, list the sequence of bases on the *messenger* RNA strand, starting at the left.

(left, 3' end) _____ (right, 5' end)

Translation occurs when a *three*-base sequence on mRNA is "read" by tRNA. This three-base sequence on mRNA is called a **codon**. Think of a codon as a three-letter word, read right (5') end to left (3') end. What is the order of the right-most (first) mRNA codon? (Remember to list the letters in the *reverse* order of that in the mRNA sequence.) The first codon on the mRNA model is

(5' end) _____ (3' end)

_____

4. Slide the mRNA strand onto the ribosome template sheet, with the first codon at the 5' end.

5. Find the tRNA-amino acid complex that complements (will fit with) the first codon. The complementary three-base sequence on the tRNA is the **anticodon.** Binding between codons and anticodons begins at the P site of the 40s subunit of the ribosome. The tRNA-amino acid complex with the correct anticodon positions itself on the P site.

6. Move the tRNA-amino acid complex onto the P site on the ribosome template sheet, and fit the codon and anticodon together. In the boxes below, indicate the codon, anticodon, and the specific amino acid attached to the tRNA.

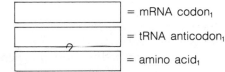

mRNA codon$_1$ = ☐

tRNA anticodon$_1$ = ☐

amino acid$_1$ = ☐

7. Now identify the second mRNA codon and fill in the boxes.

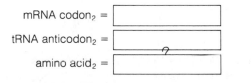

mRNA codon$_2$ = ☐   ☐ = mRNA codon$_1$

tRNA anticodon$_2$ = ☐   ☐ = tRNA anticodon$_1$

amino acid$_2$ = ☐   ☐ = amino acid$_1$

8. The second tRNA-amino acid complex moves onto the A site of the 40s subunit. Position this complex on the A site. An enzyme now catalyzes a condensation reaction, forming a **peptide bond** and linking the two amino acids into a dipeptide. (Water, HOH, is released by this condensation reaction.)

9. Separate amino acid$_1$ from its tRNA and link it to amino acid$_2$. (In reality, separation occurs somewhat later, but the puzzle does not allow this to be shown accurately; see below for correct timing.)

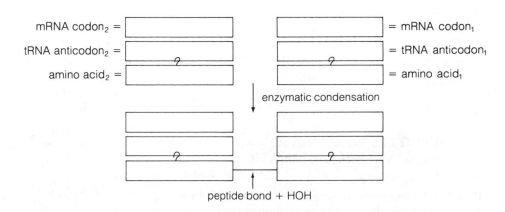

mRNA codon$_2$ = ☐   ☐ = mRNA codon$_1$

tRNA anticodon$_2$ = ☐   ☐ = tRNA anticodon$_1$

amino acid$_2$ = ☐   ☐ = amino acid$_1$

enzymatic condensation

peptide bond + HOH

One tRNA-amino acid complex remains. It must occupy the A site of the ribosome in order to bind with its codon. Consequently, the dipeptide must move to the right.

10. Slide the mRNA to the right (so that $tRNA_2$ is on the P site) and fit the third mRNA codon and tRNA anticodon to form a peptide bond, creating a model of a tripeptide. At about the same time that the second peptide bond is forming, the first tRNA is released from both the mRNA and the first amino acid. Eventually, it will pick up another specific amino acid.

$tRNA_1$ will pick up another

_____

(name the type of amino acid).

Record below the tripeptide that you have just modeled.

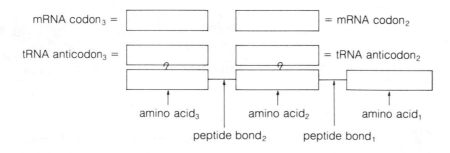

mRNA codon₃ = [   ]   [   ] = mRNA codon₂

tRNA anticodon₃ = [   ]   [   ] = tRNA anticodon₂

amino acid₃   amino acid₂   amino acid₁

peptide bond₂   peptide bond₁

You have created a short polypeptide. Polypeptides may be thousands of amino acids in length. As you see, the amino acid sequence is ultimately determined by DNA, because it was the original source of information.

Finally, let's turn our attention to the concept of a **gene**. Recall that a gene is a unit of inheritance. Our current understanding of a gene is that a gene codes for one polypeptide. This is appropriately called the **"one gene, one polypeptide hypothesis."** Given this concept, do you think a gene consists of one, several, or many deoxyribonucleotides?

A gene probably consists of _____ deoxyribonucleotides.

_____
_____

Please disassemble your models and return them to the proper location.

_____
_____

II. PRINCIPLES OF GENETIC ENGINEERING: RECOMBINATION OF DNA
(Starr, pp. 163–164)

People suffering from diabetes are unable to produce enough insulin, a hormone that is synthesized by the pancreas and that is instrumental in regulating the amount of blood sugar. Therapy for severe diabetes includes daily injections of insulin. Until recently, that insulin was extracted from the pancreas of slaughtered pigs and cows. With the advent of techniques commonly referred to as genetic engineering, human insulin is now produced by bacteria. These organisms grow and reproduce rapidly, hence producing large quantities of insulin en masse.

Genetic engineering is a convenient phrase to describe what is more properly called methods in recombinant DNA. **Recombinant DNA** is DNA into which a set of "foreign" nucleotides has been inserted. In the case of insulin production, researchers first located on human chromosomes the gene (set of nucleotides) that codes for insulin production. Once identified, the nucleotides were removed from the human DNA and inserted into the DNA of a bac-

terium. As this bacterial cell reproduced, each new generation contained the gene coding for insulin synthesis. The cells produced the hormone, which was harvested. Thus, these recombinant bacteria are "insulin factories."

Bacteria have been exchanging genes with each other for millennia. In the process, new strains of bacteria may be produced. The following experiment will familiarize you with genetic recombination in bacteria, principles of which are the basis for genetic engineering.

Two strains of the bacterium *Escherichia coli* will be used in this experiment:

- J-53R carries a chromosomal gene that causes it to be resistant to the antibiotic rifampicin; it is susceptible (killed) by another antibiotic, chloramphenicol.

- HT-99 is resistant to chloramphenicol but susceptible to rifampicin; the gene for resistance to chloramphenicol is located on a small extra-chromosomal (not on the chromosome) loop of DNA called a **plasmid.**

Like chromosomal DNA, plasmids can replicate. Insertion of a foreign DNA segment (set of nucleotides) results in the formation of a hybrid plasmid that can thereby replicate the foreign DNA as well.

Plasmids also code for the ability to transfer themselves from the host (donor) bacterium to a recipient cell by a process called **bacterial conjugation.** Thus, the plasmid acts both as a carrier of foreign DNA and

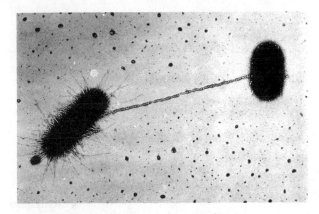

Figure 11-14 Conjugation between two bacteria. (Photo courtesy C. C. Brinton, Jr., and J. Carnahan.)

as an agent (vector) for the introduction of that DNA into the recipient cell. Once in the recipient, the plasmid replicates, and the recipient bears the genes (and hence makes the gene products) formerly in the host.

Plasmid transfer between host and recipient (in this case, two bacterial cells) occurs through a bridge formed by the host cell that connects it to the recipient. Figure 11-14 illustrates bacterial conjugation.

MATERIALS

Per student pair:

- nutrient agar plate containing chloramphenicol
- nutrient agar plate containing rifampicin
- nutrient agar plate containing both chloramphenicol and rifampicin
- culture tube containing *E. coli* J-53R
- culture tube containing *E. coli* HT-99
- culture tube containing mating mixture of both J-53R and HT-99
- test tube rack
- bottle of 10% Clorox solution
- transfer loop
- burner (Bunsen or alcohol), striker or matches
- china marker

Per lab room:

- paper toweling
- 37°C incubator
- container for discarded cultures

Your instructor will demonstrate aseptic technique before you begin. Then proceed in the manner described below.

PROCEDURE

A. Chloramphenicol Plate

1. Pour 10% Clorox disinfectant on the lab bench work area and wipe thoroughly with paper toweling.

2. Without removing the lid, turn the chloramphenicol-containing plate over and, using a china marker, draw a line on the bottom dividing the plate in half. Label one half "J-53R" and the other half "HT-99." Turn the plate right side up.

3. Light the burner.

4. Sterilize the transfer loop by flaming, as illustrated in Fig. 11-15a.

5. Allow the loop to cool for 15 seconds.

6. As shown in Fig. 11-15b, hold the sterile transfer loop and culture tube containing J-53R, and remove the cap of the tube.

Never Set the Cap on the Bench Surface.

---

7. Flame the mouth of the culture tube (Fig. 11-15c).

8. Insert the loop into the culture tube, touch the bacterial colony, and withdraw the loop. Re-flame the mouth of the culture tube and replace the cap (Fig. 11-15d).

9. Lift one edge of the chloramphenicol-containing nutrient agar plate (Fig. 11-15e). Transfer the J-53R to the appropriately labeled side of the chloramphenicol plate. Pull the loop lightly across the agar surface in a close zig-zag streaking pattern. Close the lid and set the plate aside.

Do Not Dig into the Agar.

---

10. Sterilize the transfer loop again by flaming. Allow it to cool.

11. Transfer HT-99 to the other half of the chloramphenicol plate, spreading it the same way as before.

12. Sterilize the loop.

B. Rifampicin Plate

1. With the china marker, draw a line down the center of the bottom of the rifampicin-containing culture plate, labeling one half "J-53R," the other "HT-99."

2. Using the same series of steps you used above, transfer J-53R and HT-99 to this plate.

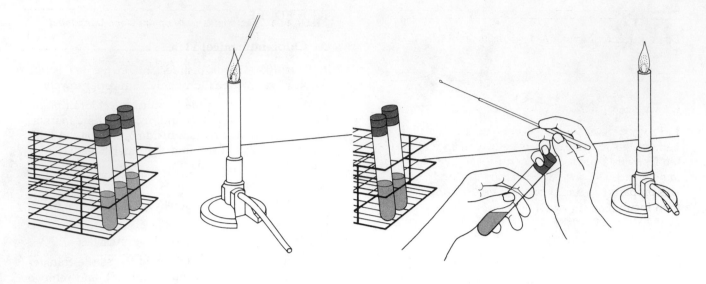

a Sterilize the loop by holding the wire in a flame until it is red hot. Allow it to cool before proceeding.

b While holding the sterile loop and the bacterial culture, remove the cap as shown.

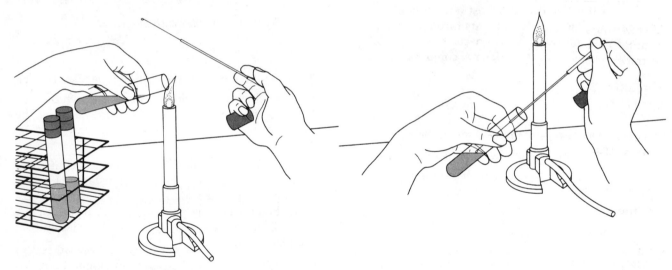

c Briefly heat the mouth of the tube in a burner flame before inserting the loop for an inoculum.

d Get a loopful of culture, withdraw the loop, heat the mouth of the tube, and replace the cap.

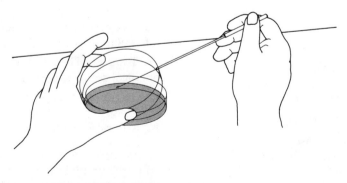

e To inoculate a solid medium in a Petri plate, place the plate on a table and lift one edge of the cover.

Figure 11-15 Procedure for inoculating a culture plate. (After Case and Johnson, 1984.)

C. Chloramphenicol Plus Rifampicin Plate

1. Proceed as above but do not divide this plate in half. Inoculate the culture plate containing both antibiotics with a drop of the culture solution that contains *both* J-53R and HT-99. This "mating solution" was prepared yesterday by mixing the two strains and then incubating it overnight.

2. Mark your name on the culture plates and place them in a 37°C incubator.

Disinfect Your Work Surface with Clorox.

After 24 hours remove the culture plates and examine them for growth. (Your instructor will provide demonstration plates for you to examine to recognize bacterial growth.)

Record your observations in Table 11-1, using a "+" to indicate the growth of bacteria, a "−" to indicate absence of growth.

Discard your plates in the container provided.

Was HT-99 susceptible or resistant to chloramphenicol?

_____

To rifampicin?

_____

Was J-53R susceptible or resistant to chloramphenicol?

_____

To rifampicin?

_____

Make a conclusion about the presence of a gene in each of the two strains of *E. coli* for resistance to each of the antibiotics.

HT-99: _____

_____

J-53R: _____

_____

Table 11-1   Bacterial Growth on Antibiotic-Containing Plates

Strain of E. coli	Growth of Bacteria on Nutrient Agar Containing:		
	Chloramphenicol	Rifampicin	Chloramphenicol plus Rifampicin
HT-99 (Donor)			
J-53R (Recipient)			
"Mating mixture"			

Make a conclusion concerning what happened when the two strains were mixed together. Incorporate your observations concerning antibiotic resistance into your conclusion.

_____

_____

_____

_____

PRE-LAB QUESTIONS

____ 1. The individuals responsible for constructing the first model of DNA structure are (a) Wallace and Watson, (b) Lamarck and Darwin, (c) Aristotle and Socrates, (d) Crick and Watson.

____ 2. Deoxyribose is (a) a five-carbon sugar, (b) present in RNA, (c) a nitrogen-containing base, (d) one type of purine.

____ 3. A nucleotide may consist of (a) deoxyribose or ribose, (b) purines or pyrimidines, (c) phosphate groups, (d) all of the above.

____ 4. Which of the following is consistent with the principle of base pairing? (a) purine-purine, (b) pyrimidine-pyrimidine, (c) adenine-thymine, (d) guanine-thymine.

____ 5. Nitrogen-containing bases between two complementary DNA strands are joined by (a) polar covalent bonds, (b) hydrogen bonds, (c) phosphate groups, (d) deoxyribose sugars.

____ 6. The difference between deoxyribose and ribose is that ribose (a) is a six-carbon sugar, (b) bonds only to thymine, not uracil, (c) has one more oxygen atom than deoxyribose has, (d) all of the above.

_____ **7.** Replication of DNA (a) takes place during interphase, (b) results in two double helices from one, (c) is semiconservative, (d) all of the above.

_____ **8.** Transcription of DNA (a) results in formation of a complementary strand of RNA, (b) produces two new strands of DNA, (c) occurs on the surface of the ribosome, (d) is semiconservative.

_____ **9.** An anticodon (a) is a three-base sequence of nucleotides on tRNA, (b) is produced by translation of RNA, (c) has the same base sequence as does the codon, (d) is the same as a gene.

_____ **10.** Bacterial conjugation (a) may result in a new strain of bacteria, (b) occurs when DNA nucleotides are transferred from one bacterial strain to another, (c) may result in genetic hybridization, (d) all of the above.

EXERCISE 11
NUCLEIC ACIDS: BLUEPRINTS FOR LIFE

POST-LAB QUESTIONS

1. How do deoxyribonucleotides differ from ribonucleotides?

2. Why is DNA often called a double helix?

3. What is the ratio of guanine to cytosine in a double-stranded DNA molecule?

   Of adenine to thymine?

4. Define:
   replication

   transcription

   translation

   codon

   anticodon

5. What does it mean to say that DNA replication is semiconservative?

6. If the base sequence on one DNA strand is ATGGCCTAG, what will the sequence be on the other strand of the helix?

   If the original strand serves as the template for transcription, what will the sequence be on the newly formed RNA strand?

7. Describe the process of translation.

8. What amino acid would be produced if *transcription* took place from a nucleotide with the three-base sequence ATA?

   Suppose a genetic mistake took place during *replication* and the new DNA strand had the sequence ATG. What would be the three-base sequence on an RNA strand transcribed from this series of nucleotides?

   Which amino acid would this codon result in?

   Explain.

9. What is a plasmid?

   How are plasmids used in genetic engineering?

10. How does bacterial conjugation differ from the process by which eukaryotic gene products are produced by bacteria?

PLANT ORGANIZATION: VEGETATIVE ORGANS OF FLOWERING PLANTS

OBJECTIVES After completing this exercise you will be able to:

1. define vegetative, morphology, dicotyledon (dicot), taproot system, node, internode, monocotyledon (monocot), adventitious root, herb (herbaceous plant), woody (woody plant), growth increment (annual ring), pore;

2. identify and give the function of the external structures of flowering plants;

3. identify and give the functions of the tissues and cell types of roots, stems, and leaves;

4. determine the age of woody branches.

INTRODUCTION The color green seems to promote a feeling of well-being among humans. Perhaps that's why we find it so relaxing to stroll through the woods on a summer day. It has been suggested that we find green so pleasant because of our own evolutionary history.

It's difficult to overstate the importance of plant life. Are you sitting upon a wooden chair? If so, you're perched upon part of a tree. No, you say? Perhaps your chair is covered with fabric. If it's natural fabric other than wool, it's a plant product. If the chair is covered with plastic, that cover was made from petroleum products derived from plant material that lived millions of years ago. The energy used to create the chair was probably derived from burning petroleum products or coal, also derived from plant material.

Obviously, plants are an important part of our lives. This exercise will introduce you to the external and internal structure of the **vegetative** (nonreproductive) organs of flowering plants—the roots, stems, and leaves.

Each organ is usually distinguished by its shape and form, its **morphology.** But there is remarkable similarity in the cells, the basic unit of life, and tissues, groups of cells functioning together, comprising these three organs. Each organ is covered by the protective dermal tissue; each organ contains ground tissue; and each also possesses vascular tissue that transports water, minerals, and the products of photosynthesis. Thus, the organs of the plant body are really much more similar than dissimilar. In fact, for this reason the differences that exist between a root, stem, or leaf are said to be *quantitative* rather than *qualitative.* That is, these differences are in the number and arrangement of cells and tissues, not the type. Consequently, the plant body is a continuous unit from one organ to the next.

NOTE *Using the photographs in your textbook while doing this exercise will help you tremendously.*

I. EXTERNAL STRUCTURE OF THE FLOWERING PLANT (Starr, pp. 292–293, 296)

MATERIALS

Per student pair:

- distilled water (dH₂O) in dropping bottles

Per student group (table):

- mature corn plant

Per lab room:

- living bean and corn plants in flats
- dishpan half filled with water

PROCEDURE

A. The Bean Plant, an Example of a Dicotyledonous Plant

The common garden bean, *Phaseolus vulgaris,* is a plant with two seed leaves and thus belongs to the large category of plants called **dicotyledons.** Other familiar examples of dicotyledons ("dicots") are sunflowers, roses, cucumbers, peas, maples, and oaks.

Obtain a bean plant by gently removing it from the medium in which it is growing. Wash the root system in the dishpan provided, not in the sink.

Label Fig. 12-1 as you study the bean plant.

1. *Root system.* The plant consists of a **root system** and a **shoot system.** Examine the root system first. The root system of the bean is an example of a **taproot system,** that is, one consisting of one large **primary root (the taproot)** from which **lateral roots** arise. Identify the taproot and lateral roots.

2. *Shoot system.* Now turn your attention to the shoot system, consisting of the stem and the leaves. What color is the stem?

Green

What structure in the cytoplasm is responsible for this color?

Chloraplst

What is the function of this structure?

Support, Energy, Photosnythses

What is one function of the stem?

~~Con~~ Conduct mAterials

Identify the points of attachment of the leaves to the stem, called **nodes;** the regions between nodes are **internodes.** Look in the upper angle created by the junction of the stem and leaf stalk for the **axillary bud.** These buds give rise to branches and/or flowers.

Find the **terminal bud** at the very tip of the shoot system. The terminal bud contains an apical meristem that accounts for increases in length of the shoot system.

If lateral branches are produced from axillary buds, each lateral branch is terminated by a terminal bud and possesses nodes, internodes, and leaves, complete with axillary buds. As you see, the shoot system can be a highly branched structure.

Look several centimeters above the soil line for the lowermost node on the stem. If the plant is relatively young, you should find the **cotyledons** attached to this node. The cotyledons shrivel as food stored in them is used for the early growth of the seedling. Eventually the cotyledons fall off.

The cotyledons are sometimes called "seed leaves" because they are fully formed (although unexpanded) in the seed. By contrast, most of the leaves you are observing on the bean plant were immature or not present at all in the seed.

Now let's examine the other component of the shoot system, the foliage leaves. Identify the **petiole** (leaf stalk) and **blade** (flattened lamina) on each leaf.

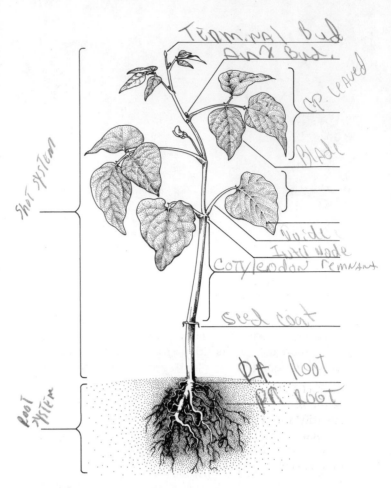

Labels: ~~root system,~~ ~~shoot system,~~ primary root (taproot), lateral root, node, internode, axillary bud, terminal bud, ~~cotyledon remnant,~~ petiole, blade, simple leaf, compound leaf, leaflet

Figure 12-1 External structure of the bean plant. (After Starr and Taggart, 1984.)

The first-formed foliage leaves are **simple leaves,** each leaf having one undivided blade. Find the simple leaves. In bean plants, subsequently formed leaves are **compound leaves,** consisting of three **leaflets** per petiole. Each leaflet has its own short stalk and blade.

Note the netted arrangement of veins in the blades. Veins contain vascular tissues, the xylem, and phloem. The **midvein** is largest and runs down the center of the blade, giving rise to numerous lateral veins.

B. The Corn Plant, a Monocotyledon

Corn (*Zea mays*) is a **monocotyledon,** or "monocot." These plants have only one cotyledon (seed leaf). You're probably familiar with a number of monocots: lilies, onions, orchids, coconuts, bananas, and the grasses. (Did you realize that corn is actually a grass?) Remove a single plant from its growing medium and wash its root system in the dishpan.

Label Fig. 12-2 as you study the corn plant.

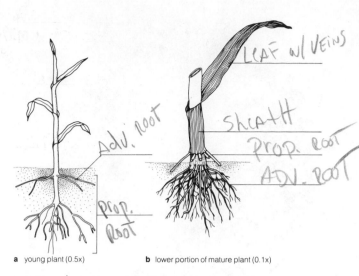

Labels: root system, adventitious root,
prop root, leaf sheath, leaf with parallel veins

Handwritten labels on figure: LEAF w/ VEINS, Adv. ROOT, Sheath, Prop. ROOT, ADV. ROOT, Prop. Root

Figure 12-2 External structure of a corn plant

1. *Root system.* The seed (grain) may still be attached to the plant if you were careful in removing the plant. Identify the root system. Note that there is no one particularly prominent root. In most monocots the primary root is short-lived and is replaced by numerous **adventitious roots.** These are roots that arise from places other than existing roots. Trace these roots back to the corn grain. Where do they originate?

About were the STOCK starts

As the roots branch, they develop into a **fibrous root system,** one particularly well suited to prevent soil erosion.

Examine the mature corn plant. Identify the large **prop roots** at the base of the plant. Where do prop roots arise from?

At the bottum of the sheath

Would you classify these as adventitious roots?

yes.

2. *Shoot system.* The shoot system of a young corn plant appears somewhat less complex than that of the bean. There seem to be no nodes or internodes. Look at the mature corn plant again. You should see that nodes and internodes do indeed exist. In your young plant, elongation of the stem has not yet taken place to any appreciable extent. Strip off the leaves of the young corn plant. Keep doing so until you find the shoot apex. (It's deeply embedded, don't you agree?)

Examine in more detail the leaves of the corn plant. Note the absence of a petiole and the presence of a **sheath** that extends down the stem. The leaf sheath adds strength to the stem. Look at the veins, which

have a parallel arrangement. (Contrast this to the petioled, netted-vein arrangement of the bean leaves.)

In the following portion of the exercise, we will study in more detail the root and shoot systems of dicotyledons.

II. THE ROOT SYSTEM (Starr, pp. 299–301)

MATERIALS

Per student:

- single-edged razor blade
- microscope slide
- coverslip
- prepared slide of buttercup (*Ranunculus*) root, cross section

Per student pair:

- distilled water (dH₂O) in dropping bottles

Per lab room:

- germinating radish seeds in large petri dishes
- demonstration slide of Casparian strip in endodermal cell walls
- demonstration slide of lateral root formation

PROCEDURE

A. Living Root Tip

Obtain a germinating radish seed. Identify the **primary root.** Its fuzzy appearance is due to the numerous tiny **root hairs.**

Using a razor blade, cut off and discard the seed, and make a wet mount of the primary root. (Add a copious amount of water so there is no air surrounding the root. *Do not* squash the root.) Examine your preparation with the low power objective of your compound microscope. (If you are having difficulty seeing the root clearly, increase its contrast by closing the microscope's diaphragm somewhat.) Locate the conical root tip. The very end of the root tip is covered by the protective **root cap.** As a root grows through the soil, the tip is thrust between soil particles. Were it not for the root cap, the apical meristem containing the dividing cells would be damaged.

Find the root hairs. Do they extend all the way down to the root cap?

yes,

Examine the root hairs carefully. Does their length increase away from the root tip?

yes,

Labels: root cap, root hairs

Figure 12-3 Drawing of living primary root (_____×)

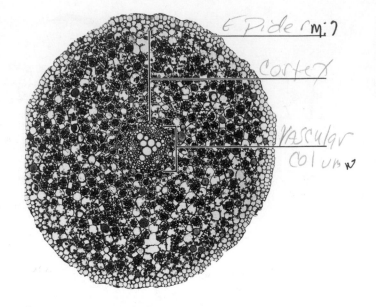

Handwritten labels: Epidermi?, Cortex, Vascular column

a

Labels: epidermis, cortex, vascular column

The youngest root hairs are the shortest. What does this imply regarding their point of origin and pattern of maturation?

they will be located towards the top of the root Tip.

Root hairs increase tremendously the absorptive surface of the root.

In Fig. 12-3 draw and label the living root.

B. Root Anatomy

Now let's see what the internal architecture of the root looks like.

Obtain a prepared slide containing a mature buttercup (*Ranunculus*) root in cross section. As you study this slide, label the described tissues and structures on the diagram of the root cross section, Fig. 12-4.

Examine the slide first with the low power objective of your compound microscope to gain an overall impression of the organization of the tissues present. Starting at the edge of the root, identify the **epidermis.** Moving inward, locate the **cortex** and the central **vascular column.** These regions represent the dermal, ground, and vascular tissue systems, respectively.

Switch to the medium power objective for further study. Look at the outermost layer of cells, the **epidermis.** What function does the epidermis serve?

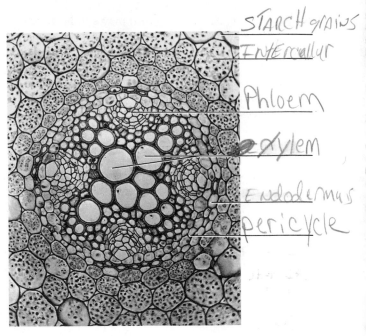

Handwritten labels: STARCH grains, Intercellur, Phloem, xylem, Endodermus, pericycle

b

Labels: starch grains, intercellular space, endodermis, pericycle, primary xylem, primary phloem

Figure 12-4 Cross section of buttercup (*Ranunculus*) root. **(a)** 70×; **(b)** portion of cortex and vascular column, 300×. (Photos courtesy Triarch, Inc.)

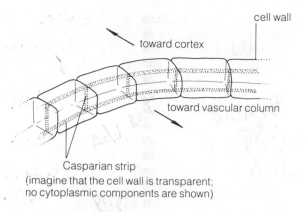

cell wall

toward cortex

toward vascular column

Casparian strip
(imagine that the cell wall is transparent;
no cytoplasmic components are shown)

Figure 12-5 Endodermal cells with Casparian strip

Beneath the epidermis find the relatively wide **cortex,** consisting of parenchyma cells that contain numerous **starch grains.** Based upon the presence of starch grains, what would you suspect one function of this root might be?

_____

_____

_____

Switch to the high-dry objective. Between the cells of the cortex find numerous **intercellular spaces.**

The innermost layer of the cortex is given a special name. This cylinder, a single cell thick, is called the **endodermis.** Locate the endodermis on your slide and label it in Fig. 12-4b. Unlike the rest of the cortical cells, endodermal cells *do not* have intercellular spaces between them. The endodermis regulates the movement of water and dissolved substances into the vascular column. Each endodermal cell possesses a **Casparian strip** within its radial and transverse walls. To visualize this arrangement, imagine a rectangular box (the endodermal cell) that has a rubber band (the Casparian strip) around its long dimension. Now imagine that the rubber band is actually part of the wall of the box. Figure 12-5 is a diagram of this arrangement.

The Casparian strip consists of waxy material that prevents water and dissolved substances from flowing through those regions of the cell wall. Instead, these substances must flow through the differentially permeable plasma membrane of the endodermal cells as they move into or out of the vascular column.

In most cases the Casparian strip is difficult to distinguish. *Examine the demonstration slide illustrating the Casparian strip.*

Return to your own slide, switch back to the medium power objective, and focus your attention on the central vascular column. The cell layer immediately beneath the endodermis is the **pericycle.** Cells of the pericycle may become meristematic and produce **lateral roots.** Examine the demonstration slide showing lateral root formation.

On your own slide find the **primary xylem,** consisting of three or four ridges of thick-walled cells. (The stain used by most slide manufacturers stains the xylem cell walls red.) The xylem is the principal water-conducting tissue of the plant. Between the "arms" of the xylem is **primary phloem,** the tissue responsible for long-distance transport of carbohydrates produced by photosynthesis (known as "photosynthates").

III. THE SHOOT SYSTEM (Starr, pp. 295–299)

MATERIALS

Per student:

- prepared slide of herbaceous dicot stem, cross section (flax, *Linum;* or alfalfa, *Medicago*)
- prepared slide of woody stem, cross section (basswood, *Tilia*)
- prepared slide of dicot leaf, cross section (lilac, *Syringa*)
- woody twig (hickory, *Carya;* or horse chestnut, *Aesculus*)

Per student pair:

- cross section of woody branch (tree trunk)

Per lab room:

- demonstration slide of lenticel

PROCEDURE

A. Primary Structure of Stems

Remember that the root and shoot are basically similar in structure; only the arrangement of tissues differs. Dicot stems have their vascular tissues (vascular tissue system) arranged in a more or less complete ring of discrete bundles of vascular tissue (called vascular bundles). Moreover, the ground tissue of dicots can be differentiated into two regions: **pith** and **cortex.** You've probably heard of the term "herb." An **herb** is an **herbaceous plant,** one that develops very little wood. Beans, flax, and alfalfa are examples of herbaceous dicots. Maples and oaks are woody dicots. Let's look at an herbaceous stem first.

Obtain a slide of an herbaceous dicot stem (flax or alfalfa). This slide may be labeled "herbaceous dicot stem." As you study this slide, label the tissues, cells, or structures indicated below in Fig. 12-6, a photograph of a partial section of an herbaceous dicot stem.

Using the low power objective, identify the single-layered **epidermis** covering the stem, a multilayered **cortex** between the epidermis and **vascular bundles,** and the **pith** in the center of the stem.

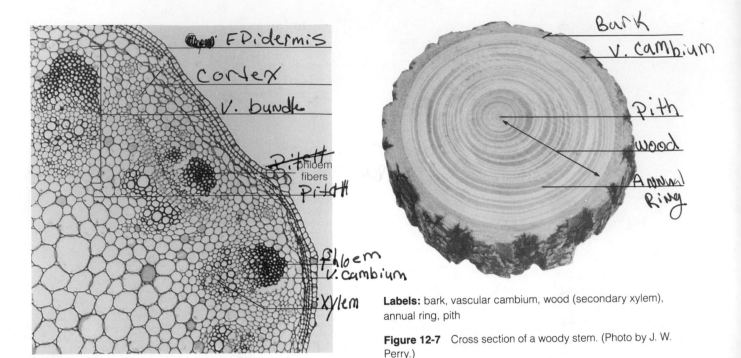

Handwritten labels on Figure 12-6: EPidermis, Cortex, V. bundle, Pith, Pith, Phloem, V. Cambium, Xylem

Handwritten labels on Figure 12-7: Bark, V. Cambium, Pith, wood, Annual Ring

Labels: epidermis, cortex, vascular bundle, pith, primary phloem, vascular cambium, primary xylem

Figure 12-6 Portion of a cross section of an herbaceous dicot stem (66×). (Photo by J. W. Perry.)

Labels: bark, vascular cambium, wood (secondary xylem), annual ring, pith

Figure 12-7 Cross section of a woody stem. (Photo by J. W. Perry.)

Observe a vascular bundle with the medium power objective. Adjacent to the cortex, find the **primary phloem.** Just to the inside of the primary phloem will be located a **vascular cambium,** which for the most part is inactive. (The vascular cambium is that lateral meristem which produces wood and secondary phloem, both secondary tissues. Despite your specimen being an herbaceous plant, a vascular cambium may be present. Generally, this meristem does not produce enough secondary tissue to result in the plant being considered woody.) Locate thick-walled vessel members of the **primary xylem.** The primary xylem is that vascular tissue closest to the pith. (The cell wall of the vessel members is probably stained red.)

B. Secondary Growth: Woody Stems

Woody plants are those that undergo secondary growth. Both roots and shoots may have secondary growth. This growth occurs because of activity of the two lateral meristems—the vascular cambium and the **cork cambium.** The vascular cambium produces **secondary xylem (wood)** and **secondary phloem,** while the cork cambium produces the **periderm.**

Examine a cross section of a tree trunk. A tiny region in the center of the stem is the **pith.** Most of the trunk is made up of *wood* (also called secondary xylem). Count the **growth increments (annual rings)** within the wood to estimate the age of the stem when the section was cut. (In most woods, a growth increment includes *both* a light and a dark layer of cells.) How old would you estimate your section to be?

_____ years

The vascular cambium is located *between* the most recently formed wood (secondary xylem) and the **bark;** thus, the bark is everything *external* to the vascular cambium. It consists of secondary phloem and periderm. The periderm performs the same function as did the epidermis before the epidermis ruptured as a result of increase in girth of the stem. What is the function of the periderm?

_____

Within the wood, find the **rays** that appear as lines running from the center toward the edge of the stem. Examine the wood with a dissecting microscope and locate the numerous holes in the wood. These are the cut ends of the vessel members, often called **"pores."**

In Fig. 12-7, label the features you have found on the woody stem section you examined.

Now that you've got an idea of the composition of a woody stem, let's examine one with the microscope.

Obtain a prepared slide of a cross section of a woody stem (basswood, *Tilia,* or another stem). Examine it with the various magnifications available on the *dissecting* microscope.

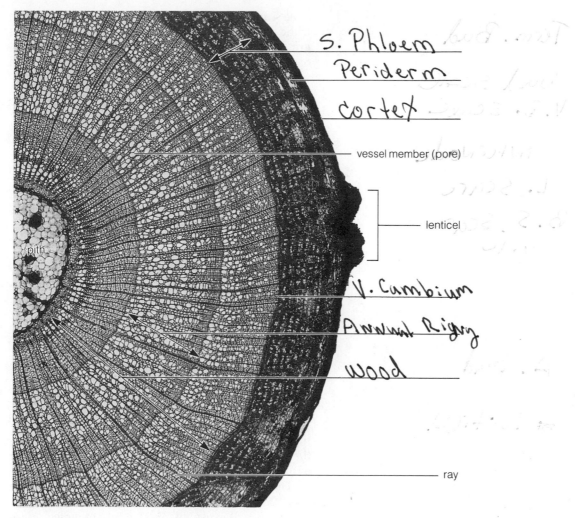

Handwritten labels on figure: S. Phloem, Periderm, Cortex, V. Cambium, Annual Ring, Wood

Printed labels on figure: vessel member (pore), lenticel, pith, ray

Labels: periderm, cortex, secondary phloem, vascular cambium, secondary xylem (wood), annual ring

Figure 12-8 Cross section of a woody dicot stem (25×). (Photo courtesy Ripon Microslides, Inc.)

Starting at the edge, identify the **periderm** (darkly stained cells). Depending upon the age of the stem, **cortex** (thin-walled cells with few contents) may be present just beneath the periderm. Now find the broad band of **secondary phloem** (consisting of cells in pie-shaped wedges). Identify the **vascular cambium** (a narrow band of cells separating the secondary phloem from secondary xylem), **secondary xylem (wood)**, and pith (large, thin-walled cells in the center). Count the number of growth increments (annual rings). How old is this section?

_____ years

Note the largest, thick-walled cells in the wood. These are the pores (vessel members). Find the **rays** running through the wood.

Recall that the periderm replaces the epidermis as secondary growth takes place. The epidermis had stomata that allowed the exchanging of gases between the plant and the environment. When the epidermis was shed, so were the stomata. But the need for exchange of gases still exists because the living cells require oxygen for carbohydrate metabolism (Exercise 7). The plant has solved this problem by having special regions, **lenticels,** in the periderm. Lenticels are groups of cells with lots of intercellular space, in contrast to the tightly packed cells in the rest of the periderm.

Label Fig. 12-8.

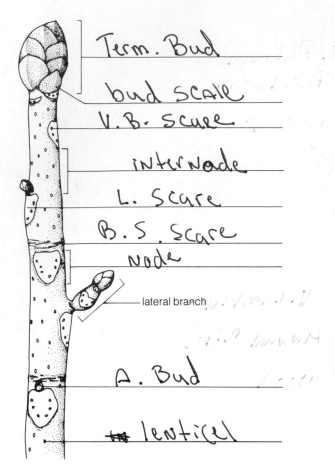

Term. Bud

bud scale

V.B. scare

internode

L. Scare

B.S. Scare

node

— lateral branch

A. Bud

lenticel

Labels: terminal bud, leaf scar, node, vascular bundle scar, axillary bud, internode, lenticel, bud scale, terminal bud scale scars

Figure 12-9 External structure of a woody stem

C. External Features of Woody Stems

Examine a twig of hickory (*Carya*) or buckeye (*Aesculus*) that has lost its leaves. Label Fig. 12-9 as you study the twig.

Find the large **terminal bud** at the tip of the twig. If your twig is branched, each branch has its own terminal bud. Identify the shield-shaped **leaf scars** at each **node.** (Remember, a node is the region where a leaf attaches to the stem.) Leaf scars represent the point at which the leaf petiole was attached on the stem. Within each leaf scar, note the numerous dots. These are the **vascular bundle scars.** Immediately above and adjacent to most leaf scars should be an **axillary bud.** In the **internode** regions of the twig locate the small raised bumps on the surface; these are **lenticels,** the regions of the periderm that allow for exchange of gases.

Return to the terminal bud. Note that the bud is surrounded by **bud scales.** When these scales fall off during spring growth, they leave **terminal bud scale scars.** Because a terminal bud is produced at the end of each growing season, the groups of terminal bud

scale scars can be used to determine the age of a twig. If the most recent growth took place during the last growing season (summer), when was the portion of the twig immediately adjacent to the cut end produced?

_____

(Remember to date all regions between successive bud scale scars.)

D. Leaf Anatomy

Obtain a prepared slide of a cross section of a dicot leaf (lilac or other). Label Fig. 12-10 as you examine the leaf.

Examine the leaf first with the low power objective to gain an overall impression of its morphology. Note the size and orientation of the **veins** within the leaf. The veins contain the xylem and phloem. Find the centrally located **midvein** within the **midrib,** the midvein-supporting tissue. You might think of the midvein as the major artery of the leaf, carrying water and minerals to the leaf and materials produced during photosynthesis to sites where they will be used during carbohydrate metabolism.

Use the high-dry objective to examine a portion of the blade to one side of the midvein.

Starting at the top surface of the leaf, find the **cuticle,** a waxy, water-impervious substance covering the **upper epidermis.** The epidermis is a single layer of tightly appressed cells.

The ground tissue of the leaf is represented by the **mesophyll** (literally "middle leaf"). In dicot leaves the mesophyll is usually divided into two distinct regions; immediately below the upper epidermis find the two layers of **palisade mesophyll.** These columnar-shaped cells are rich in chloroplasts. Below the palisade mesophyll find the loosely arranged **spongy mesophyll.** Note the large volume of intercellular space within the spongy mesophyll. Does the spongy mesophyll contain any chloroplasts?

_____

What then is one function that occurs within the spongy mesophyll?

_____

In the **lower epidermis** find a **stoma** (plural: stomata) with its **guard cells** and the **pore** (inset, Fig. 12-10).

Large **epidermal hairs** shaped somewhat like mushrooms are usually found on the lower epidermis. Is the lower epidermal layer covered by a cuticle?

_____

Compare the abundance of stomata within the lower epidermis with that in the upper epidermis. Which epidermal surface has more stomata?

_____

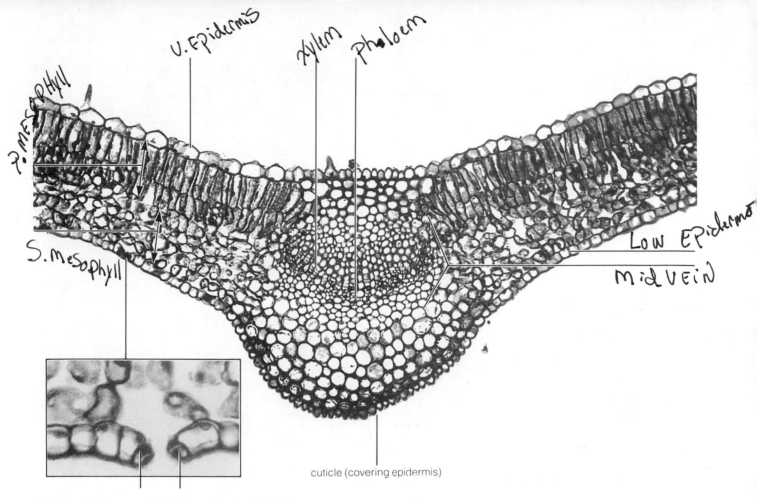

Labels: midvein, upper epidermis, palisade mesophyll, spongy mesophyll, lower epidermis, xylem, phloem, guard cell (2)

Figure 12-10 Cross section of a dicot leaf with midrib (100×; inset of a stoma, 250×). (Photos courtesy Ripon Microslides, Inc.)

Examine the midvein in greater detail. The thick-walled cells (often stained red) are part of the xylem tissue. Below the xylem locate the phloem (usually stained green). Now identify and examine the smaller veins within the lamina (blade). Note that these, too, contain both xylem and phloem.

Label Fig. 12-10.

PRE-LAB QUESTIONS

_____ **1.** The study of a plant's structure is (a) physiology, (b) morphology, (c) taxonomy, (d) botany.

_____ **2.** A plant with two seed leaves is (a) a monocotyledon, (b) a dicotyledon, (c) exemplified by corn, (d) a dihybrid.

_____ **3.** A taproot system lacks (a) lateral roots, (b) a taproot, (c) both of the above, (d) none of the above.

_____ **4.** Which of the following is not part of the shoot system? (a) stems, (b) leaves, (c) lateral roots, (d) axillary buds.

_____ **5.** An axillary bud (a) would be found along internodes, (b) produces new roots, (c) is the structure from which branches and flowers arise, (d) is the same as a terminal bud.

_____ **6.** Corn (_Zea mays_) (a) is a dicotyledon, (b) produces adventitious roots and is a monocotyledon, (c) has leaves with netted veins, (d) all of the above.

_____ **7.** The endodermis (a) is the outer covering of the root, (b) is part of the vascular tissue, (c) contains the Casparian strip, which regulates the movement of substances, (d) none of the above.

_____ **8.** Meristems (a) are located at the tips of stems, (b) are located at the tips of roots, (c) are regions of active growth, (d) all of the above.

_____ **9.** To determine the age of a woody twig, you would count the number of (a) nodes, (b) leaf scars, (c) lenticels, (d) regions between sets of terminal bud scale scars.

_____ **10.** The midrib of a leaf (a) contains the midvein, (b) contains only xylem, (c) is part of the spongy mesophyll, (d) contains only phloem.

ANIMAL ORGANIZATION I: MAMMALIAN HISTOLOGY

OBJECTIVES After completing this exercise you will be able to:

1. define tissue, organ, system, organism, histology, basement membrane, goblet cell, cilia, brush border of microvilli, keratinization, keratin, collagen fiber, elastic fiber, fibroblast, fat cell, lumen, chondrocyte, lacuna, Haversian canal, lamella, osteocyte, actin filaments, myosin filaments, intercalated disks, neuron, cell body, dendrites, axon, neuroglia;

2. discuss the high degree of organization present in animal structure and explain its significance;

3. recognize the four basic tissues and their common mammalian subtypes;

4. list the functions of the four basic tissues and their common mammalian subtypes;

5. explain how the four basic tissues are combined to make organs.

INTRODUCTION Unicellular eukaryotic organisms contain *organelles* that carry on all the functions vital to life. In multicellular organisms, each cell is specialized to emphasize certain functions and relies on the other cells for the remainder.

Collections of similar cells that interact as a structural, functional unit are called **tissues.** A tissue includes any extracellular material produced and maintained by its cells. There are four basic tissue types in animals, and they are combined much like a quadruple-decker sandwich to make all of the **organs.** Organs perform specific functions and are strung together functionally, and usually structurally, to form **systems.** Each system carries on a vital function, and considered together, the systems equal the animal **organism.** Tissues, organs, and systems are present in most, but not all, animals.

I. TISSUES

(Starr, pp. 337–339)

The study of tissues is called **histology.** The four basic tissues types are **epithelial tissue, connective tissue, muscle tissue,** and **nervous tissue.** Each has subtypes.

In order to study these subtypes, you will examine a variety of tissues and organs, all permanently mounted on glass microscope slides. Most slides have sections of tissues and organs, usually 6 to 10 μm thick. Other slides have whole pieces of organs that are either transparent or are teased (gently pulled apart) and spread on the surface of the slide until they are thin enough to see through. Some organs are simply smeared onto the surface of slides.

These prepared slides have been chosen to show not only the characteristics of each tissue but also how variations in the basic structure of each allows for the related but distinct functions of its subtypes. For the sake of simplicity, the following exercise will consider only the common subtypes of adult mammalian tissues.

MATERIALS

Per student:

- compound microscope, lens paper, a bottle of lens cleaning solution (optional), a lint-free cloth (optional), a dropper bottle of immersion oil (optional)

- prepared slides of the following:
  —whole mount of mesentery (simple squamous epithelium)
  —section of the cortex of the mammalian kidney
  —section of trachea
  —transverse section of small intestine (preferably of the ileum)
  —section of esophagus
  —section of mammalian skin
  —sections of contracted and distended urinary bladders
  —teased spread of loose (areolar) connective tissue
  —section of white adipose tissue
  —longitudinal section of tendon
  —ground transverse section of compact bone
  —sections of the 3 muscle types
  —smear of the spinal cord of an ox (neurons)

Per lab room:

- demonstration of intercalated disks in cardiac muscle tissue
- 50-mL beaker three-fourths full of water
- 50-mL beaker three-fourths full of immersion oil
- 2 small glass rods

PROCEDURE

A. Epithelial Tissues

Epithelial tissues are widespread throughout the body, covering both the body's outer (epidermis of skin) and inner surfaces (ventral body cavities), and lining the inner surfaces of tubular organs (the small intestine, for example). Their main functions are *protection* and *transport* (secretion and absorption, for example). Specialized functions include sensory reception and the maintenance of the body's gametes (egg and developing sperm).

Epithelial cells carry on rapid cell division in the adult, and various stages of mitosis (see Chapter 8) are often seen in this tissue type. As a consequence, epithelia have the highest rates of cell turnover among tissues. For example, the lining of the small intestine replaces itself every three to four days. Epithelial tissues don't have blood vessels and are attached to the underlying connective tissue by an extracellular **basement membrane** that is difficult to see if it is not specially stained. The following rules are used to name the subtypes of epithelial tissue:

RULE 1 What is the shape of the outermost cells? There are three choices: *squamous* (scalelike or flat), *cuboidal*, or *columnar*. These choices describe the shape of the cells when viewed in a section perpendicular to the surface. In surface view, all three types are polygonal.

RULE 2 How many layers of cells are there? Epithelial tissues are *simple* if there is one layer of cells; they are *stratified* if there are two or more.

These principles are illustrated for squamous epithelia in Fig. 13-1.

There is a further complication. If an epithelium appears stratified because it has more than one layer of nuclei, but electron microscopic examination shows that all cells reach the basement membrane, it is called *pseudostratified* (Fig. 13-2).

Also, the epithelium lining the inside of the urinary bladder and of some of the other excretory system organs changes its subtype as it fills with urine. This epithelium is called *transitional* to avoid confusion.

There are six common subtypes of epithelia: *simple squamous, simple cuboidal, simple columnar, pseudostratified columnar, stratified squamous,* and *transitional.* Find and examine these subtypes in the following slides:

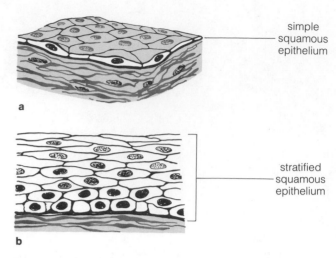

Figure 13-1 Squamous epithelia. **(a)** simple, **(b)** stratified.

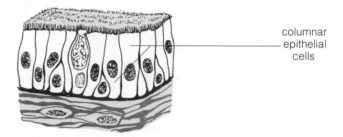

Figure 13-2 Pseudostratified columnar epithelium

1. *Whole mount of mesentery.* The mesentery is a fold of the abdominal wall and holds the intestines in place. This slide provides a surface view of the **simple squamous epithelium** that lines the ventral body cavities. What is the shape of the surface cells?

Draw in Fig. 13-3 what you see. In the space provided at the end of the figure title, note the total magnification of the compound microscope you used to make this drawing. Repeat this procedure for each subsequent drawing.

2. *Kidney.* With the help of Fig. 13-4, find a *renal corpuscle.* A renal corpuscle is composed of a tuft of capillaries called the *glomerulus,* which lies in *Bowman's capsule,* the cup-shaped end of one of the kidney's functional units, the nephron. The walls of Bowman's capsule and the capillaries of the glomerulus are *simple squamous epithelia.* This is easiest to see in the outer wall of Bowman's capsule. But if you look carefully and use high power, you may be able to see transverse sections of capillary walls made up of one or two circular or C-shaped simple squamous epithelial cells. Around the renal corpuscle you can see a number of transverse and oblique sections of the tubular portion of the nephron. Their walls are composed of **simple cuboidal epithelium.** Label Fig. 13-4.

Figure 13-3 Drawing of a surface view of simple squamous epithelium (_____×)

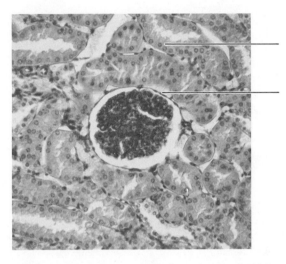

Labels: simple squamous epithelium, simple cuboidal epithelium

Figure 13-4 Photomicrograph of a section of the cortex of the kidney (285×). (Photo courtesy Ripon Microslides, Inc.)

3. *Trachea.* The trachea is a tubular organ that conveys air to and from the lungs. Note that its inner surface is lined by **pseudostratified columnar epithelium.** Locate unicellular glands called **goblet cells** because of their shape. They secrete mucus that is difficult to see unless it is specifically stained. Using high power, do you see the numerous motile hairs that project from the surface of the columnar epithelial cells? These are **cilia,** which in life move synchronously, sweeping mucus and trapped bacteria and debris up the trachea to the throat. When you clear your throat, you collect this mucus and swallow it. Label Fig. 13-5.

4. *Small intestine.* The small intestine is a tubular organ that connects the stomach to the large intes-

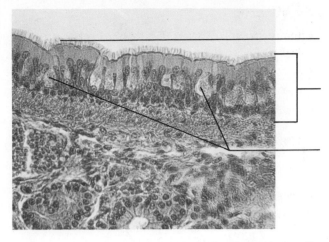

Labels: pseudostratified columnar epithelium, goblet cells, cilia

Figure 13-5 Photomicrograph of a section of the inner surface of the trachea (280×). (Photo courtesy Ripon Microslides, Inc.)

tine. Its inner surface is lined with **simple columnar epithelium** (Fig. 13-6). Note that *goblet cells* are present. Don't confuse the **brush border of microvilli** with cilia. Compare the height of this border with that of the cilia from the previous slide. The size of the nucleus is a good reference point. Label Fig. 13-6.

The microvilli are primarily responsible for the large surface area of the small intestine. Individual microvilli can best be seen at the electron microscopic level (see Fig. 13-7).

5. *Esophagus.* The tubular esophagus connects the throat to the stomach. Examine the epithelium lining the inner surface of the esophagus. Is the shape of the outermost cells squamous, cuboidal, or columnar?

_____

Are there one or many layers of cells in this tissue?

_____

Name this subtype of epithelial tissue.

_____

Draw in Fig. 13-8 what you have observed.

6. *Skin.* The skin is divided into three layers (Fig. 13-9). The *epidermis* is composed of stratified squamous epithelium, while the *dermis* and *hypodermis* (or subcutaneous layer) are connective tissue and will be studied later. The epidermis is the most extreme example of a protective epithelium. Strata in the epidermis are caused by the process of **keratinization,** whereby the cells transform themselves into "bags" of **keratin.** It is this protein that gives skin its tough, flexible, and water-resistant surface.

At one of the free edges of your section of mammalian skin, locate the **keratinized stratified squamous epithelium.** It will be stained bluer than the predominately pink connective tissue layers. Hair follicles and

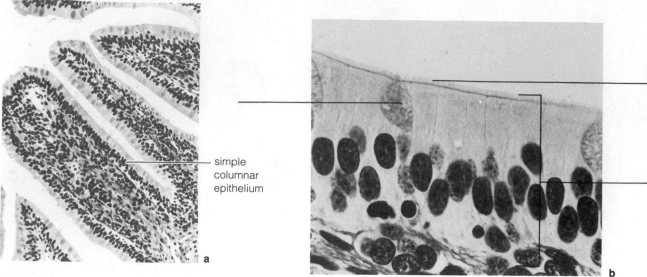

simple
columnar
epithelium

a

b

Figure 13-6 Photomicrographs of a section of the inner surface of the small intestine. **(a)** 10×; **(b)** 1000×. (Photos courtesy Triarch, Inc.)

Labels: simple columnar epithelium, goblet cells, brush border of microvilli

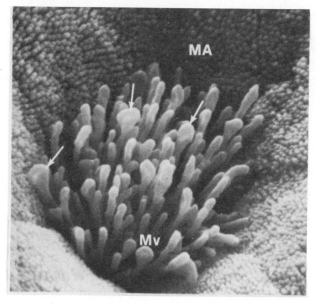

MA

Mv

Figure 13-7 Scanning electron microscopic view of the surface of the small intestine; MA—Small microvilli on surface of absorptive cells and Mv—Longer, larger microvilli on surface of a goblet cell (12,220×). (Photo from R. Kessel and R. Kardon, *Tissues and Organs*. Copyright © 1979 W. H. Freeman and Company. Used by permission.)

Figure 13-8 Drawing of a section of the epithelial tissue lining the inner surface of the esophagus (_____×)

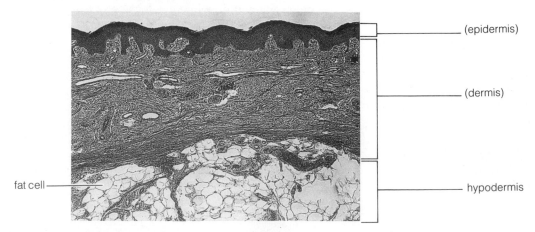

(epidermis)

(dermis)

hypodermis

fat cell

Figure 13-9 Photomicrograph of a section of skin (50×). (Photo courtesy Ripon Microslides, Inc.)

Labels: keratinized stratified squamous epithelium, dense fibrous irregular connective tissue

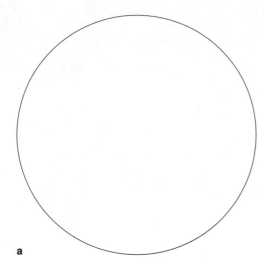

a

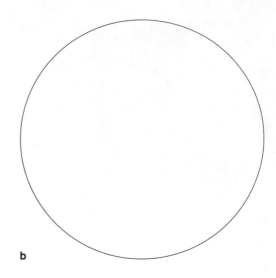

b

Figure 13-10 Drawings of transitional epithelia from **(a)** contracted and **(b)** distended urinary bladders

multicellular sweat glands may be present in the connective tissue layers. These structures grow into the connective tissue layers from the epidermis during the development of the skin. Label the epithelium in Fig. 13-9.

7. *Urinary bladder.* There are two sections on this slide. One is from a contracted bladder, the other from a distended bladder. Locate the **transitional epithelium** at the surface of one of these sections. The transitional epithelium from the contracted bladder looks like stratified cuboidal epithelium. The transitional epithelium of the distended bladder is thinner and looks like stratified squamous epithelium. Draw these two extremes in Fig. 13-10.

B. Connective Tissue

Connective tissues occur in all parts of the body. They contain a large amount of material external to the cells called the *extracellular matrix*. This matrix consists of *fibers* embedded in *ground substance*. There are two basic kinds of fibers: collagen and elastic. Collagen fibers are tough, flexible, and inelastic, whereas elastic fibers stretch when pulled, returning to their original length when the pull is removed. Except for cartilage, connective tissues contain blood vessels. Similar to epithelia, connective tissue cells are capable of cell division in the adult, but at a reduced rate. Connective tissue subtypes are classified as follows:

GROUP 1 Soft (connective tissue proper)

a. loose (few fibers that run in all directions; also called areolar)

b. dense (many fibers)
   i . regular (fibers run in the same direction)
   ii. irregular (fibers run in all directions)

c. special (for example, white adipose)

GROUP 2 Hard

a. cartilage (ground substance is polymerized or jellylike)

b. bone (ground substance is mineralized)

GROUP 3 Blood (although blood is a connective tissue, we will cover it in a later exercise)

Find and examine these subtypes in the following slides.

1. *Teased spread of loose connective tissue.* **Loose connective tissue** forms much of the packing material of the body and fills in the spaces between other tissues. Many other kinds of cells live in this tissue and play important roles in the body's immune defense system.

Find **collagen fibers** and **elastic fibers** in your slide of loose connective tissue (Fig. 13-11). Collagen fibers are stained light pink and are variable in diameter, but they are wider than the elastic fibers. Elastic fibers are darkly stained, thin, and branched. Areolar connective tissue contains a number of different cell types. To see a **fibroblast**—the cell that produces the matrix of loose and other soft connective tissues—look for an elongated, oval-shaped nucleus associated with a fiber. The amorphous ground substance of all soft connective tissues consists of a viscous soup of carbohydrate-protein molecules and is usually extracted from sections of these tissues during processing. Label Fig. 13-11.

2. *Skin.* Look at the second layer of the skin, the dermis, which is primarily composed of **dense fibrous irregular connective tissue.** The term "fibrous" refers to the high concentration of collagen fibers produced by the resident fibroblasts. As the name of this subtype indicates, there is a large number of apparently randomly oriented collagen fibers in the matrix. The dermis cushions the body from everyday stresses and strains. Label the connective tissue in Fig. 13-9.

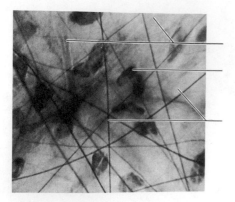

Labels: collagen fibers, elastic fibers, fibroblast

Figure 13-11 Photomicrograph of a spread of loose connective tissue (196×). (Photo by D. Morton.)

Figure 13-12 Drawing of a section of dense regular fibrous connective tissue (_____×)

Note that the looser fibrous irregular connective tissue of the hypodermis has a lower concentration of collagen fibers and islands of fat cells. It functions as a shock absorber, as an insulating layer, and as a site for storing water and energy (white adipose tissue). In animals that move their skin independently of the rest of the body, skeletal muscle tissue is found in the hypodermis.

3. *Tendon.* A tendon connects a skeletal muscle organ to a bone organ. Tendons are composed predominately of **dense regular fibrous connective tissue.** Examine a longitudinal section of a tendon. How are the fibers arranged?

_____

_____

_____

This design makes tendons very strong, much like a rope composed of braided strings, which in turn are made of even smaller fibers. How are the fibroblasts oriented relative to the arrangement of the fibers?

_____

_____

_____

Draw in Fig. 13-12 what you have observed.

4. *White adipose tissue.* As you see, **white adipose tissue** consists mainly of **fat cells.** However, careful examination of the section using high power shows them to be surrounded by delicate collagen fibers, fibroblasts, and capillaries. Because the fat has been lost during the slide preparation, the fat cells look empty. The primary function of this tissue is energy storage. Label Fig. 13-13.

5. *Trachea.* Locate a portion of one of the rings of **hyaline cartilage** in the wall of the trachea. The rings

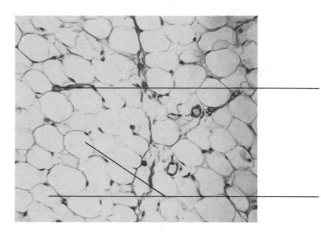

Labels: fat cells, capillary

Figure 13-13 Photomicrograph of a section of white adipose tissue (134×). (Photo courtesy Ripon Microslides, Inc.)

of cartilage prevent the lumen of the trachea from collapsing. A **lumen** is the space within a hollow organ. Look for **chondrocytes** (cartilage cells) within the matrix. Chondrocytes are located in **lacunae** (the singular is *lacuna*), small holes in the matrix.

Although invisible, many collagen fibers are embedded in the polymerized ground substance. They cannot be seen because the indices of refraction of these two matrix components are similar. Your instructor has set up a demonstration of this phenomenon. Observe the two labeled beakers, one filled with immersion oil and the other with water. Look at the glass rod that has been placed in each of them. The index of refraction of glass is about 1.58, that of water about 1.33, and that of immersion oil about 1.52. In which fluid is it easier to see the glass rod?

_____

Label Fig. 13-14.

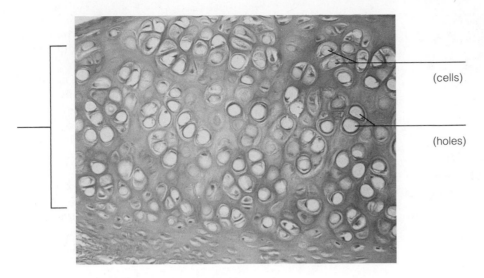

(cells)

(holes)

Labels: hyaline cartilage, chondrocytes, lacunae

Figure 13-14 Photomicrograph of hyaline cartilage in the wall of the trachea (290×). (Photo courtesy Ripon Microslides, Inc.)

In locations where cartilage has to be more durable (intervertebral disks, for example), the collagen content is higher and the fibers are visible. In other sites (such as outer ear flaps), large numbers of elastic fibers are present. Elastic cartilage is deformed by a small force and returns to its original shape when the force is removed.

6. *Bone.* Living bones are amazingly strong. **Bone** is a hard, yet flexible, tissue. Its hardness is due to a mineral (predominantly a calcium-phosphate salt called hydroxyapatite) deposited in the matrix. Its flexibility comes from having the highest collagen content of all connective tissues.

Examine a transverse section of compact bone tissue (Fig. 13-15). This preparation has been made by grinding a piece of the shaft of a long bone with coarse and then finer stones until a thin wafer remains. Although only the mineral part of the matrix is present, the basic architecture has been preserved. In living bone, blood vessels and nerves are present in the large **Haversian canals.** These are surrounded by concentric layers of matrix called **lamellae** (the singular is *lamella*). There are intervening rings of **lacunae** (smaller holes), which in living bone contain cells called **osteocytes.**

In young living bone, the lines that you see connecting lacunae with each other and with the Haversian canal are little canals that contain the cytoplasmic processes of osteocytes. Thus, the osteocytes are able to exchange nutrients, wastes, and other molecules with the blood. By comparison, substances in cartilage have to diffuse across the matrix between chondrocytes and blood vessels in the surrounding soft connective tissue. Label Fig. 13-15.

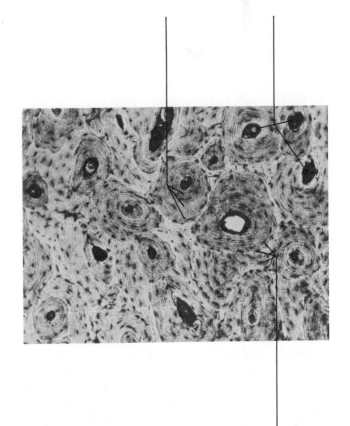

Labels: Haversian canals, lamellae, lacunae

Figure 13-15 Transverse section of ground compact bone (62×). (Photo courtesy Triarch, Inc.)

C. Muscle Tissue

Muscle tissue is contractile. Its cells (fibers) can shorten and produce changes in the position of body parts, or they try to shorten and produce changes in tension. Contraction results from interactions between two types of protein filaments: **actin** and **myosin**. Like epithelial tissue, muscle tissue is primarily cellular; but unlike both epithelial and connective tissues, its cells do not normally divide in the adult. Therefore, dead fibers usually cannot be replaced. There are three subtypes of muscle tissue: *skeletal, cardiac,* and *smooth.* The main characteristics of their fibers are summarized in Table 13-1.

Table 13-1 Characteristics of Muscle Fibers				
Muscle Tissue	Nucleus		Transverse Striations	Special Features
	Number	Position		
skeletal	many	peripheral	yes	——
cardiac	one	central	yes	intercalated disks
smooth	one	central	no	——

Skeletal muscle tissue is found in skeletal muscle organs and is under voluntary control. This means that your conscious mind can order it to contract, but not all of its contractions are voluntary. In fact, most are not.

Cardiac muscle (located in the heart) and smooth muscle (found in the walls of tubular organs like the small intestine) are both involuntary. Involuntary means that these muscle types are normally controlled at the unconscious level and cannot be directly controlled by the conscious mind.

Examine your slide with sections of all three muscle subtypes. Identify their characteristics (see Table 13-1). In skeletal and cardiac muscle fibers, actin and myosin filaments overlap to produce the alternating pattern of light and dark bands (transverse striations) seen in these tissues. Only cardiac muscle cells are branched. Where the branch of one fiber joins another, the cells are stuck together and in direct communication through a complex of cell-to-cell junctions. The complex is called an **intercalated disk.** They are present in your section, but if you have trouble seeing them, look at the demonstration of intercalated disks set up by your instructor. Draw in Fig. 13-16 several fibers from each subtype of muscle tissue.

D. Smear of the Spinal Cord of an Ox

Nervous tissue is found in the brain, spinal cord, nerves, and all of the body's organs. Its function is the point-to-point transmission of information. A message is carried by impulses that travel along the functional unit of the nervous system, the **neuron.** Like muscle cells, neurons do not divide in the adult, and therefore replacement is impossible.

The largest cells in Fig. 13-17 are *motor neurons,* which connect the spinal cord to muscle fibers or glands. Find a similar cell in your smear of an ox spinal cord. The motor neuron has a **cell body** and a number of slender cytoplasmic extensions called neuron "processes," including one long **axon** and several shorter **dendrites.** The dendrites and cell body are stimulated within the spinal cord, and the axon conducts impulses out of it. The cells with smaller nuclei are accessory cells of the spinal cord and brain and are called **neuroglia.** Accessory cells help neurons function and make up about half the mass of nervous tissue. Label Fig. 13-17.

II. ANALYSIS OF AN ORGAN (Starr, pp. 364, 365)

As part of your study of tissues, you already have examined some aspects of the microscopic structure of a number of organs, including the kidney, trachea, small intestine, and skin. To repeat an important point, each of these organs is composed of various subtypes of the four basic tissues. The subtypes present in any organ contribute to its specific function. Let us examine more closely the function of the tissues of the small intestine (Fig. 13-18).

The specific functions of the small intestine are digestion of food, absorption of the end products of digestion, and transportation of indigestible material (fiber, etc.) to the large intestine.

MATERIALS

Per student:

- compound microscope, lens paper, a bottle of lens cleaning solution (optional), a lint-free cloth (optional), a dropper bottle of immersion oil (optional)
- prepared slide with a transverse section of small intestine (preferably of the ileum)

PROCEDURE Find and examine the following tissue subtypes in your section of the small intestine.

A. Simple Columnar Epithelium

This tissue is ideally suited for absorption of molecules from a hostile environment. The epithelium is one cell thick to facilitate transport and contains goblet cells that secrete mucus, which is thought to protect its surface from being digested. The simple columnar epithelium is continuous with the ducts of multicellular glands that deliver enzymes and other molecules important in digestion to the lumen of the small intestine.

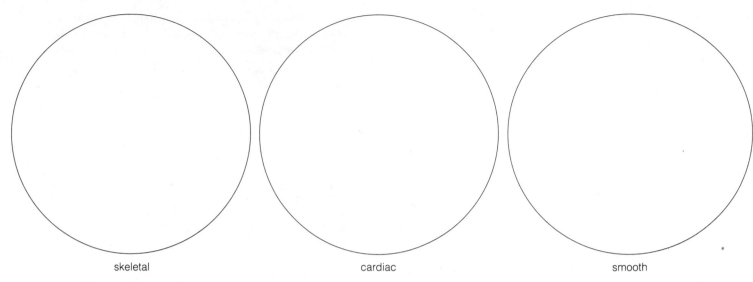

skeletal   cardiac   smooth

Figure 13-16 Drawing of sections of muscle tissue
(———×)

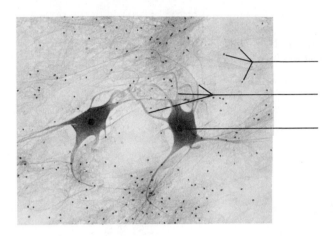

Labels: cell body, processes, neuroglia

Figure 13-17 Motor neurons in smear of spinal cord (134×).
(Photo courtesy Ripon Microslides, Inc.)

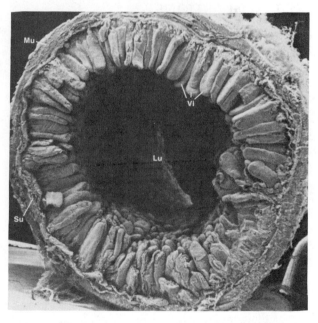

Figure 13-18 Scanning electron micrograph of a section of
the small intestine; Lu—Lumen, Mu—Muscle layer, Su—
Submucosa, and Vi—Villi (14×). (Photo from R. Kessel and R.
Kardon, *Tissues and Organs*. Copyright © 1979 W. H. Freeman
and Company. Used by permission.)

B. Loose Connective Tissue

The simple columnar epithelium combines with loose
connective tissue, which in turn combines with a thin
layer of smooth muscle called the *muscularis mucosae*
to form the *mucosa*. The muscularis mucosae separates the mucosa from a denser connective tissue
layer, the *submucosa*.

The epithelial and connective tissue layers of the
mucosa form fingerlike projections that protrude
into the lumen. These projections are called *villi*.
After the microvilli discussed earlier in this exercise,
the villi are the second most important contributor to
the large surface area of the small intestine. The connective tissue core of a villus contains a rich network
of capillaries and a lymphatic capillary to receive the
absorbed molecules. The walls of these blood vessels
are porous, simple squamous epithelium.

C. Smooth Muscle of the Muscle Layer

The muscle layer contains two sublayers of smooth
muscle arranged perpendicular to each other. In the
inner sublayer the fibers are circular, while in the
outer sublayer they are parallel to the longitudinal
axis of the small intestine. These sublayers produce
both *segmental contractions*, which mix the contents of
the small intestine, and waves of contraction, known
as *peristalsis*, that sweep the contents of the small intestine along its length.

D. Nervous Tissue

Two nets of neurons in the wall of the small intestine control the contractions of the smooth muscle. One is between the submucosa and the muscle layer, while the other is between the circular and longitudinal smooth muscle sublayers. The latter one is easier to find. Look for neuron cell bodies between the sublayers of the muscle layer.

E. Simple Squamous Epithelium of Serosa

The outer wall of the small intestine is lined with simple squamous epithelium that is continuous with and similar to the lining of the body cavity. This serosal membrane provides a moist, low-friction surface that allows for the free movement of abdominal organs.

As you see, the specific function of an organ is the sum of the individual functions of its tissues. Each organ in an animal can be analyzed in a similar fashion.

PRE-LAB QUESTIONS

____ 1. Histology is the study of (a) cells, (b) organelles, (c) tissues, (d) none of the above.

____ 2. Collections of similar cells and their products that interact as a structural, functional whole are called (a) organs, (b) systems, (c) tissues, (d) organelles.

____ 3. Organs strung together functionally and usually structurally form (a) organs, (b) systems, (c) tissues, (d) organelles.

____ 4. The structures that carry on all the functions vital to life in unicellular eukaryotic organisms are called (a) organs, (b) systems, (c) tissues, (d) organelles.

____ 5. Structures formed when the four tissues are combined are called (a) organs, (b) systems, (c) tissues, (d) organelles.

____ 6. An epithelial tissue formed by more than one layer of cells and with columnlike cells at the surface would be called (a) simple squamous, (b) stratified squamous, (c) simple columnar, (d) stratified columnar.

____ 7. The middle layer of the skin (a) is called the dermis, (b) is connective tissue, (c) contains collagen fibers, (d) is all of the above.

____ 8. Connective tissues are composed of (a) cells, (b) extracellular matrix, (c) axons, (d) a and b.

____ 9. To which subtype of muscle tissue does a fiber with transverse striations and many peripherally located nuclei belong? (a) skeletal, (b) cardiac, (c) smooth, (d) none of the above.

____ 10. In which tissue would you look for cells that function in point-to-point communication? (a) connective, (b) epithelial, (c) muscle, (d) nervous.

EXERCISE 13
ANIMAL ORGANIZATION I: MAMMALIAN HISTOLOGY

POST-LAB QUESTIONS

1. In the correct order from smallest to largest, list the levels of organization present in most animals.

2. Describe the main structural characteristics of the four basic tissues.
   a. epithelial tissue

   b. connective tissue

   c. muscle tissue

   d. nervous tissue

3. Describe the main functions of the four basic tissues.
   a. epithelial tissue

   b. connective tissue

   c. muscle tissue

   d. nervous tissue

4. Choose any organ. What is its specific function? Describe its functional histology.

ANIMAL ORGANIZATION II: MOUSE DISSECTION

OBJECTIVES   After completing this exercise you will be able to:

1. define head, neck, trunk, tail, thorax, abdomen, appendages, scrotum, diaphragm, ventral body cavities, thoracic cavity, abdominopelvic cavity, serosal membrane, pleural sacs, pericardial sac, mesentery, dorsal body cavities, cranial cavity, vertebral cavity;

2. list each system of a mammal and its vital functions;

3. describe the basic plan of the mammalian body;

4. locate the openings on the surface of a mammal's body;

5. locate the major organs in a mammal's body.

INTRODUCTION   Exercise 13 described in general the organization of an animal. Specifically, it introduced the four basic *tissues* and how they are combined in mammals to make *organs*. Organs with related functions form *systems* that carry on the vital functions of life and that collectively comprise the *organism*.

Health and the institutions and individuals that help us maintain it are an essential segment of our society. It is to everyone's benefit that each person be well-informed in this area. How many times have you or someone else asked where a particular organ is located? We should all know the location of the major organs in the body and what the insides of the body look like. To accomplish this, you will dissect a mouse. This animal has been chosen because, like you, it is a mammal, and because along with the rat, mice belong to the most common group of mammals used in biomedical research.

Mice and humans are both members of the class Mammalia. The young of most mammals develop in the uterus of the mother and are nourished until birth by a shared maternal/fetal organ, the *placenta*. The exceptions are the egg-laying duck-billed platypus and spiny anteater. After birth, young mammals drink milk from the mother's *mammary glands*. Other mammalian characteristics include *hair*, a *permanent set of teeth* in the adult, *sweat and sebaceous glands in the skin, red blood cells without nuclei*, a skeletal muscular *diaphragm* separating the body cavities of the upper and lower trunk, and a relatively *constant, high body temperature*.

Because the similarities between the different species of mammals far outnumber the ways in which they differ, the locations and functions of their organs and systems are essentially the same in all. Thus, we can study the mouse as representative of all mammals, even though it is specialized for its own way of life (by having gnawing teeth, for example).

There are 11 organ systems in mammals. Before continuing with this exercise, use your text to complete Table 14-1.   (Starr, pp. 340, 341)

MATERIALS

Per student:

- colored pencils—green, yellow, blue, red, black, brown, and pink

Per group (4):

- basic dissecting kit containing a scalpel, scissors, 4 pins, and a dissecting needle
- mouse, euthanized immediately prior to lab
- squeeze bottle containing 0.9% saline solution
- cotton balls
- dissecting tray
- small plastic bag for disposal of organs and soiled cotton balls during the dissection

Per lab section:

- large plastic bag for disposal of the small plastic bags and the carcasses at the end of the exercise

Per lab room:

- demonstration dissection of the nervous system of the mouse

Table 14-1 The Organ Systems of Mammals	
Systems	Vital Functions
Integumentary	
Nervous	
Endocrine	
Skeletal	
Muscular	
Circulatory	
Defense	
Respiratory	
Digestive	
Excretory	
Reproductive	

PROCEDURE

A. External Features

1. One member of each group obtains a fresh dead mouse and places it in a dissecting tray.

2. Identify the four major body regions—the **head, neck, trunk,** and **tail**—and the two major surfaces—the *dorsal surface* (back) and the *ventral surface* (belly). *See Appendix for the definitions of terms of orientation around and in the animal body.*

3. The trunk is divided into a front (anterior) portion, the **thorax** (supported by the rib cage), and a rear (posterior) portion, the **abdomen.** Feel the rib cage and the soft abdomen.

4. Observe the four **appendages** attached to the trunk. There are two *arms* attached to the thorax and two *legs* attached to the abdomen. Each appendage has three segments. The arm is divided into the *upper arm, forearm,* and *front foot,* and the leg is divided into the *thigh, shank,* and *hind foot.* There are five *digits* at the end of each appendage, although the thumb is reduced in size and has a *nail* rather than a *claw.*

5. Note that most of the mouse's body is covered by hair, which traps air and forms an insulating layer. Most of the human body surface is sparsely covered by hair and has sweat glands in the skin that secrete sweat onto its surface for cooling via evaporation. Mice have sweat glands only on the pads of their feet.

6. Find the following openings on the body surface:

a. mouth (entrance to the digestive system)

b. two **external nostrils** (openings to the respiratory system)

c. two **external auditory canals** (in ears)

d. anus (exit from the digestive system)

e. If you have a male mouse (see Fig. 14-1a), note the *preputial opening.* A portion of the copulatory organ, the **penis,** protrudes from this depression. Feel the rest of the penis under the skin of the preputial opening. At the tip of the penis is the **urethral opening,** which is shared by the reproductive and excretory systems.

f. If you have a female mouse (see Fig. 14-1b), note the **vulva.** This opening of the reproductive system is in front of the anus on the ventral surface. The **urethral opening** is the exit of the excretory system and is located underneath the protruding **clitoris,** which is immediately in front of the vulva. The clitoris has a common developmental origin with the penis.

7. If you have a male, locate the sac-like **scrotum,** which covers the anus ventrally. Feel for the paired testes in the scrotum. Each **testis** produces the male sex hormone and sperm. Unlike human males, the testes of the mouse may be withdrawn into the abdomen. The function of the scrotum and the muscles that attach to each testis is temperature regulation. The ideal temperature for sperm production is slightly less than body temperature.

If you have a mature female, find the well-developed teats of the **mammary glands.** How many do you count?

Exchange your mouse for one of the opposite sex from another group and examine the differences in external features.

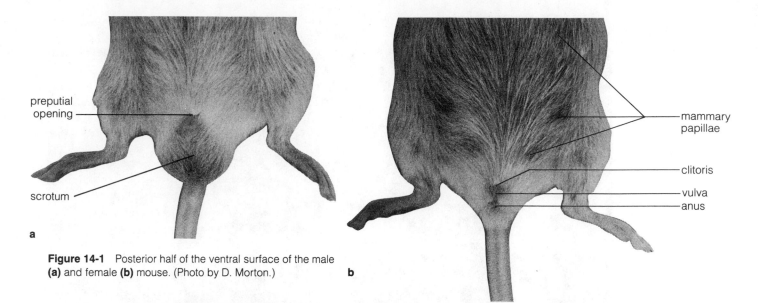

Figure 14-1 Posterior half of the ventral surface of the male **(a)** and female **(b)** mouse. (Photo by D. Morton.)

Labels for (a): preputial opening, scrotum

Labels for (b): mammary papillae, clitoris, vulva, anus

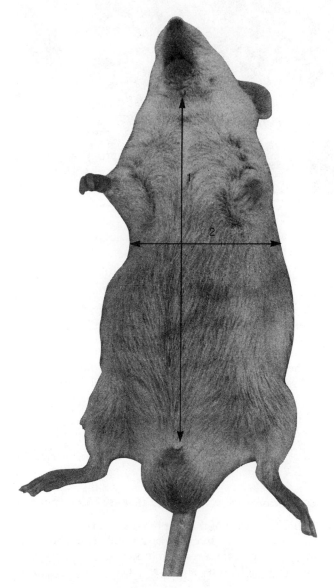

Figure 14-2 Opening the ventral body cavities. (Photo by D. Morton.)

B. Opening the Ventral Body Cavities

Figure 14-2 illustrates the incisions (cuts) necessary to open the ventral body cavities.

1. With a scalpel, make a longitudinal incision through the skin just to the right (the mouse's right) of the midline. This cut should extend from the neck to the preputial opening or clitoris, depending on the sex of your mouse. Then with a pair of scissors pierce the body wall just below the ribs. With the blades pointing up to prevent damage to the internal organs, cut along the incision through the rib cage and collar bone. Turn the scissors around and likewise cut through the abdominal muscles.

2. Pull the sides of the longitudinal incision apart and look for the **diaphragm.** This is a thin partition of skeletal muscle that separates the body cavities of the thorax and abdomen. Again use the scissors to make two perpendicular cuts through the body wall just below the diaphragm. Extend these cuts around and to the back of the mouse. Now cut the diaphragm away from its attachment to the body wall and separate the organs of the thoracic cavity from their attachment to the ventral body wall.

3. You should at this point be able to fold back four triangular flaps of body wall, opening up the **ventral body cavities,** the **thoracic cavity** in the thorax, and the **abdominopelvic cavity** in the abdomen (Fig. 14-3). To fold back the two upper flaps, you must break the ribs near their attachment to the *vertebral column* (backbone). You can easily do this with your hands. Pin the flaps to the dissecting tray. Rinse any coagulated blood from the body cavities with 0.9% saline solution and remove any excess fluid with cotton balls. Put the soiled cotton balls in the small plastic bag. Note that each body cavity is lined by a thin, wet **serosal membrane** that reduces friction between the inner body wall and the organs contained therein.

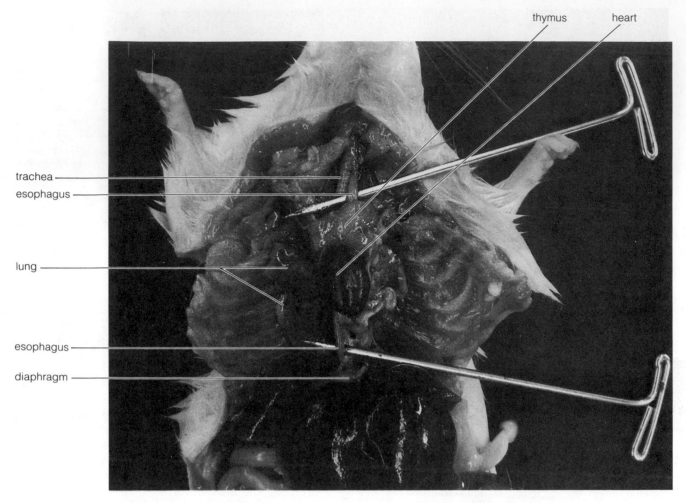

thymus heart

trachea

esophagus

lung

esophagus

diaphragm

Figure 14-3 Ventral view of organs in thoracic cavity. (Photo by D. Morton.)

C. Thoracic Cavity

The thoracic cavity (Fig. 14-3) is comprised of two lateral (away from the midline) **pleural sacs** and a medial (at or near the midline) **pericardial sac.** At the top, the medial walls of the pleural sacs join to form a septum, the *mediastinum.* Below, the mediastinum expands to surround the pericardial sac.

1. Locate the **lungs** in the pleural cavities. The net diffusion of oxygen into the blood and of carbon dioxide out of the blood occurs in the lungs.

2. Between the pleural cavities, find the **trachea** (windpipe), **esophagus** (foodpipe), and blood vessels traveling to and from the head in the mediastinum. Especially in young mice, you must remove the **thymus,** which covers the mediastinum. The thymus is important in the development of the immune defenses. The size and function of the thymus decrease with age. Dispose of the thymus in the small plastic bag.

3. Open the pericardial sac and examine the pump of the circulatory system, the **heart.** Just above and behind the heart, find where the trachea divides into a right and left **bronchus.** During the process of inspiration, the bronchi carry air from the trachea toward the lungs. Under the trachea and pericardial sac is the esophagus. Do you see where it pierces the diaphragm, passing into the abdominopelvic cavity?

4. In Fig. 14-3, underline the lung, thymus, and heart, using a different color for each organ system. Underline the organs of the respiratory system green, the defense system blue, and the circulatory system red.

D. Abdominopelvic Cavity (Fig. 14-4)

1. Just posterior to the diaphragm observe the body's largest gland, the **liver.** It has a number of lobes. Lift them and look for a reddish brown sac, the **gallbladder.** The gallbladder stores *bile* secreted by the liver.

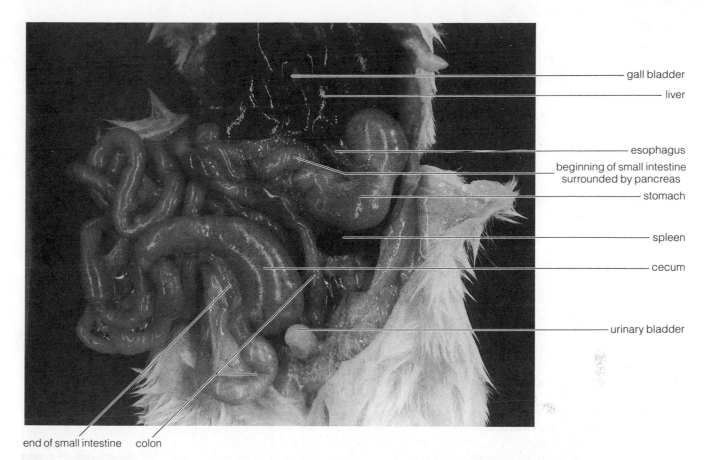

gall bladder
liver
esophagus
beginning of small intestine surrounded by pancreas
stomach
spleen
cecum
urinary bladder

end of small intestine   colon

Figure 14-4 Ventral view of organs of abdominopelvic cavity. (Photo by D. Morton.)

When the gallbladder contracts, bile is squirted into the small intestine to aid in the digestion of fats.

2. Locate the **stomach** in the left side of the abdominopelvic cavity, under the liver. The stomach helps process swallowed food. Can you discover where the esophagus joins the stomach at its anterior end? Examine the **spleen,** a dark red organ located just to the left of the stomach. The spleen filters the blood, removes old blood cells, and plays a role in immune defense.

3. Note where the **small intestine** arises from the lower end of the stomach. The small intestine completes the digestion of food and is the main site for absorption of the end products of digestion. It is the longest organ in the body, and its convolutions fill most of the abdominopelvic cavity. The small intestine is attached by a double membrane, the **mesentery,** to the dorsal body wall.

4. To observe the **pancreas,** look for a hard to see organ in the first loop of the small intestine. This gland produces enzymes for the digestion of food and, as part of the endocrine system, secretes *insulin* and other hormones into the blood.

5. Find the lower end of the small intestine, the place where it empties into the **large intestine.** The large intestine is shaped like an upside-down *U*. To see it, lift the small intestine. Follow the large intestine until it joins the short **rectum,** which terminates at the anus. Return to the junction between the small and large intestines. Note the blind sac extending in the direction opposite from the large intestine. This is the **caecum,** which in humans is a relatively small structure. In humans, a blind twisted tube, the *vermiform appendix,* is attached to the caecum.

6. Carefully remove the abdominal portion of the digestive tract by cutting (a) the esophagus between the diaphragm and stomach, (b) the dorsal mesentery, and (c) the descending portion of the large intestine (the third arm of the upside-down *U*). Do not disturb the urinary bladder, which lies in front of the caecum. Also, if you have a female, do not disturb the reproductive system, which is situated under the small intestine and in front of the rectum.

7. Stretch out the length of the small intestine and measure it.

length of small intestine = _____ cm

Now measure the mouse from the tip of the nose to the base of the tail.

length of head and body of mouse = _____ cm

How many times the length of the mouse is its small intestine?

_____ times

Dispose of the stomach and intestines in the small plastic bag.

8. In Figs. 14-3 and 14-4, underline the liver, gall bladder, stomach, spleen, small intestine, large intestine, and caecum. Underline the digestive system yellow; otherwise, the color code is the same as used for Fig. 14-3.

9. The **urinary bladder** stores urine delivered via two **ureters** from the paired **kidneys** (Fig. 14-5). Find the kidneys, which are located on the dorsal body wall under the serosal membrane, which in this location is called the *peritoneum*. With a dissecting needle, tear away the serosal membrane above one kidney. Each kidney is like a thick *C*, with the indentation pointing toward the midline of the body. The ureter arises from the center of this concave surface. Trace the ureter to the urinary bladder. At the lower end of the urinary bladder find the **urethra,** the duct that conveys urine out of the body when the bladder contracts.

10. If you have a female mouse, examine the organs of the reproductive system. Find the **body of the uterus** immediately behind the urinary bladder. Trace the body of the uterus upward until it branches into two **uterine horns,** which pass under the ureters and extend almost to the kidneys. Here each uterine horn joins a short **oviduct,** which expands to help form a sac surrounding the **ovary**. The paired ovaries are the source of female sex hormones and eggs. If the mouse is pregnant, the uterus will be quite large. Follow the uterus in the other direction and see where it joins the **vagina,** which in turn opens into the vulva.

If you have a male mouse, make sure you see a female before you finish this exercise.

11. In Fig. 14-5, underline the kidney, ureter, urinary bladder, and urethra brown; color the body of uterus, uterine horns, oviduct, ovary, and vagina pink.

E. Muscular and Skeletal Systems

1. Skin one of the mouse's legs. Individual skeletal muscle organs are held together by dense fibrous connective tissue. Separate one **skeletal muscle organ** by tearing this connective tissue with a dissecting needle. Describe the general appearance and consistency of a skeletal muscle.

_____

_____

_____

_____

_____

_____

Functions of skeletal muscles include movement and posture (in concert with the skeletal system) and heat production (e.g., shivering).

2. Remove all of the skeletal muscles surrounding the bone that supports the upper leg. Describe the general appearance and consistency of a **bone organ.**

_____

_____

_____

_____

_____

_____

Functions of bones include support, protection (for example, the cranium protects the brain), movement, and posture (in concert with skeletal muscles). Bones also store minerals (calcium, for example) and produce blood cells (in the *bone marrow*).

3. Dispose of the carcass, cut-out tissues and organs, and any soiled cotton balls into the small plastic bag. After closing it, place the small plastic bag in the large plastic bag provided by your instructor.

F. Nervous System

The nervous system (Fig. 14-6) is divided into the *central nervous system* (CNS) and *peripheral nervous system* (PNS). The PNS connects the organs and tissues of the body to the CNS. *Receptors* on the surface and inside of the body receive *stimuli* and send information to the CNS via *sensory neurons*. The CNS integrates this information and may react by sending instructions via *motor neurons* to *effectors* (muscle tissue or gland) to respond to the stimuli.

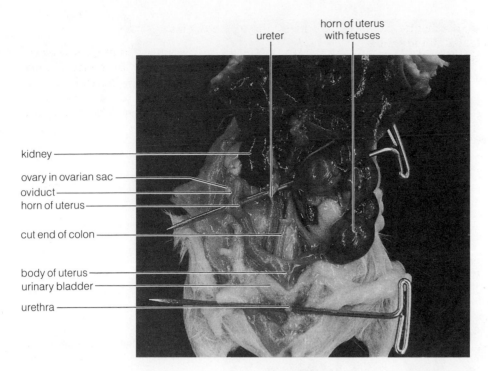

ureter

horn of uterus
with fetuses

kidney

ovary in ovarian sac

oviduct

horn of uterus

cut end of colon

body of uterus

urinary bladder

urethra

a

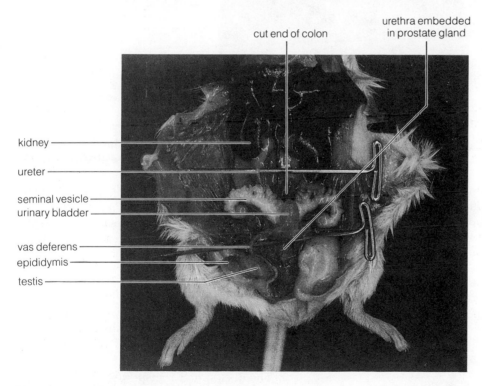

cut end of colon

urethra embedded
in prostate gland

kidney

ureter

seminal vesicle

urinary bladder

vas deferens

epididymis

testis

b

Figure 14-5 Ventral view of the excretory and reproductive systems of a female (**a**) and a male (**b**) mouse. (Photo by D. Morton.)

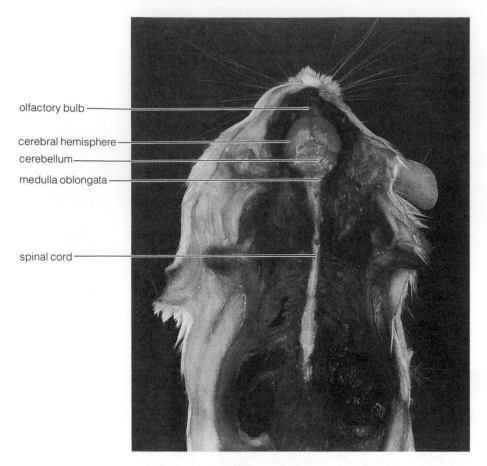

olfactory bulb

cerebral hemisphere

cerebellum

medulla oblongata

spinal cord

Figure 14-6 Dorsal view of brain and spinal cord. (Photo by D. Morton.)

Examine the demonstration dissection of the nervous system of a mouse. Identify the **brain** in the **cranial cavity** and the **spinal cord** in the **vertebral cavity.** The cranial and vertebral cavities together comprise the **dorsal body cavities.** Try to find the roots of *cranial nerves* and *spinal nerves* located on either side of the brain and spinal cord. These nerves along with nerves in general contain the axons of many sensory and motor neurons.

PRE-LAB QUESTIONS

_____ **1.** Most mammals have (a) hair, (b) mammary glands, (c) a set of permanent teeth in the adult, (d) all of the above.

_____ **2.** Which series of numbers places the levels of organization in mammals in order from smallest to largest? (1) tissue, (2) organ, (3) system, (4) cell; (a) 1,2,3,4; (b) 2,1,3,4; (c) 4,1,2,3; (d) 4,2,1,3.

_____ **3.** The front (anterior) portion of the trunk of a mouse's body is called the (a) tail, (b) appendages, (c) thorax, (d) abdomen.

_____ **4.** The anus is the exit from the (a) respiratory system, (b) excretory system, (c) digestive system, (d) circulatory system.

_____ **5.** The vulva is the opening into the (a) respiratory system, (b) excretory system, (c) digestive system, (d) female reproductive system.

_____ **6.** The testes are located in the (a) scrotum, (b) thoracic cavity, (c) penis, (d) urethra.

_____ **7.** The largest gland in the body is (a) the liver, (b) the thymus, (c) the pancreas, (d) none of the above.

_____ **8.** The blind sac to which the vermiform appendix is attached in humans is (a) the small intestine, (b) the large intestine, (c) the rectum, (d) none of the above.

_____ **9.** The duct(s) that transport(s) urine from the kidneys to the urinary bladder is (are) (a) oviducts, (b) the urethra, (c) ureters, (d) uterine horns.

_____ **10.** The brain is found in the (a) cranial cavity, (b) spinal cavity, (c) dorsal body cavities, (d) a and c.

EXERCISE 14
ANIMAL ORGANIZATION II: MOUSE DISSECTION

POST-LAB QUESTIONS

1. Matching: (one best answer)

_____ penis

_____ serosal membranes

_____ pancreas

_____ pleural sac

_____ pericardial sac

_____ abdominopelvic cavity

_____ uterus

_____ diaphragm

_____ clitoris

_____ spleen

a. organ of both endocrine and digestive systems

b. organ of female reproductive tract

c. contains the heart

d. same developmental origin as penis

e. male copulatory organ

f. divides the thoracic and abdominopelvic cavities

g. filters blood

h. contains a lung

i. lines body cavities

j. contains the liver

2. Describe the basic plan of the mammalian body.

3. Name the two systems that control the functions of all the organs of the body.

a. _____

b. _____

4. Describe the vital functions of the circulatory system.

5. Describe how the vital functions of the digestive and excretory systems complement each other.

6. Individual organs are usually part of a system in which their specific functions are combined to perform one of the vital functions necessary for the life of an organism. Usually, the organs of a system are physically in contact (e.g., the digestive system), but not always. List two systems where the individual organs are not in direct contact with each other (a and b).

   a. _____

   b. _____

7. Express your thoughts on the following questions. Do you think that the organism is the highest level of organization for animals? Can you name any insect species where individual members specialize to perform certain functions necessary to maintain life (e.g., gathering of food, protection, waste disposal, reproduction) and rely on the other individuals for the remainder? How about the human family? How about human society?

HUMAN SENSATIONS, REFLEXES, AND REACTIONS

OBJECTIVES After completing this exercise you will be able to:

1. define consciousness, sensory neurons, receptors, stimulus, motor neurons, somatic motor neurons, autonomic motor neurons, chemical synapse, effectors, interneurons, integration, sensations, proprioception, modality, projection, adaptation, free neuron endings, encapsulated neuron endings, phantom pain, reflex, reflex arc, stretch reflexes, patella reflex, swallowing reflex, pupillary reflex, reaction, reaction time;

2. describe the flow of information through the nervous system;

3. state the nature and function of sensations;

4. describe a stretch reflex;

5. describe the pupillary reflex;

6. distinguish between a reflex and a reaction;

7. measure visual reaction time.

INTRODUCTION How do you interact with the external environment? To answer this question, you first have to be able to analyze your interactions. This means you have to be conscious. **Consciousness** is the state of being aware of the things around you, your responses, and your own thoughts. Being conscious allows you to learn, remember, and to show emotion. Second, you have to understand the flow of information through the nervous system (Fig. 15-1).

Sensory neurons carry messages from **receptors** to the central nervous system (CNS). Receptors are located either within the body or on its surface. Receptors within the body receive information from the internal environment, while those on the surface of the body receive information from the external environment. Each piece of information received by a receptor is called a **stimulus** (the plural is *stimuli*).

Motor neurons carry messages from the brain and spinal cord to **effectors**. Effectors are muscles or glands that respond to stimuli. **Somatic motor neurons** control skeletal muscles, and **autonomic motor neurons** control smooth muscles, cardiac muscle, and glands.

In the CNS, a sensory neuron can directly stimulate a motor neuron across a **chemical synapse;** more frequently, though, one or more **interneurons** con-

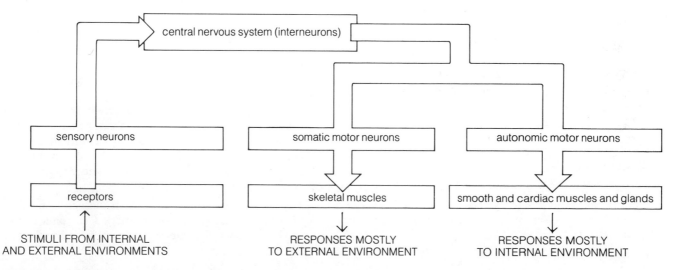

Figure 15-1 The flow of information through the nervous system

nect the sensory and motor neurons.

A chemical synapse is a junction between two neurons, or between a neuron and an effector, that is separated by a small gap. A chemical transmitter substance released from the first neuron binds to, and produces changes in, the receiving cell.

The function of the neurons within the CNS is **integration.** At this level, integration is the processing of messages received from receptors and the activation of the appropriate motor neurons, if any, to initiate responses by effectors. Your conscious mind is located in the cerebral cortex of the brain and is aware, and indeed is a part, of some of this activity.

I. SENSATIONS (Starr, pp. 441–447)

A receptor is the smallest part of a sense organ (such as skin) that can respond to a stimulus. The receptor is linked to the CNS by a single sensory neuron. Our bodies have receptors for light, sound waves, chemicals, heat, cold, tissue damage, and mechanical displacement. Senses for which we have sensations include sight, hearing, taste, smell, pain, touch, pressure, temperature, vibration, equilibrium, and **proprioception** (knowledge of the position and movement of the various body parts). **Sensations** are that portion of the sensory input to the CNS that is perceived by the conscious mind. There are also a number of complex sensations such as thirst, hunger, and nausea.

Most sensations inform the conscious mind about the state of the external environment. Sensations from the internal environment inform the conscious mind about problems such as dehydration. If you are thirsty, you will make a conscious decision to find and drink water.

Receptors and the sensations they produce have three characteristics: modality, projection, and adaptation. These characteristics can be easily demonstrated by investigating the skin's receptors.

MATERIALS

Per student pair:

- compound microscope
- prepared slide of mammalian skin stained with hematoxylin and eosin
- felt-tip, nonpermanent pen
- bristle
- dissecting needle
- 2 blunt probes in a 250-mL beaker of ice water
- 2 blunt probes in a 250-mL beaker of hot tap water (the hot water will have to be changed every 5 minutes)
- ice bag
- camel-hair brush
- reflex hammer
- 1,000-mL beaker containing ice water
- 1,000-mL beaker containing 45°C water
- 1,000-mL beaker containing room temperature water
- tissue paper

PROCEDURE

A. Modality

Although the nature of the impulses passing along all sensory neurons is the same, the stimulation of a specific receptor produces a specific sensation. This is modality. For example, stimulating a receptor for light in the eye causes a visual sensation and not sound. Different sensations result from where in the CNS, specifically the brain, these sensory neurons, or the interneurons to which they connect, terminate.

1. Examine a prepared section of skin (Fig. 15-2). There are two categories of receptors present: **free neuron endings** and **encapsulated neuron endings.**

Free neuron endings are almost impossible to see in typically stained sections, but note their distribution in Fig. 15-2. Stimulating different free neuron endings produces sensations of pain, crude touch, and perhaps cold and hot.

Encapsulated neuron endings consist of neuron endings surrounded by a connective tissue capsule. You can see the connective tissue capsule in typically stained sections. Look for Meissner's corpuscles in the dermal papillae. Meissner's corpuscles are receptors for fine touch and low-frequency vibration. Now find Pacinian corpuscles between the dermis and hypodermis. Pacinian corpuscles look like a cut onion and are receptors for pressure and high-frequency vibration.

2. With a felt-tip, nonpermanent ink pen, have your lab partner draw a 25-cell, 0.5-cm grid (Fig. 15-3) on the inside of your forearm, just above the wrist.

3. You are now the subject, and your lab partner is the investigator. At this point the investigator asks the subject to close his or her eyes. Using a bristle, the investigator touches the center of each box in the grid. If the bristle bends, you are pressing too hard. Ask the subjects to announce when they feel the touch. Do not count those responses that are given when you remove the bristle. Just count those that coincide with the initial touch. Mark each positive response with a *T* in the upper left-hand corner of the corresponding box in Fig. 15-3.

4. Repeat the above with a clean dissecting needle. This time, if you feel a prick, mark *P* for pain in the upper right-hand corner of the corresponding box in Fig. 15-3.

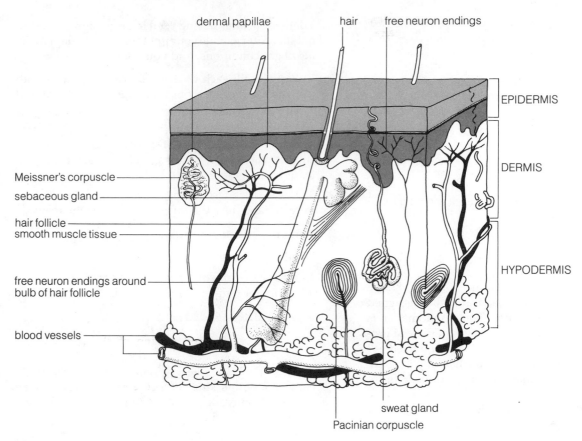

Figure 15-2 Vertical section of skin. (After Fowler, 1984.)

Labels in figure:
dermal papillae, hair, free neuron endings
EPIDERMIS
DERMIS
HYPODERMIS
Meissner's corpuscle
sebaceous gland
hair follicle
smooth muscle tissue
free neuron endings around bulb of hair follicle
blood vessels
sweat gland
Pacinian corpuscle

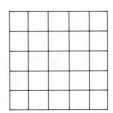

Figure 15-3 Grid for testing skin stimuli and recording modality data

_____

CAUTION *Do Not Press; Simply Let the Tip of the Dissecting Needle Rest on the Surface of the Skin*

_____

5. Repeat the above with a chilled blunt probe. Before using the blunt probe, dry it with tissue paper. The blunt probe will warm up over time, so switch it with the second chilled blunt probe every five trials. This time mark each positive response with a *C* for cold in the lower left-hand corner of the corresponding box in Fig. 15-3.

6. Repeat the above with a heated blunt probe. Before using the blunt probe, dry it with tissue paper. Use the two blunt probes alternately every five trials. This time mark each positive response with an *H* for hot in the lower right-hand corner of the corresponding box in Fig. 15-3.

7. What is the total number of positive responses for each stimulus?

touch _____/25 trials

pain _____/25 trials

cold _____/25 trials

hot _____/25 trials

8. Can you see a pattern in the distribution of positive responses marked in Fig. 15-3? (yes or no)

_____

9. What can you conclude about the modality of skin receptors?

_____

_____

_____

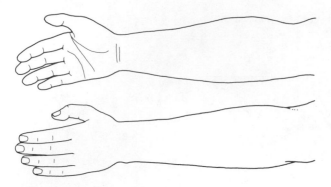

Figure 15-4 Front and back views of forearm and hand for recording projection data

B. Projection

All sensations are felt in the brain. However, before the conscious mind receives a sensation it is assigned back to its source, the receptor. This phenomenon is called projection. You have probably experienced projection. A common example is the "pins and needles" you feel in your hand and forearm when you accidentally jar the nerve that passes over the outside of the elbow. The sensory neurons in the nerve are stimulated, and your brain projects the sensation back to the receptors. Another example is the **phantom pain** that amputees feel in missing limbs.

1. Obtain an ice bag from the freezer.

2. The investigator holds the ice bag against the inside of the subject's elbow for 2 to 5 minutes.

3. The subject describes any sensations felt in the hand or forearm and notes them on Fig. 15-4.

4. While the ice bag is applied to the elbow, check for any loss of sensation by gently stroking the arm with a camel-hair brush.

5. Sensations may also be felt after the ice bag is removed.

6. If no results are obtained, try tapping the inside of the elbow with the reflex hammer.

7. What can you conclude about projection and the receptors on the surface of the hand and forearm?

_____

_____

_____

C. Adaptation

The intensity of the signal produced by a receptor depends in part on the strength of the stimulus and in part on the degree to which the receptor was stimulated before the stimulus. Receptors undergo adaptation to a constant stimulus over time. When you enter

a dark room after having been in bright light, you cannot see. After a while your photoreceptors adapt to the new light conditions, and your vision improves.

1. Partially fill each of three 1,000-mL beakers with ice water, water at room temperature, and water at 45°C.

2. The subject places one hand in the ice water and the other in the warm water. After one minute, the subject places both hands simultaneously in the water at room temperature.

3. The subject describes the sensation of temperature in each hand, and the investigator notes the results.

Hand pre-adapted in ice water _____

Hand pre-adapted in warm water _____

4. What can you conclude about adaptation and the receptors for temperature in the skin of the hand?

_____

_____

_____

How about other kinds of receptors? Can you give an example from your own experience? Do you feel the touch of your clothes?

_____

_____

_____

II. REFLEXES (Starr, pp. 433, 434)

A **reflex** is an involuntary response to the reception of a stimulus. A **reflex arc** consists of the nervous system components activated during the reflex. The simplest reflex arc consists of a receptor, sensory neuron, motor neuron, and effector.

Involuntary means your conscious mind does not help decide the response to the stimulus. However, the conscious mind may be aware after the fact that the reflex has taken place. Reflexes of which we are not aware occur most often in the internal environment (for example, reflexes involved in adjustments of blood pressure).

MATERIALS

Per student pair:

- reflex hammer
- penlight

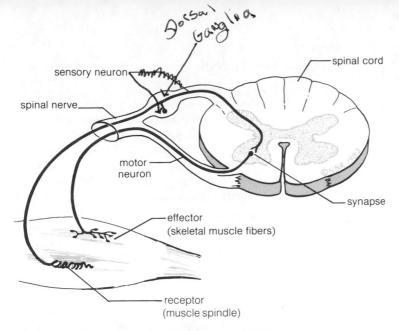

Dorsal Ganglia

Figure 15-5   The stretch reflex. (After Creager, 1983.)

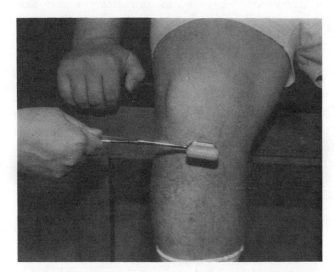

Figure 15-6   Area to tap to produce patella reflex. (Photo by) D. Morton.)

PROCEDURE

A. Stretch Reflexes

Stretch reflexes are the simplest type of reflex because there is no direct involvement of interneurons (Fig. 15-5). The sensory neuron synapses directly with the motor neuron in the spinal cord. Stretch reflexes are important in controlling balance and complex skeletal muscular movements such as walking. They are often tested by physicians during a physical as a check for spinal nerve damage.

1. The subject sits on a clean lab bench.

2. The investigator taps the patella ligament just below the patella bone (kneecap) with a reflex hammer (Fig. 15-6).

3. If you have trouble producing a response, ask the subject to shut both eyes and count backward from 10.

4. Describe the **patella reflex** in the space below. The receptor is the **muscle spindle** in the quadriceps femoris muscle of the thigh that is attached to the patella ligament. The muscle spindle detects any stretching of the muscle. The effector is the muscle itself.

_____

_____

_____

5. Why is a stretch reflex a somatic reflex? (Check the definition of a somatic motor neuron.)

_____

_____

_____

6. Even with your eyes shut, are you aware of the stimulus or the effect, or both?

stimulus (yes or no) _____

effect (yes or no) _____

This is because of pressure receptors that sense the tap and because of proprioceptors that sense movement of the leg.

B. Pupillary Reflex

1. The investigator shines the penlight into one of the subject's eyes. Does the size of the pupil (the opening into the eye that is surrounded by the iris, the pigmented part of the eye) get larger or smaller?

_____

2. Now turn off the penlight. Does the size of the pupil get larger or smaller?

_____

3. Repeat steps 1 and 2 and note which is faster, constriction of the iris (which makes the pupil smaller) or dilation of the iris (which makes the pupil larger)?

_____ is faster.

4. Ask if the subject is aware of the pupil's changing diameter.

(yes or no) _____

5. Why is the pupillary reflex an autonomic reflex?

_____

_____

_____

C. Complex Reflexes

Complex reflexes involve many reflex arcs and interneurons. A good example is swallowing. The stimulus in the **swallowing reflex** is the movement of saliva, food, or drink into the posterior oral cavity. The response is swallowing.

1. Cup your hand around your neck and swallow. Feel the complex skeletal muscular movements involved in swallowing. Do you consciously control all these muscles?

(yes or no) _____

2. Is it possible to swallow several times in quick succession?

(yes or no) _____

3. Explain this result. It has something to do with the stimulus.

_____

_____

_____

4. What part of swallowing does your conscious mind control, and what part is a reflex?

_____

_____

_____

III. REACTIONS

A **reaction** is a voluntary response to the reception of a stimulus. Voluntary means your conscious mind initiates the reaction. An example is swatting a fly once it has landed in an accessible spot. Because neurons must carry the sensory message to the cerebral cortex and the message to the motor neuron to react, a reaction takes more time than a reflex. **Reaction time** has the following components:

1. The time it takes for the stimulus to reach the receptive unit.

2. The time it takes for the receptor to process the message.

3. The time it takes for a sensory neuron to carry the message to the integration center.

4. The time it takes for the integration center to process the information.

5. The time it takes for a motor neuron to carry the response to the effector.

6. The time it takes for the effector to respond.

Visual reaction time can easily be measured with a reaction time ruler. This device makes use of the principle of progressive acceleration of a falling object.

MATERIALS

Per student group (4):

- Reaction Time Kit (Carolina Biological Supply Company)
- chair or stool
- calculator (optional)

PROCEDURE The following instructions are modified from the *Reaction Time Kit Instructions* booklet (Fig. 15-7).

1. The subject sits on a chair or stool.

2. The investigator stands facing the subject and holds the "release end" of the reaction time ruler with the thumb and forefinger of the dominant hand, at eye level or higher.

3. The subject positions the thumb and forefinger of the dominant hand around the "thumb line" on the ruler. The space between the subject's thumb and forefinger should be about 1 inch.

4. The subject tells the investigator when he or she is ready to be tested.

5. Once the investigator is told the subject is ready, at any time during the next 10 seconds, the investigator lets go of the ruler.

6. The subject catches the ruler between the thumb and forefinger as soon as it starts to fall. The line under his or her thumb represents visual reaction time in milliseconds.

Figure 15-7 Two students measuring visual reaction time. (Photo by D. Morton.)

Table 15-1 Record of Your Reaction Time Data	
Trial #	Reaction Time (milliseconds)
1	150
2	0
3	160
4	175
5	100
6	140
7	130
8	158
9	140
10	145
Total	1290
Average (Total/10)	129

7. The subject reads the reaction time from the ruler out loud, and the investigator records the data in Table 15-1.

8. Repeat steps 1 through 7 ten times and calculate the average reaction time from ten trials.

9. Repeat steps 1 through 8 for each member of the group.

10. The reaction time of most of the 10 trials should be similar, but perhaps the first few or one at random may be relatively different from the others. If this is true for your data, suggest some reasons for this variability.

_____

_____

_____

11. If opportunity and interest allow, the *Reaction Time Kit Instructions* booklet has a number of suggestions for other experiments you can easily do with the reaction time ruler.

PRE-LAB QUESTIONS

____ **1.** Neurons that carry messages from receptors to the CNS are (a) sensory, (b) motor, (c) interneurons, (d) none of the above.

____ **2.** Neurons that carry messages from the CNS to effectors are (a) sensory, (b) motor, (c) interneurons, (d) none of the above.

____ **3.** Neurons that carry messages within the CNS are (a) sensory, (b) motor, (c) interneurons, (d) none of the above.

____ **4.** Knowledge of the position and movement of the various body parts is (a) modality, (b) projection, (c) adaptation, (d) none of the above.

____ **5.** Skin contains (a) free neuron endings, (b) encapsulated neuron endings, (c) a and b, (d) no nervous tissue.

____ **6.** Which characteristic of receptors does phantom pain illustrate? (a) modality, (b) projection, (c) adaptation, (d) none of the above.

____ **7.** A simple reflex arc is made up of a receptor and (a) a sensory neuron, (b) a motor neuron, (c) an effector, (d) all of the above.

____ **8.** A stretch reflex is (a) somatic, (b) autonomic, (c) a and b, (d) none of the above.

____ **9.** A pupillary reflex is (a) somatic, (b) autonomic, (c) a and b, (d) none of the above.

____ **10.** A reaction is (a) a reflex, (b) involuntary, (c) voluntary, (d) a and b.

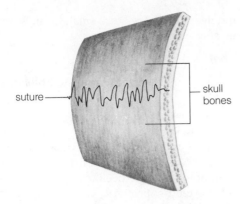

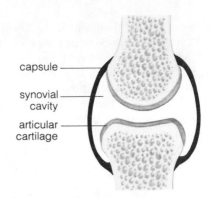

Figure 16-2 A suture (**a**) and a diagrammatic synovial joint (**b**). (After Fowler, 1984.)

Per lab room:

- labeled chart and illustrations of the adult human skeleton

PROCEDURE

A. Identification of Some Bones

There are 206 separate bones in the adult human skeleton. Using the labeled chart and illustrations of the human skeleton, identify the following bones on the articulated human skeleton and label them in Fig. 16-1.

1. Axial skeleton
   a. **skull** (28 separate bones, including middle ear bones);
   b. **vertebrae** (singular is *vertebra;* 26 separate bones, including the **sacrum,** which is composed of five fused vertebrae, and the **coccyx,** which is usually composed of four fused vertebrae)
   c. **ribs** (24 separate bones)
   d. **sternum** (three fused bones)
   e. **hyoid** (only bone that does not form a joint with another bone)

2. Appendicular skeleton (these bones are all found on each side of the body)

   pectoral girdle (shoulder)
   a. **scapula**
   b. **clavicle**

   upper appendage (arm)
   c. **humerus**
   d. **radius**
   e. **ulna**
   f. **carpals** (8)
   g. **metacarpals** (5)
   h. **phalanges** (14)

   pelvic girdle (hip)
   i. **coxal bone** (three fused bones—pubis, ischium, ilium)

   lower appendage (leg)
   j. **femur**
   k. **tibia**
   l. **fibula**
   m. **patella**
   n. **tarsals** (7)
   o. **metatarsals** (5)
   p. **phalanges** (14)

B. Joints

Joints can be immobile, like the **sutures** that connect the bones of the roof of the skull of young adults; they can be freely movable, like **synovial joints** (such as the elbow and knee); or they can fall between these two extremes in the degree of motion they allow (Fig. 16-2). The capsule of synovial joints is lined by a *synovial membrane* that secretes lubricating *synovial fluid.* Identify the synovial joints listed in Fig. 16-1 and list the participating bones that form them in Table 16-1.

Table 16-1 Bones That Form the Major Synovial Joints	
Joint	Bones
wrist	
elbow	
shoulder	
hip	
knee	
ankle	

C. Surface Features

There are many places on your body surface where bones can be felt. However, it is often difficult to tell specifically which bone you are feeling. Some are easy. For example, feel the bone of the lower jaw, the *mandible*. This is the only bone of the skull that forms a synovial joint with another skull bone.

Let us try a harder example, the process that projects from the point of the elbow joint. Touch it and alternately extend and flex the forearm, increasing and decreasing the angle between the forearm and upper arm, respectively. Which part of the arm does the process move with, forearm or upper arm?

_____

While still touching this process, alternately turn the hand palm down and up. Does the process move?

(yes or no) _____

Identify the bone to which this process belongs.

_____

In general, to identify a portion of a bone near a joint, move the body parts adjacent to the joint while touching the bone.

Identify the bones to which the surface features listed in Table 16-2 belong.

Table 16-2 Surface Features and Bones	
Surface Feature	Bone
knuckles	_____
bump next to the wrist and on the same side of the upper appendage as the little finger	_____
smaller bump next to the wrist and on the same side of the upper appendage as the thumb	_____
bump next to and outside the ankle	_____
bump next to and inside the ankle	_____

D. Structure of a Bone

1. Look at a femur, the longest bone of the skeleton. It consists of a shaft, or **diaphysis,** with two knobby ends, or **epiphyses** (the singular is *epiphysis*). One of the ends has a narrow neck and a round head. To which bone does it join?

PAtella

To which bone of the skeleton does the other end join?

Pelvic Girdle

Note that there are other surface features on the femur such as projections of various sizes and lines. These surface features are attachment sites for ten-

dons and ligaments. Are there small tunnels opening onto the surface of the femur?

(yes or no) _yes_

In life, these tunnels serve as routes for blood vessels and nerves.

2. Examine a femur that has been sawed in half lengthwise. There are two kinds of adult bone tissue: **compact bone** and **spongy bone.** Compact bone is solid and dense and is found on the surface of the femur. Spongy bone is latticelike and is found on the inside of the femur, primarily in the epiphyses and surrounding the **marrow cavity.** Which kind of bone tissue looks denser?

_____

Comparing pieces of equal size, which kind of bone tissue looks lighter?

_____

E. Structure of a Skeletal Muscle

Like bones, skeletal muscles are composed of all four basic tissue types. Skeletal muscles are mostly skeletal muscle tissue with the individual skeletal muscle fibers arranged parallel to the axis along which the muscle shortens when contracting. There is also a lot of connective tissue that surrounds the fibers and connects them to the tendons.

II. LEVERAGE AND MOVEMENT

(Starr, pp. 353–358)

Much of the skeletal system can be considered a system of levers where each bone is a lever and the joints are fulcrums. **Levers** are simple machines. When a pulling force or effort is applied to a lever, it moves about its fulcrum, overcoming a resistance or moving a load.

During a typical movement, one end of a skeletal muscle, the **origin,** remains stationary. The other end, the **insertion,** moves along with the bone and surrounding body part. The movement produced by the contraction is the **action** of the skeletal muscle. Most insertions are close to their joints, and the advantage gained by this is that the muscle has to shorten a small distance to produce a corresponding large movement of a body part.

MATERIALS

Per student pair:

- pair of scissors
- toggle switch mounted on a board (Alternatively, you can use any light switches present in the room.)
- pair of forceps
- pencil
- textbook

PROCEDURE

A. Classes of Levers

There are three classes of levers:

1. Class I. The fulcrum is located between the effort and the load (Fig. 16-3).

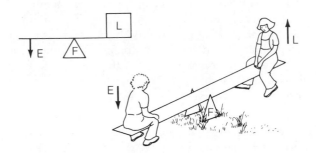

Figure 16-3 A seesaw is an example of a first-class lever; E—effort, F—fulcrum, and L—load

2. Class II. The load is located between the fulcrum and the effort (Fig. 16-4).

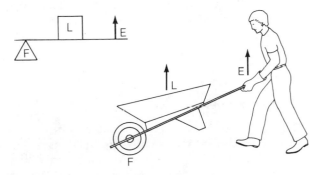

Figure 16-4 A wheelbarrow is an example of a second-class lever; E—effort, F—fulcrum, and L—load

3. Class III. The effort is located between the fulcrum and the load (Fig. 16-5). Third-class levers are the most common in the skeletal system.

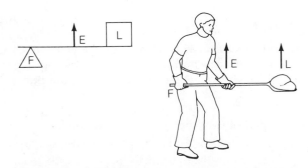

Figure 16-5 Lifting a spade with one hand while holding the handle stationary with the other hand is an example of a third-class lever; E—effort, F—fulcrum, and L—load

4. Test your understanding of the three classes of levers by examining the following objects and completing the following matching question.

class of lever	*objects*
_____ **1.** class I	**a.** scissors
_____ **2.** class II	**b.** toggle switch or light switch
_____ **3.** class III	**c.** forceps

B. Analysis of Simple Movements

Let us analyze three simple movements: flexion of the forearm, extension of the forearm, and plantar flexion of the foot (Fig. 16-6).

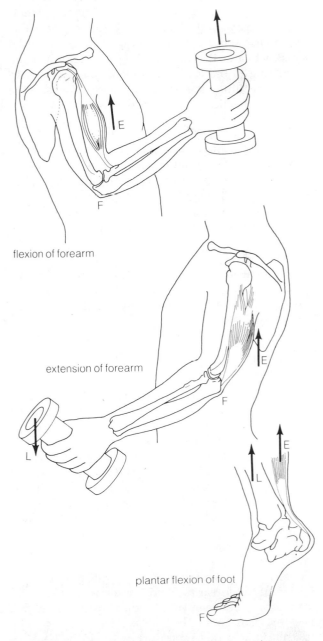

flexion of forearm

extension of forearm

plantar flexion of foot

Figure 16-6 Some simple actions of skeletal muscles

1. *Flexion of forearm.* While sitting, turn your hand so the palm is up and place it under the lab bench. Try to flex the forearm (decrease the angle between the forearm and upper arm). Because the skeletal muscle that is attempting to flex the forearm cannot shorten, the tension in it will increase. A contraction of a skeletal muscle where tension increases but no movement results is called an **isometric contraction.** Feel with your other hand the front surface of the upper arm. The large tense muscle is the biceps brachii. Its origin is the scapula and its insertion the radius. Which joint is the fulcrum?

_____

Now hold a pencil in the palm of your hand and flex the forearm. A contraction of a skeletal muscle that results in movement is called an **isotonic contraction.** There is no increase in tension during the movement. Feel the tension in the biceps brachii as you make this movement. Repeat this procedure, only replace the pencil with a textbook. Both the pencil and the book are adding to the load being lifted, the forearm. In which case, lifting the pencil or the textbook, was the tension in the biceps brachii the greatest?

lifting the _____

When you lift any object, the tension in the muscle must equal the weight of that object before movement can occur. Therefore, normal movements have an isometric phase followed by an isotonic phase.

Where is the pulling force applied? (insertion, origin, or both the insertion and origin)

_____

Even simple movements require the coordination of a group of muscles. For example, the origin does not move because other skeletal muscles hold the scapula stationary.

What class of lever is illustrated by this example?

(I, II, or III) _____

2. *Extension of forearm.* Place the hand, still palm up, on the top of the lab bench and try to extend the forearm (increase the angle between the forearm and the upper arm). Feel for a tense muscle on the back surface of the upper arm. This is the triceps brachii. The origin of the triceps brachii is the scapula and the upper humerus; its insertion is the *olecranon process* of the ulna. The fulcrum is the same as the previous example, only it has shifted its position relative to the effort and the load. What class of lever is illustrated by this example?

_____

Extension of the forearm is the opposite movement to flexion of the forearm. Hold the textbook, palm still up, halfway between full flexion and full extension. Feel the tension in the biceps brachii and triceps brachii. Repeat this procedure without the book. Is

the tension in the biceps brachii greater with or without the book?

_____

Is the tension in the triceps brachii greater with or without the book?

_____

The state of contraction of a group of skeletal muscles has to be coordinated to accomplish a particular movement or element of posture. Both the tendons of the biceps brachii and the triceps brachii are pulling on their insertions on the bones of the forearm to keep the forearm stationary. Other muscles are keeping the shoulder stationary.

3. *Plantar flexion of foot.* You need to stand up for the last example. A lab partner should stand behind you and watch that you do not fall during this procedure. With one hand on the lab bench to steady your balance, stand on the tips of your toes. With your other hand feel one of the very large tense muscles on the back of each calf. This is the gastrocnemius. The origin of the gastrocnemius is the femur, and its insertion is a tarsal—the calcaneus (the so-called heel bone). The fulcrum is the metatarsal-phalangeal joints, and the weight is the weight of the body transmitted through the tibia. Of what class of lever is this an example?

_____

III. WALKING

Walking is a complex activity that requires many movements and the coordinated contractions of several groups of skeletal muscles. Walking for each leg involves two phases that together make up the *step cycle.* The *stance phase* is the time when the leg is bearing weight, and the *swing phase* is when the leg is in the air.

MATERIALS

Per lab room:

▪ safe place to walk

PROCEDURE

1. Your instructor will tell you where you can walk safely. Walk a few normal steps, concentrating on one leg. What part of the foot strikes the ground first?

(toe or heel) _____

What part of your foot leaves the ground last?

_____

Does it leave passively, or does it push off?

_____

2. Now put your hands on your hips and concentrate on what your pelvic girdle is doing while you walk. First take short strides and then long ones. Does the pelvic girdle rotate more during short or long strides?

_____

This can be demonstrated in a different way. Find a lab partner of about equal height. Walk right next to each other but out of step, i.e., with opposite feet leading. First take short steps and then long ones. What happens?

_____

_____

This is called _lateral displacement_. Incidentally, females in general have to rotate their pelvic girdles a little more than males for a given length of stride. This is due to differences in the proportions of the female and male pelvic girdles.

3. _Vertical displacement_ also occurs during walking. From the side observe two individuals of equal height walking out of step and next to each other. Do their heads remain at the same level, or do they bob up and down?

_____

PRE-LAB QUESTIONS

____ **1.** Ligaments connect (a) bones to bones, (b) skeletal muscles to bones, (c) tendons to bones, (d) none of the above.

____ **2.** Tendons connect (a) bones to bones, (b) skeletal muscles to bones, (c) ligaments to bones, (d) none of the above.

____ **3.** Which of the following bones is part of the axial skeleton? (a) clavicle, (b) radius, (c) coxal bone, (d) none of the above.

____ **4.** The two kinds of bone tissue are (a) compact and porous, (b) compact and spongy, (c) porous and spongy, (d) none of the above.

____ **5.** There are _____ classes of levers. (a) two, (b) three, (c) four, (d) more than four.

____ **6.** The class of lever where the effort is located between the fulcrum and the load is called (a) first class, (b) second class, (c) third class, (d) fourth class.

____ **7.** The end of the skeletal muscle that remains stationary during a movement is the (a) action, (b) origin, (c) insertion, (d) none of the above.

____ **8.** In an isotonic contraction of a skeletal muscle (a) the tension in the muscle increases, (b) movement occurs, (c) no movement occurs, (d) a and c.

____ **9.** In an isometric contraction of a skeletal muscle (a) the tension in the muscle increases, (b) movement occurs, (c) no movement occurs, (d) a and c.

____ **10.** The step cycle of walking consists of a (a) stance phase, (b) swing phase, (c) a and b, (d) none of the above.

HUMAN BLOOD AND CIRCULATION

OBJECTIVES After completing this exercise you will be able to:

1. define blood vessels (different types), heart, blood, plasma, extracellular fluid, intracellular fluid, interstitial fluid, homeostasis, agglutination, blood typing system, intrinsic clotting mechanism, serum, hemoglobin, pulmonary circuit, systemic circuit, blood pressure, elastic membranes, valves, sinoatrial node;

2. identify and give the functions of the different types of blood cells;

3. describe the ABO and D(Rh) blood typing systems;

4. outline how a clot forms;

5. distinguish between an artery, capillary, and vein;

6. describe how blood flows through capillaries;

7. name the four chambers of the heart and describe the route blood takes through them;

8. describe how the heart contracts.

INTRODUCTION Circulatory systems, whether they be simple channels for water or closed networks of vessels containing complex fluids, provide the link between the cells of multicellular organisms. The evolution of circulation goes hand in hand with the evolution of cell specialization and paves the way for the evolution of tissues, organs, systems, and larger organisms.

Vertebrates and some invertebrates have a closed *circulatory system*, or a connected system of pipes called **blood vessels**, usually with a pump, or **heart.** The blood vessels convey fast-flowing rivers of blood that connect every system, organ, tissue, and cell in the body (Fig. 17-1).

The blood is piped around the circulatory system in blood vessels. In mammals and birds, there are two completely separate routes leading to and from the heart: the pulmonary and systemic circuits. In each circuit, branching **arteries** convey blood to

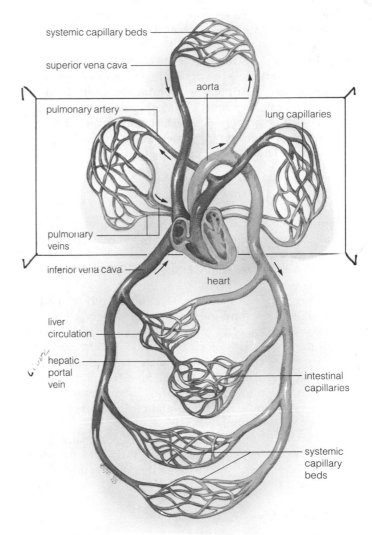

Figure 17-1 Schematic of the human circulatory system. The pulmonary circuit is enclosed by the box. Darker shading indicates oxygen-depleted blood. (After Starr and Taggart, 1987.)

smaller and more numerous **arterioles,** which in turn deliver the blood to beds of **capillaries.** It is across the capillary walls that the exchange of dissolved gases, nutrients, wastes, etc., takes place. Within the beds the capillaries branch and merge, finally merging into **venules.** Venules merge into a smaller number of **veins,** which continue merging and which drain blood back toward the heart.

The **pulmonary circuit** carries oxygen-depleted blood to the capillary beds of the lungs, where oxygen is loaded and where excess carbon dioxide is unloaded. Pulmonary veins drain the oxygen-rich blood back to the heart. The **systemic circuit** takes the oxygen-rich blood from the heart and conveys it to the rest of the body's capillary beds, where oxygen is unloaded and excess carbon dioxide is picked up. Systemic veins drain the oxygen-depleted blood back to the heart.

In a few cases this pattern of blood flow—heart → arteries → arterioles → capillaries → venules → veins → heart is interrupted by a **portal vein,** which connects two capillary beds. The most prominent example is the *hepatic portal vein,* which transports blood from capillary beds in the intestines, stomach, and spleen to beds of large capillaries in the liver.

I. BLOOD (Starr, pp. 376–378, 385–387)

Human **blood** is about 45% cells by volume. Most of the cells are erythrocytes, or red blood cells. They are what gives blood its red color. Blood cells are suspended in a straw-colored fluid called **plasma** (55% of the blood).

Plasma is part of the **extracellular fluid,** which is all of the body fluid outside of cells (45% of all body fluid). Fluid in cells is collectively called the **intracellular fluid** (55% of all body fluid). Fluid directly surrounding cells is also part of the extracellular fluid and is called **interstitial fluid.** The extracellular fluid is the same thing as the internal environment, and body functions strive to maintain its constancy. This process is called **homeostasis.**

Plasma is mostly water but contains many dissolved substances including gases, nutrients, wastes, ions, hormones, enzymes, antibodies, and other proteins.

MATERIALS

Per student:

- a compound microscope, lens paper, a bottle of lens cleaning solution (optional), a lint-free cloth (optional), a dropper bottle of immersion oil (optional)
- a prepared slide of a Wright-stained smear of human blood

- 4 disposable sterile alcohol wipes (optional)
- 4 disposable sterile blood lancets (optional)
- 3 microscope slides (optional)
- 4 sterile absorbent pads (optional)

Per student pair:

- 6 microscope slides
- box of mixing sticks
- wax pencil
- dissecting needle

Per lab group (4):

- 3 dropper bottles, each one containing anti-A, anti-B; or anti-D(Rh) sera
- 4 dropper bottles, each one containing type A⁻, B⁻, AB⁺, or O⁺ erythrocytes
- dropper bottle containing erythrocytes of unknown type (U1)
- dropper bottle containing erythrocytes of unknown type (U2)
- dropper bottle of Wright stain (optional)
- dropper bottle of Wright buffer (optional)
- staining rack (optional)
- squirt bottle of distilled water (optional)
- Tallquist booklet (optional)

Per lab room:

- eosinophil on demonstration (compound microscope)
- basophil on demonstration (compound microscope)
- container labeled *Disposal of Materials Contaminated with Blood*
- clock with a second hand

PROCEDURE

A. Formed Elements (Cells and Platelets) of Blood

1. Use your compound microscope to examine the prepared slide of Wright-stained smear of blood with medium power. Note the numerous pink-stained red blood cells, or **erythrocytes** (Fig. 17-2a). Each erythrocyte is a biconcave disk without a nucleus. Scattered among them are a much smaller number of blue/purple-stained cells. These are white blood cells, or **leukocytes.** Center a leukocyte and rotate the nosepiece to the high-dry objective. What part of the cell is stained blue/purple?

(nucleus or cytoplasm) _____

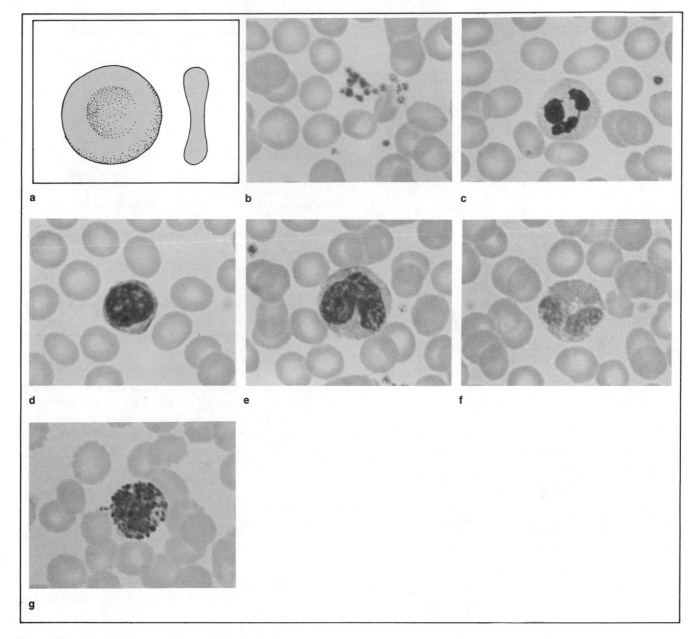

Figure 17-2 Formed elements of blood. **(a)** diagram of a top and side view of an erythrocyte, **(b)** platelets, **(c)** neutrophil, **(d)** lymphocyte, **(e)** monocyte, **(f)** eosinophil, **(g)** basophil. (Photos by D. Morton.)

Using high power, preferably the oil immersion objective, move the slide slowly and look for fragments of cells between the erythrocytes and leukocytes. They usually have one small blue-stained granule in them and are often clumped together. These cell fragments are **platelets,** or thrombocytes (Fig. 17-2b).

2. Identify at least three of the five leukocytes—**neutrophils, lymphocytes,** and **monocytes** (Figs. 17-2c, 2d, and 2e). You may also find **eosinophils** and **basophils** (Figs. 17-2f and 2g), which are usually the rarest leukocyte types.

The three types of leukocytes with the suffix -*phil* (for *philic,* meaning to like) have large *specific gran-*ules. The prefix in their names refers to the staining characteristics of the specific granules: *neutro-* for neutral (i.e., little staining by either of the two dyes in Wright stain); *eosino-* because the specific granules stain with the pink dye eosin; and *baso-* because the specific granules stain with the basic dye methylene blue. Eosin and methylene blue are the two dyes present in Wright stain.

3. If you have not found an eosinophil or a basophil by the time you have identified the three common leukocyte types, look at one or both of the demonstrations of these cells that have been set up by your instructor.

4. The abundance and size of the various blood cells are presented in Table 17-1.

Table 17-1	Characteristics of Formed Elements of Blood		
Cell or Fragment	Number/mm³ in Peripheral Blood	Percent of Leukocytes	Size (μm)
erythrocyte	4.5 to 5.5 million	—	7 by 2
platelets	250,000 to 300,000	—	2 to 5
neutrophils	3,000 to 6,750	65	10 to 12
eosinophils	100 to 360	3	10 to 12
basophils	25 to 90	1	8 to 10
lymphocytes	1,000 to 2,700	25	5 to 8
monocytes	150 to 750	6	9 to 15

5. Table 17-2 lists the functions of the blood cells.

Table 17-2	Functions of Formed Elements of Blood
Cell or Fragment	Functions
erythrocytes	contain hemoglobin, which transports oxygen, and carbonic anhydrase, which promotes transport of carbon dioxide by the blood
platelets	source of substances that aid in blood clotting
neutrophils	leave the blood early in an inflammation to become phagocytes (cells that eat bacteria and debris)
eosinophils	phagocytosis of antigen-antibody complexes; numbers are elevated during allergic reactions
basophils	granules contain a substance (histamine) that makes blood vessels leaky and a substance (heparin) that inhibits blood clotting
lymphocytes	perform many functions central to immunity
monocytes	leave the blood to form phagocytic cells called macrophages

The following procedure is optional.

6. It is easy to make a blood smear of your own blood and stain it with Wright stain.

CAUTION *The sight of blood is a psychological shock for some people. Therefore, do not be a subject or a helper unless you feel comfortable doing this procedure. You can still be an observer. Everyone should be seated and ready with the materials necessary to do this procedure before anyone starts.*

---

a. Work in pairs. Collect all the materials necessary to make a blood smear: sterile alcohol wipe, disposable sterile blood lancet, sterile absorbent pad, and two microscope slides. In addition, each group will need dropper bottles of Wright stain and buffer, a staining rack, and a squirt bottle of distilled water.

b. Read all of this section before you start.

CAUTION *Blood is a potential source of disease-causing organisms that can be transmitted with a puncture. If you have been exposed to serum hepatitis or AIDS, or have hemophilia, do not do this procedure. Do not touch any material contaminated by another person's blood. Never reuse a lancet or use a contaminated lancet.*

---

c. Alternate being the subject and the helper. Thoroughly wash and dry your hands. With an alcohol wipe, the subject then cleanses the terminal pad of the middle finger of either his or her left hand (if right-handed) or his or her right hand (if left-handed).

d. The helper opens an unopened lancet by peeling off the paper covering the blunt end of the lancet (the orientation of the lancet is pictured on the side of its package). Leave the pointed end covered until just before use.

CAUTION *Do not touch the pointed end of the lancet with any object before pricking the finger.*

---

e. The subject places his or her finger (nail side down) on the surface of the lab bench or table. Depending on who is to prick the finger, one of you places the index finger of the dominant hand on the left side of the cleansed pad and the thumb on its right side. **Do not touch the pad just to the left of center, because this is the area to be pricked.** Press gently so as to make the pad bulge.

f. The other partner removes the lancet from its package and pricks the pad just to the left of center. Note that the lancet has a stop to prevent too deep a prick.

g. Place a drop of blood on a clean slide lying on the lab bench. While holding the slide with the drop of blood still, make a smear as illustrated in Fig. 17-3.

h. Put a sterile absorbent pad over the wound.

i. Put the smear on a staining rack and let it air dry.

j. When the bleeding has stopped, discard the soiled alcohol wipe, blood lancet, absorbent pad, and second microscope slide in the container labeled *Disposal of Materials Contaminated with Blood.*

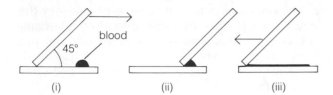

Figure 17-3 How to make a blood smear

(i)	(ii)	(iii)

k. Drop Wright stain onto the slide. Count and record the number of drops needed to cover the smear.

Stain for 2 minutes.

l. Add double the number of drops of Wright buffer. After 4 minutes, flood the slide with distilled water. Rinse the smear with more distilled water at the sink and stand the slide on end to air dry.

m. When the slide is thoroughly dry, examine it with high power.

You do not need a coverslip to use an oil immersion objective. When you are finished, remove the oil from the slide by placing a piece of lens paper on the oil, letting it absorb the oil, and gently sliding it off to the side. If oil is still present, repeat this procedure. After the oil is removed, the slide may be stored for future examination.

B. Blood Typing (Grouping)

Your erythrocytes may have on their surfaces one or more *antigens* that will cause their agglutination if exposed to the complementary *antibodies*. **Agglutination** is the clumping together of erythrocytes. This could theoretically occur during a blood transfusion. The transfusion of incompatible blood causes the destruction of donor erythrocytes and perhaps the death of the patient when the clumped cells block blood vessels. For these reasons, blood used for transfusion is very carefully matched for compatibility with the patient's blood.

In the ABO **blood typing system,** erythrocytes can have the A or B antigen, both, or neither. If one or both antigens are not present, the plasma contains the antibody or antibodies for the missing antigen.

For example, if an individual's blood is type A, then his or her plasma contains anti-B antibodies. A person with type A blood cannot safely receive blood from type B and AB donors, because the donor erythrocytes will be agglutinated by the anti-A antibodies in the plasma. This information is summarized in Table 17-3.

Which blood type can theoretically give blood to any other type and therefore has been called the *universal donor?*

Table 17-3 ABO Blood Types

Blood Type	Antigen Present on Erythrocytes	Antibody Present in Plasma	Plasma Agglutinates
A	A	anti-B	B and AB
B	B	anti-A	A and AB
AB	A and B	none	none
O	none	anti-A and B	A, B, and AB

Explain why this is so.

Which blood type can theoretically receive blood from any other type and therefore has been called the *universal recipient?*

Explain why this is so.

Usually a standard blood typing procedure includes a test for the Rh factor or D antigen. For example, a person with A⁺ blood has both the A and D antigens on the surface of his or her erythrocytes. About 86% of the population in the United States is RH⁺. However, Rh⁻ individuals do not have the anti-D antibody unless they have been exposed to Rh⁺ erythrocytes. This could happen during a transfusion of Rh⁺ blood or during the birth of an Rh⁺ child.

1. Work in pairs. Gather together six microscope slides, a box of mixing sticks, and a wax pencil.

2. As in Fig. 17-4, with the wax pencil divide each microscope slide into thirds with two lines perpendicular to the long axis of the slide. Label the upper left-hand corner of the left-hand box with an *A*. Similarly, label the next box to the right with a *B* and the last box with a *D*. In the upper right-hand corner, label each slide with a number (1 to 6).

3. Drip one drop of type A⁻ erythrocyte suspension in the middle of each box of slide 1. Then drip one

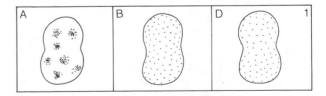

Figure 17-4 Results of ABO/D antisera testing on type A⁻ erythrocytes

drop of anti-A serum next to the drop of blood cells in the A box, one drop of anti-B serum next to the drop of blood cells in the B box, and one drop of anti-D serum next to the drop of blood in the D box. Mix together the two drops in each box with a mixing stick. Use an unused end of a mixing stick for each box.

If the antigen is present, the erythrocytes will clump together in the appropriate box. If it is not present, the mixed solution will remain cloudy (Fig. 17-4).

4. Repeat step 3 for B⁻, AB⁺, and O⁺ erythrocytes using slides 2, 3, and 4, respectively. If you have trouble deciding whether a result is positive or negative, check the slide with the compound microscope.

5. Using slides 5 and 6, identify the unknowns. Summarize all the results in Table 17-4.

Table 17-4	Results of ABO/D Blood Typing			
Slide No.	A	B	D	Blood Type
1	+	–	–	A⁻
2	–	+	–	B⁻
3	+	+	+	AB⁺
4	–	–	+	O⁺
5(U1)	___	___	___	___
6(U2)	___	___	___	___

The following procedure is optional.

6. It is easy to type your own blood.

a. Work in pairs. Collect all the materials necessary to obtain blood from the finger: sterile alcohol wipe, disposable sterile blood lancet, and sterile absorbent pad. You also will need a microscope slide marked with a waxed pencil as illustrated in Fig. 17-5 and mixing sticks.

b. Read all of this section before you start.

c. Read the cautions in Section A.6. Obtain blood as described in sections A.6c to f, except use the index finger instead of the middle finger.

d. Drip or touch the center of each box to deposit one drop of blood in the middle of each box.

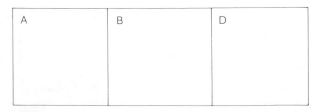

A	B	D

Figure 17-5 Results of ABO/D antisera testing on your blood

e. The subject puts a sterile absorbent pad over the wound. The partner drips one drop of anti-A serum next to the drop of blood cells in the A box, one drop of anti-B serum next to the drop of blood cells in the B box, and one drop of anti-D serum next to the drop of blood in the D box.

f. Mix together the two drops in each box with a mixing stick. Use an unused end of a mixing stick for each box.

If the antigen is present, the erythrocytes will clump together in the appropriate box. If it is not present, the mixed solution will remain cloudy. Record your results in Fig. 17-5.

Record your blood type.

If you have trouble deciding whether a result is positive or negative, check the slide with the compound microscope.

g. Discard the soiled alcohol wipe, blood lancet, absorbent pad, and microscope slide in the container labeled *Disposal of Materials Contaminated with Blood.*

h. Write your ABO/Rh blood type on the board. It is not necessary to include your name. In Table 17-5 record the total number of students in your lab section for each blood type. Calculate the percent of each blood type (number/total × 100) and record it in Table 17-5.

Table 17-5	Results of ABO/Rh Blood Typing for Lab Section	
Blood Type	Number	Percent
A⁻	___	___
A⁺	___	___
B⁻	___	___
B⁺	___	___
AB⁻	___	___
AB⁺	___	___
O⁻	___	___
O⁺	___	___
Total	___	100

C. Blood Clotting

When a blood vessel is torn, platelets stick to the damaged vessel and to each other to form a platelet plug. Substances are released that cause the ves-

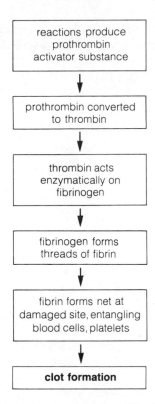

reactions produce
prothrombin
activator substance

↓

prothrombin converted
to thrombin

↓

thrombin acts
enzymatically on
fibrinogen

↓

fibrinogen forms
threads of fibrin

↓

fibrin forms net at
damaged site, entangling
blood cells, platelets

↓

clot formation

Figure 17-6 Blood clotting mechanism. (After Starr and Taggart, 1987.)

sel walls to spasm. Both of these mechanisms help limit blood loss from a wound. The platelets as well as damaged cells in the vessel wall release other substances that activate the **intrinsic blood clotting mechanism** (Fig. 17-6).

After the net of fibrin is produced, platelets and a substance they release cause the net to contract and form a clot. Blood cells also become trapped as the clot forms. The fluid that is squeezed out of a forming clot is called **serum** (the plural is *sera*).

The following procedure is optional.

1. Work in pairs. Collect all the materials necessary to obtain blood from the finger: sterile alcohol wipe, disposable sterile blood lancet, and sterile absorbent pad. You also will need a microscope slide.

2. Read all of this section before you start.

3. Read the cautions in section A.6. Obtain blood as described in Sections A.6c to f, except use the little finger instead of the middle finger.

4. Note the time blood first appears on the surface of your finger.

5. Drip a drop of blood on the center of the microscope slide.

6. The subject puts a sterile absorbent pad over the wound.

7. Every 30 seconds from the time noted in step 7 draw a dissecting needle through the drop of blood. As the needle leaves the blood observe carefully whether clotted strands of blood adhere to its tip. Note the time these threads are first seen.

8. The difference between the two times is clotting time and is a relative measure of the efficiency of the intrinsic clotting mechanism.

Record your clotting time.

_____ minutes

When tested with this method, normal clotting time ranges from 2 to 8 minutes.

9. Discard the soiled alcohol wipe, blood lancet, absorbent pad, and microscope slide in the container labeled *Disposal of Materials Contaminated with Blood.*

D. Erythrocyte Function

Erythrocytes are essentially bags of hemoglobin. **Hemoglobin** is an iron-containing pigment that transports oxygen and carbon dioxide.

Erythrocytes also contain an enzyme, *carbonic anhydrase*, which promotes the rapid conversion of carbon dioxide to bicarbonate ion. This ion then diffuses out of the erythrocyte and is transported in the plasma. About 23% of carbon dioxide is transported bound to hemoglobin, and 70% is converted to bicarbonate ion. The rest is dissolved in the blood.

The following procedure is optional.

1. Work in pairs. Collect all the materials necessary to obtain blood from the finger: sterile alcohol wipe, disposable sterile blood lancet, and sterile absorbent pad. You will also need a Tallquist test booklet, which contains filter paper sections, a scale, and instructions.

2. Read all of this section before you start.

3. Read the cautions in section A.6. Obtain blood as described in Sections A.6c to f, except use the ring finger instead of the middle finger.

4. Wipe away the first drop of blood with the sterile absorbent pad and place the second drop on the Tallquist filter paper. Hold the absorbent pad over the wound until bleeding stops.

5. After the blood loses its glossy appearance, match its color with that of the Tallquist scale and record the value.

_____ percent hemoglobin

The ratio 15.6 gm/100 mL blood is taken as 100%. Values from 70% and up are normal for adults, with females having slightly lower values than males.

6. Discard the soiled alcohol wipe, blood lancet, absorbent pad, and filter paper in the container labeled *Disposal of Materials Contaminated with Blood.*

II. BLOOD VESSELS

(Starr, p. 384)

The basic structure of blood vessels is illustrated in Fig. 17-7.

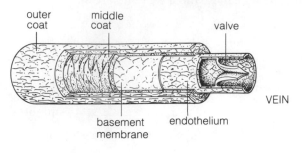

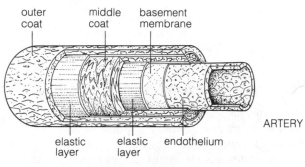

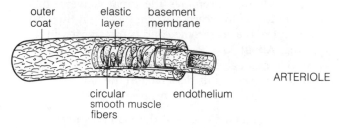

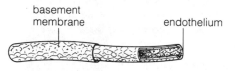

Figure 17-7 Structure of blood vessels. (After Spence, 1982, and Starr and Taggart, 1987.)

MATERIALS

Per student:

- compound microscope, lens paper, bottle of lens cleaning solution (optional), lint-free cloth (optional)
- prepared slide of a transverse section of a companion artery and vein

Per student pair:

- fish net
- small fish (3–4 cm long) in an aquarium
- 3-by-7-cm piece of absorbent cotton
- half a Petri dish
- coverslip
- dissecting needle

Per student group (4):

- container of anesthetic dissolved in dechlorinated water
- squeeze bottle of dechlorinated water

Per lab room:

- safe place to run in place
- several meter sticks taped vertically to the walls
- clock with a second hand

PROCEDURE

A. Arteries

Blood flows from the large arteries leaving the heart to the capillary beds because of the pressure gradient set up by the contraction of the heart. **Blood pressure** is the hydrostatic pressure of the blood at any part of the circulatory system. Average blood pressure decreases down the pressure gradient.

The walls of the largest arteries contain many **elastic membranes,** which are stretched during contraction (*systole*) of the heart. When the heart is relaxing (*diastole*), these membranes rebound and squeeze the blood, maintaining blood pressure and flow. Valves at the point where the aorta and pulmonary arteries leave the heart prevent the backflow of blood.

When a physician takes your blood pressure, it is usually of the brachial artery of the upper arm and with the body at rest. A blood pressure of 120/80 means that the pressure during systole is 120 mm of mercury (Hg) and that the diastolic pressure is 80 mm Hg. The difference between systolic and diastolic pressures (*pulse pressure*) produces a pulse that you can feel in arteries that pass close to the skin.

Blood pressure is not the same in all parts of the circulatory system. If it were, blood would not flow. Table 17-6 lists the blood pressures in various parts of the circulatory system.

Table 17-6 Blood Pressure in the Circulatory System of a Young Man at Rest	
Location	Blood Pressure (mm/Hg)
right atrium of heart	5/0 (systolic/diastolic)
right ventricle of heart	25/5
pulmonary arteries	20
arterioles and capillaries of lung	20 to 10
pulmonary veins	10
left atrium of heart	10/0
left ventricle of heart	120/10
brachial artery	120/80
arterioles	100 to 50
capillaries	50 to 20
veins	20 to 0

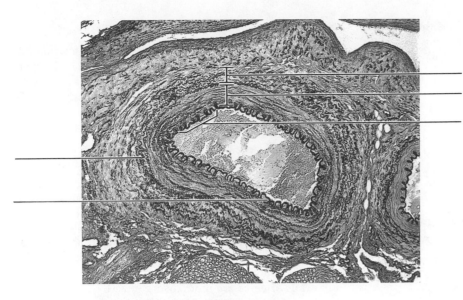

Labels: outer coat, middle coat, endothelium, inner elastic membrane, outer elastic membrane

Figure 17-8 Photomicrograph of a transverse section of an artery (136×). (Photo courtesy Ripon Microslides, Inc.)

Blood pressure also changes with health, emotional state, activity, and other factors.

1. Get a prepared slide of a companion artery and vein. Find and examine the transverse section of an artery.

Arteries have thick walls compared to other blood vessels. They have an outer coat of connective tissue, a middle coat of smooth muscle tissue, and an inner coat of simple squamous epithelium (*endothelium*). The three coats are separated by elastic membranes. The middle coat is the thickest layer of the three coats. Label Fig. 17-8.

2. Find your *radial pulse* in the radial artery (Fig. 17-9). Use the index and middle fingers of your other hand. A pulse occurs every time the heart contracts. The strength of the pulse is a measure of the difference between the systolic and diastolic blood pressure.

3. Determine and record your heart rate by counting the number of pulses in 15 seconds and multiplying by four.

_____/minute

_____

CAUTION *Do not do the following procedures if you have any medical problems with your lungs or heart. All subjects should be seated and should stop immediately if they feel faint.*

_____

4. Hold your breath. After 10 seconds have passed, determine your heart rate as in step 2.

_____/minute

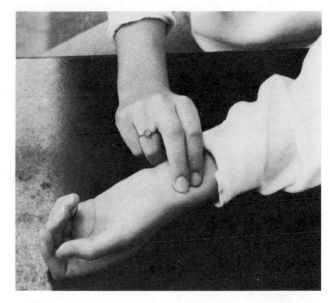

Figure 17-9 Feeling the radial pulse. (Photo by D. Morton.)

Compared to when you were breathing normally, does the strength of the pulse increase, decrease, or remain the same when holding your breath?

_____

When you hold your breath, you decrease the return of blood to the heart. This reduces pulse pressure. Homeostatic mechanisms increase heart rate to compensate for reduced blood pressure.

5. Now run in place for 2 minutes in the area designated by your lab instructor. Immediately after sitting down, again measure your heart rate as in step 2.

_____/minute

Compared to when you are at rest, does the strength of the pulse increase, decrease, or remain the same immediately after exercise?

_____

Explain these results.

_____

_____

_____

B. Capillaries

Capillaries have a very thin wall consisting of endothelium.

1. Use the net to catch a small fish from the aquarium and place it in the anesthetic fluid. Treat the fish gently and it will not be harmed by this procedure.

2. After the fish turns belly up, wrap its body in cotton made soaking wet with dechlorinated water. Place the fish in half a Petri dish so that the tail is in the center.

3. Using dechlorinated water, make a wet mount of the posterior two-thirds of the fish's tail and examine it with the low power, medium power, and high-dry objectives of the compound microscope. Use the lowest illumination that still allows you to see the blood flowing in the vessels. If necessary, you can temporarily close the condenser iris diaphragm to create more contrast.

Can you see erythrocytes?

(yes or no) _____

What vessels can you identify?

_____

_____

Is the blood flowing at the same speed in all of the capillaries?

(yes or no) _____

Describe blood flow.

_____

_____

_____

_____

4. Return the fish to the aquarium, wash the half Petri dish, and squeeze out the cotton in the sink before dropping it into the trash can.

C. Veins

For the blood to return to the heart after passing through capillary beds below the heart, it must overcome the force of gravity. Veins have **valves** to prevent the backflow of blood away from the heart. Blood is moved from one segment between valves to another primarily by muscular and breathing movements.

1. Get a prepared slide of a companion artery and vein. Find and examine the transverse section of a vein.

Veins have thin walls compared to their companion artery. They have an outer coat of connective tissue, a middle coat of smooth muscle tissue, and an inner coat of endothelium. Elastic membranes may be present. The outer coat is the thickest layer of the three coats. Compared to arteries and their walls, the walls of veins are more disorganized.

Label Fig. 17-10.

2. Work in pairs. Notice the veins as the subject's arm hangs down at the side of the body. You can easily see the veins because they are full of blood. This is usually best seen on the back of the hand. Now raise the arm above the head. Describe and explain any changes that take place.

_____

_____

_____

3. Using one of the meter sticks vertically taped to the wall, determine the venous pressure in the veins of the hand. Hold the subject's arm straight out at the level of the heart. Record the reading in millimeters where the hand crosses the meter stick.

measurement 1 = _____ mm

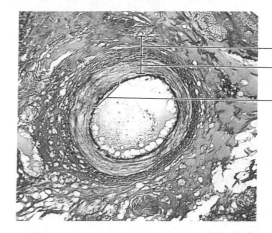

Labels: outer coat, middle coat, endothelium

Figure 17-10 Photomicrograph of a transverse section of a vein (107×). (Photo courtesy Ripon Microslides, Inc.)

Raise the arm slowly until the veins in the hand collapse (be sure that most of the muscles in the arm are relaxed). Record the height to which the arm has been raised in millimeters.

measurement 2 = _____ mm

The difference between the two readings gives you the venous pressure expressed in millimeters of water.

_____ mm (measurement 1) − _____ mm
(measurement 2) = _____ mm H$_2$O

Use the following formula to convert this into millimeters of mercury.

_____ mm H$_2$O × 0.074 mm Hg/mm H$_2$O = _____ mm Hg

How does this compare to arterial pressures (Table 17-6)?

_____

_____

4. Look at the veins of the subject's forearm and hand. There are swellings at various intervals. The swellings are valves. Choose a section between two swellings that does not have any side branches. Place one finger on the swelling away from the heart and with another finger press the blood forward beyond the next swelling. Does the vein fill up with blood again?

(yes or no) _____

Now remove the finger and observe what happens. Try this again, but press blood in the opposite direction. Discuss your observations.

_____

_____

_____

_____

III. THE HEART (Starr, pp. 379, 380)

Normally the heart beats over 100,000 times a day, pumping the blood around the circulatory system. The hearts of birds and mammals have four chambers (Fig. 17-11).

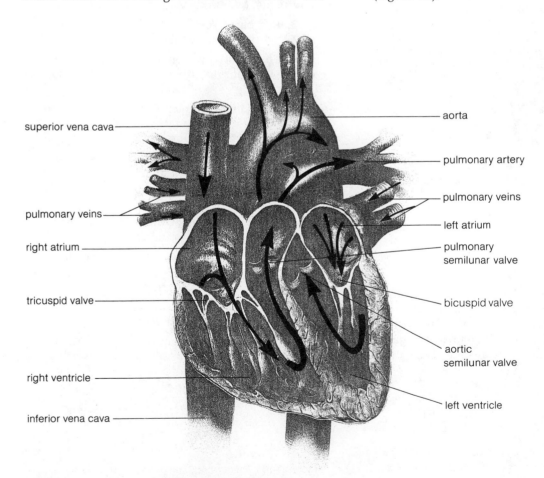

Figure 17-11 Diagram of a human heart. The arrows indicate the direction of blood flow, and the chambers and vessels carrying blood that is rich in oxygen are shaded. (After Fowler, 1984.)

The **right atrium** receives blood from the *superior vena cava, inferior vena cava* (Fig. 17-1), and *coronary sinus* (which drains blood from capillary beds in the heart itself). When the right atrium contracts, blood is pushed through the *tricuspid valve* into the **right ventricle.** Contraction of the right ventricle pushes blood into the trunk of the *pulmonary arteries.* The **left atrium** receives blood from the *pulmonary veins,* and its contraction pushes blood through the *bicuspid valve* into the **left ventricle.** Contraction of the left ventricle pushes blood into the *aorta.* The *semilunar valves* prevent the backflow of blood from the pulmonary trunk and aorta.

MATERIALS

Per lab section:

- demonstration of the beating heart of a doubly pithed frog kept moist with amphibian Ringer's solution (balanced salts solution)

PROCEDURE   Your instructor has set up a demonstration of a frog heart in place in the opened thorax of a frog. Although the frog has a three-chambered heart—two atria and one ventricle—the heart's function and control are essentially the same as those of humans. The nervous system of the frog has been destroyed so it does not feel pain or control heart action. Is the heart beating?

(yes or no) _____

Are the contractions of the heart organized or disorganized?

_____

If you observe carefully, you can see the order in which the chambers contract. Record your observations.

_____

_____

_____

_____

The primary pacemaker of the heart, the **sinoatrial node,** is located in the right atrium. It can function independently of the nervous system, firing rhythmically. Each time the sinoatrial node fires, it initiates a message to contract. This message spreads over the atria. Then special heart cells amplify and conduct the message throughout the ventricle.

Count and record how many times the heart contracts in one minute (heart rate).

_____ beats/minute

In an intact frog, the heart rate is modified by input from the central nervous system. The heart rate is affected by the amount of *acetylcholine* and *norepinephrine* secreted by neurons around the sinoatrial node. During restful activities, acetylcholine slows the heart rate and thus acts as a "brake" on the sinoatrial node. Pain, strong emotions, the anticipation of exercise, and the "fight or flight" response all can increase the secretion of norepinephrine. Norepinephrine speeds the heart rate and thus acts as an "accelerator" on the sinoatrial node and can override the parasympathetic "brake."

PRE-LAB QUESTIONS

____ **1.** The extracellular fluid consists of (a) plasma, (b) interstitial fluid, (c) intracellular fluid, (d) a and b.

____ **2.** Blood functions in the transport of (a) gases, (b) nutrients, (c) hormones, (d) all of the above.

____ **3.** Red blood cells are (a) erythrocytes, (b) leukocytes, (c) platelets, (d) all of the above.

____ **4.** The most common leukocyte in the blood is a (a) lymphocyte, (b) eosinophil, (c) basophil, (d) neutrophil.

____ **5.** The cellular fragments in the blood that function in blood clotting are (a) erythrocytes, (b) leukocytes, (c) platelets, (d) none of the above.

____ **6.** The clumping together of erythrocytes due to type A cells being exposed to anti-A serum is called (a) formation of a platelet plug, (b) agglutination, (c) clotting, (d) none of the above.

____ **7.** Blood vessels that connect capillary beds are (a) arteries, (b) veins, (c) portal veins, (d) b and c.

____ **8.** From which chamber of the heart does the right ventricle receive blood? (a) right atrium, (b) left atrium, (c) left ventricle, (d) none of the above.

____ **9.** How many chambers does the frog heart have? (a) one, (b) two, (c) three, (d) four.

____ **10.** The primary pacemaker of the heart is the (a) bicuspid valve, (b) tricuspid valve, (c) aorta, (d) sinoatrial node.

HUMAN RESPIRATION

OBJECTIVES After completing this exercise you will be able to:

1. define breathing, inspiration, expiration, ventilation, negative pressure inhalation, cohesion, positive pressure exhalation, positive pressure inhalation, tidal volume, inspiratory reserve volume, expiratory reserve volume, residual volume, vital capacity, chemoreceptor;

2. list the skeletal muscles used in breathing and give the specific function of each;

3. explain how air moves in and out of the lungs during respiration in the human;

4. explain how air moves in and out of the lungs during respiration in the frog;

5. describe the relationship between vital capacity and lung volumes and the interrelationships between lung volumes;

6. explain the importance of CO_2 concentration in the blood and other body fluids to the control of respiration.

INTRODUCTION Exercise 10 investigated carbohydrate metabolism and cellular respiration. Oxygen (O_2) is consumed and carbon dioxide (CO_2) and water (H_2O) are produced during the breakdown of glucose to provide the energy (adenosine triphosphate, or ATP) to fuel cellular activities.

In land vertebrates, O_2 uptake and CO_2 elimination occur by diffusion across moistened thin membranes located in the lungs, the main organs of the respiratory system. Animals are protected from excessive water loss via evaporation from moist respiratory surfaces by having the lungs inside the body. O_2 is carried from the lungs to the body's cells and CO_2 is delivered to the lungs from the body's cells by the blood and circulatory system.

The main function of the rest of the respiratory system is **ventilation**—the exchange of gases between the lungs and the atmosphere. The movement of gases in (inhalation) and out (exhalation) of the respiratory system requires the contraction of skeletal muscles. The rhythm of these muscular contractions and resulting ventilation is called respiration, or **breathing.**

Before we go on with this exercise, we need to review some terms from Exercise 14. The trunk of the body is divided into an upper *thorax*, which is supported by the rib cage and contains the *thoracic cavity*, and a lower *abdomen*. The thoracic cavity contains two *pleural sacs*, which contain the lungs and the *pericardial sac* around the heart. The thoracic cavity and the cavity of the abdomen are separated by a partition of skeletal muscle called the **diaphragm.**

The muscles of breathing include two sets of skeletal muscles between the ribs (**external** and **internal intercostal muscles**), the diaphragm, and skeletal muscles in the abdominal wall (**abdominal muscles**).

Breathing alternates between **inspiration** (breathing in) and **expiration** (breathing out). The mechanisms of breathing are as follows:

A. Inspiration—size of thorax is increased

1. *thoracic*—contract *external intercostal muscles* and relax *internal intercostal muscles* to raise rib cage

2. *abdominal*—contract *diaphragm* to lower floor of thoracic cavity

3. *forced*—both 1 and 2, plus other muscles in the anterior regions of the neck and shoulders

B. Expiration—size of thorax is decreased

1. *passive*—relax inspiratory muscles

2. *forced*—contract *internal intercostal muscles* and contract *abdominal muscles*

MATERIALS

Per student:

- 2 pieces of paper (half letter size)

Per group (4):

- functional model of lung
- metric tape measure

- large caliper with linear scale (e.g., Collyer pelvimeter)
- noseclip (optional)
- simple spirometer or lung volume bags

Per lab room:

- frogs in terrarium
- clock with second hand that is visible to all
- designated safe area for running in place

PROCEDURE

(Starr, pp. 410–416)

A. Ventilation

All flow occurs down a pressure gradient. When you let go of an untied inflated balloon, it flies away, propelled by the jet of air flowing out of it. The air flows out because the pressure is higher inside than outside the balloon. The high pressure inside the balloon is maintained by the energy stored in its stretched elastic wall.

When the thoracic cavity expands during inspiration, first the pressure in the pleural sacs decreases, and then the pressure within the lungs decreases. Because the pressure outside the body is now higher than that in the lungs, and assuming the connecting *ventilatory ducts* (trachea, etc.) are not blocked, air flows into the lungs (Fig. 18-1). This is called **negative pressure inhalation.**

The opposite occurs during expiration. The size of the thorax and pleural sacs decreases, the pressure in the lungs increases, and air flows out of the body down its concentration gradient. This is called **positive pressure exhalation.**

The pressure in the pleural sacs is actually always below atmospheric pressure. Thus, a hole in a pleural sac or lung will result in a collapsed lung. Inspiration is aided by **cohesion** (sticking together) of the wet serosal membranes lining the lungs and outer walls of the pleural sacs. Expiration depends in part on the

elastic recoil (like letting go of a stretched rubber band) of lung tissue.

Figure 18-2 illustrates inspiration and expiration.

1. Work in groups of four. Look at the functional lung model. The "Y" tube is analogous to the ventilatory ducts. The balloons represent the lungs. The space within the transparent chamber represents the thoracic spaces and its rubber floor (rubber "diaphragm"), the muscular diaphragm.

2. Pull down the rubber diaphragm. Describe what happens to the balloons.

_____

As you pull down the rubber diaphragm, does the volume of the space in the container increase or decrease?

_____

As the volume changes, is the pressure in the container increased or decreased?

_____

As the balloons inflate, does the volume of air in the balloons increase or decrease?

_____

Why do the balloons inflate?

_____

_____

_____

3. Push up on the rubber diaphragm. Describe what happens to the balloons and why it happens.

_____

_____

_____

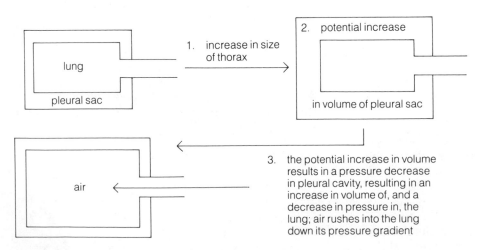

Figure 18-1 Pressure relationships during inspiration

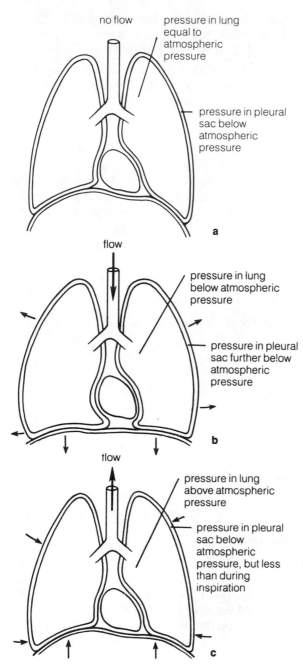

no flow

pressure in lung equal to atmospheric pressure

pressure in pleural sac below atmospheric pressure

a

flow

pressure in lung below atmospheric pressure

pressure in pleural sac further below atmospheric pressure

b

flow

pressure in lung above atmospheric pressure

pressure in pleural sac below atmospheric pressure, but less than during inspiration

c

Figure 18-2 Changes in the thoracic cavity during inspiration (**b**) and expiration (**c**) and corresponding movements of air. (**a**) shows the situation at the end of an expiration. (After Weller and Wiley, 1985.)

4. Pull the rubber diaphragm down and push it up several times in succession to simulate breathing.

5. Pucker up your lips and inhale. As you inhale, place one of the pieces of paper directly over your lips. What occurs?

_____

_____

This suction is caused by the negative pressure created in your lungs by the contraction of the muscles of inspiration.

6. Fold the narrow ends of the two pieces of paper to produce 2–3 cm flaps. Open the flaps and use them as handles. Hold a piece of paper with each hand and touch their flat surfaces together in front of you. Pull them apart.

Now, thoroughly wet both pieces of paper with water and again touch their flat surfaces together in front of you. Pull them apart. What difference did the water make?

_____

_____

Inflation of the lungs of vertebrates that inhale using negative pressure is aided by the cohesion (sticking together) of the wet serosal membranes that line the lungs and the outer walls of the pleural sacs.

B. Positive Ventilation

Some vertebrates such as the frog inhale by pushing air into the lungs. This is called **positive pressure inhalation.**

Observe a frog out of water. The frog inhales by sucking in air through the nostrils by lowering the floor of the mouth. Valves in the nostrils are then closed and the floor of the mouth raised, thus increasing the pressure and forcing the air into the lungs. The upper portion of the ventilatory duct can be closed to keep the air in the lungs. Exhalation occurs by elastic recoil of the lungs with the ventilatory duct open. In the frog, both inhalation and exhalation are the result of positive pressure.

What is the frog's respiratory rate (breaths per minute)? Count and record how many times the frog lowers and raises the floor of the mouth (one breath) in 3 minutes.

_____ breaths

Divide by three to calculate the average respiratory rate.

_____ /minute

C. Breathing Movements

1. Place your hands on your abdomen and take three deep breaths—three forced inspirations followed by three forced expirations. Describe and explain what you feel during:

each inspiration—

_____

_____

each expiration—

_____

_____

2. Place your hands on your chest and repeat step 1. Describe and explain what you feel during:

each inspiration—

_____

_____

each expiration—

_____

_____

D. Measurements of the Thorax

The size of the thorax can be described by three so-called diameters: the lateral diameter (LD), the anterioposterior diameter (A/PD), and the vertical diameter (Fig. 18-3). The vertical diameter is the only one that cannot be measured easily.

Make the following observations and record them in Table 18-1.

1. Work in groups of four. Measure the circumference of the chest with a tape measure at two levels, under the armpits (axillae—C_{AX}) and the lower tip of the sternum (xiphoid process—C_{XP}).

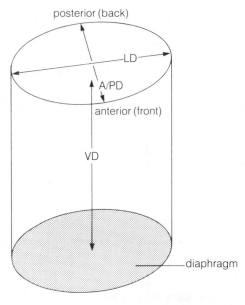

Figure 18-3 Thoracic diameters; LD—lateral diameter, A/PD—anterioposterior diameter, VD—vertical diameter

Table 18-1	Chest Measurements (cm)				
Condition	C_{AX}	C_{XP}	A/PD	LD	
a	_____	_____	_____	_____	
b	_____	_____	_____	_____	
c	_____	_____	_____	_____	
d	_____	_____	_____	_____	

2. With calipers measure the A/PD and LD at the nipple line. Read the distance between the tips of the calipers off the scale in centimeters.

3. Record data in Table 18-1 for the following conditions: **(a)** at the end of a restful inspiration; **(b)** at the end of a passive expiration; **(c)** at the end of a forced inspiration; and **(d)** at the end of a forced expiration. While the measurements are being taken, it is extremely important not to tense muscles other than those used for respiration.

4. About two-thirds of the air inhaled during a restful inspiration is due to contraction of the diaphragm. Interpret the data in Table 18-1, and in your own words describe changes in the size of the thorax during:

a. a restful inspiration—

_____

_____

b. a passive expiration—

_____

_____

c. a forced inspiration—

_____

_____

d. a forced expiration—

_____

_____

E. Spirometry

Air in the lungs is divided into four mutually exclusive volumes: tidal volume (TV), inspiratory reserve volume (IRV), expiratory reserve volume (ERV), and residual volume (RV).

Tidal volume is the volume of air inhaled or exhaled during breathing. It normally varies from a minimum at rest to a maximum during strenuous exercise.

Inspiratory reserve volume is the volume of air you can voluntarily inhale after inhalation of the tidal volume. **Expiratory reserve volume** is the volume of air you can voluntarily exhale after an exhalation of the tidal volume. IRV and ERV both decrease as TV increases.

Residual volume is the volume of air that cannot be exhaled from the lungs. That is, normal lungs are always partially inflated.

There are four capacities derived from the four volumes:

inspiratory capacity (IC) = TV + IRV
functional residual capacity (FRC) = ERV + RV
vital capacity (VC) = TV + IRV + ERV
total lung capacity (TLC) = total of all four lung volumes

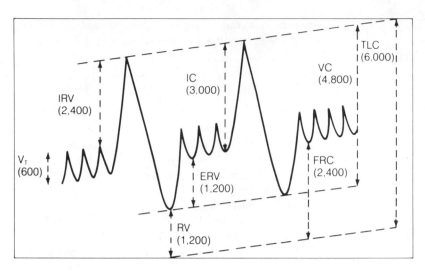

Figure 18-4 A spirogram showing the defined lung volumes and capacities. The numbers in parentheses are average values in milliliters. (After Weller and Wiley, 1985.)

All the lung volumes except the residual volume can be measured or calculated from measurements obtained using a simple spirometer or lung volume bag. A more sophisticated recording spirometer makes a trace of respiration over time called a spirogram. Figure 18-4 illustrates a spirogram and the relationships of the lung volumes and capacities.

CAUTION *Always use a sterile mouthpiece and do not inhale air from either a simple spirometer or a lung volume bag.*

1. Work in groups of four. Sit quietly and breathe restfully. Use a noseclip or hold your nose. After you feel comfortable, start counting as you inhale. After the fourth inhalation, exhale normally into the spirometer or lung volume bag. Read the volume indicated by the spirometer or squeeze the air to the end of the lung volume bag and read the volume from the wall of the bag. Record the volume below (trial 1). Reset the spirometer or squeeze the air out of the lung volume bag. Repeat this procedure two more times (trials 2 and 3) and calculate the total and average tidal volume at rest.

trial 1	_____ mL
trial 2	_____ mL
trial 3	_____ mL
total =	_____ mL

Divide the total by three = _____ mL to calculate the average TV at rest.

2. Determine the volume of air you can forcibly exhale after a restful inspiration (average of three trials).

trial 1	_____ mL
trial 2	_____ mL
trial 3	_____ mL
total =	_____ mL

Divide the total by three = _____ mL. This is the average sum of expiratory reserve volume and tidal volume at rest.

3. Determine the volume of air you can forcibly exhale after a forceful inspiration (average of three trials).

trial 1	_____ mL
trial 2	_____ mL
trial 3	_____ mL
total =	_____ mL

Divide the total by three = _____ mL to calculate the average vital capacity.

4. Calculate the ERV at rest by subtracting the result of step 1 from the result of step 2.

_____ mL (step 2) – _____ mL (step 1) = _____ mL.

5. Calculate the IRV at rest by subtracting the result of step 2 from the result of step 3.

_____ mL (step 3) – _____ mL (step 2) = _____ mL.

6. Summarize your results in Table 18-2.

Table 18-2   Vital Capacity and Lung Volumes at Rest (mL)

Measure	Volume (mL)
Tidal volume	_____
Inspiratory reserve volume	_____
Expiratory reserve volume	_____
Vital capacity	_____

Does vital capacity change as tidal volume increases or decreases?

(yes or no) _____

Measure and record your height in centimeters.

_____ cm

Write your vital capacity/height on the board—your name is not necessary.

Plot the vital capacity of each of the students in your lab section on the graph provided in question 6 of the Post-lab Questions.

F. Control of Respiration

The control of respiration, both the rate and depth of breathing, is very complex. Simply stated, **chemoreceptors** (receptors for chemicals such as O_2, CO_2, and hydrogen ions, or H^+), stretch receptors in the ventilatory ducts, and centers in the brain stem (part of brain that connects to the spinal cord) control respiration. By far the most important stimulus is the CO_2 concentration in the blood and other body fluids.

Our own experience has taught us that respiration is to some extent under the control of the conscious mind. We can decide to stop breathing or to breathe more rapidly and deeply. However, the unconscious mind can override voluntary control. The classical example of this is the inability to hold one's breath for more than a few minutes. Once the CO_2 concentration rises above a specific point, you are forced to breathe.

_____

_____

CAUTION *Do not do the following procedures if you have any medical problems with your lungs or heart. All subjects should be seated and should stop immediately if they feel faint.*

_____

1. Work in pairs. The subject sits down unless instructed to do otherwise. After the subject feels comfortable, determine the respiratory rate at rest. The investigator counts and records the number of times the subject breathes in 3 minutes.

_____ breaths

Divide by three to calculate the average respiratory rate.

_____ /minute

2. The subject breathes deeply, as rapidly as possible. Try to take at least 10 breaths but stop as soon as you can answer the following question. In any case do not continue for more than 20 breaths. As time goes on, does it become easier or more difficult to continue rapid deep breathing?

Forced deep breathing results in overventilation of the lungs, or **hyperventilation.** How does hyperventilation affect the CO_2 concentration of the blood? (increases CO_2, decreases CO_2, or no effect on CO_2)

_____

3. When fully recovered from step 2, measure how long the subject can hold his or her breath after a restful inspiration.

_____ seconds

Now, the subject carefully runs in place for 2 minutes in the area designated by your lab instructor. Immediately after sitting down, again measure how long the subject can hold his or her breath after a restful inspiration.

_____ seconds

How does running in place affect the CO_2 concentration of the blood? (increases CO_2, decreases CO_2)

_____

What causes the CO_2 concentration to change while you are running in place?

PRE-LAB QUESTIONS

____ **1.** Which of the following muscles contract during inspiration? (a) external intercostals, (b) internal intercostals, (c) abdominal, (d) b and c.

____ **2.** Which of the following muscles contract during a passive expiration? (a) external intercostals, (b) internal intercostals, (c) diaphragm, (d) none of the above.

____ **3.** Which of the following muscles contract during a forced expiration? (a) external intercostals, (b) diaphragm, (c) abdominal, (d) b and c.

____ **4.** An untied inflated balloon flies because (a) the pressure is higher inside than outside the balloon, (b) the pressure is lower inside than outside the balloon, (c) air flows down its pressure gradient, (d) a and c.

_____ 5. Human ventilation is (a) negative pressure inhalation, (b) positive pressure inhalation, (c) negative pressure exhalation, (d) b and c.

_____ 6. Frog ventilation is (a) negative pressure inhalation, (b) positive pressure inhalation, (c) positive pressure exhalation, (d) b and c.

_____ 7. Vital capacity is always equal to (a) tidal volume, (b) inspiratory reserve volume, (c) expiratory reserve volume, (d) a + b + c.

_____ 8. An instrument that measures lung volumes is (a) a caliper, (b) a spirometer, (c) a barometer, (d) none of the above.

_____ 9. Respiration is controlled by (a) chemoreceptors, (b) stretch receptors, (c) centers in the brain stem, (d) all of the above.

_____ 10. The most important stimulus in the control of respiration is the concentration in the blood and other body fluids of (a) oxygen (O_2), (b) carbon dioxide (CO_2), (c) hydrogen ions (H^+), (d) none of the above.

EXERCISE 18
HUMAN RESPIRATION

POST-LAB QUESTIONS

1. Which skeletal muscles are contracted during:

    a. restful inspiration—

    b. forced inspiration—

    c. passive expiration—

    d. forced expiration—

2. Describe changes in the size of the thorax during:

    a. inspiration—

    b. expiration—

3. Describe changes in the potential volume of the pleural sacs during:

    a. inspiration—

    b. expiration—

4. Define for humans:

    a. negative pressure inhalation—

    b. positive pressure exhalation—

5. How does breathing in a human differ from that in a frog?

6.

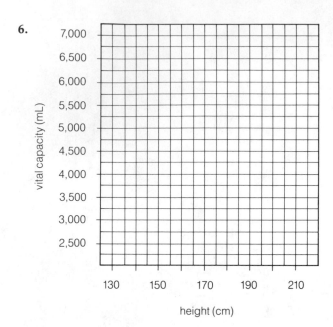

Is there a relationship between vital capacity and height? If so, describe it mathematically or with words.

7. Explain why hyperventilation can prolong the time you can hold your breath. Can this be dangerous (for example, hyperventilation followed by swimming under water)?

ANIMAL DEVELOPMENT: GAMETOGENESIS AND FERTILIZATION

OBJECTIVES   After completing this exercise you will be able to:

1. define sexual reproduction, fertilization, gametes, gonads, ovum, sperm, dioecious, zygote, monoecious, asexual reproduction, parthenogenesis, yolk, acrosome, blastodisc, vegetal pole, animal pole, gametogenesis, testis, ovary, seminiferous tubules, interstitial cells, testosterone, Sertoli cells, follicle, ovulation, estrogen, progesterone, corpus luteum, external fertilization, external development, internal fertilization, internal development;

2. draw and label a diagram of a mammalian sperm;

3. describe the structure of chicken and frog ova, and tell how they differ from a typical mammalian ovum;

4. recognize interstitial cells, seminiferous tubules, Sertoli cells, and sperm in a prepared section of a mammalian testis;

5. recognize follicles, primary oocytes, and a corpus luteum in a prepared slide of a mammalian ovary;

6. describe spermatogenesis and oogenesis in mammals;

7. describe the events and consequences of sperm penetration and fertilization.

INTRODUCTION   Most animal species reproduce sexually. **Sexual reproduction** usually involves the fusion of the nuclei of two gametes, called the ovum (the plural is *ova*) and sperm. This fusion is referred to as **fertilization.**

Gametes are produced by *meiosis* in reproductive organs called **gonads,** usually in individuals of two separate sexes. A female's gametes are **ova,** while those of a male are **sperm.** Species with male and female gonads in separate individuals are said to be **dioecious.**

Each gamete is haploid, and fertilization creates a new diploid cell, the **zygote,** whose combination of genes is unlike those of either parent. It is also very unlikely that the genes of one zygote will be identical to those of any other zygote, even those derived from the same parents. What two Mendelian principles largely account for this?

_____

_____

_____

_____

This variation is an advantage to a species in a changing and unpredictable environment. Why?

_____

_____

_____

Some animals (earthworms and snails, for example) produce both ova and sperm in the same individual, but self-fertilization is rare, occurring only in parasites with constant and predictable environments (tapeworms, for example). A species with both male and female reproductive organs in the same individual is referred to as **monoecious,** or *hermaphroditic.*

Except for a few invertebrate groups, animals reproduce either sexually or asexually, but not in both ways. **Asexual reproduction** is the production of new individuals by any mechanism that does not involve gametes (fission in prokaryotes, for example). Also, the ova of many animals, either naturally or in the laboratory, are capable of development without fertilization. This is called **parthenogenesis.**

(Starr, Chapter 32)

I. GAMETES

Most sperm have at least one flagellum. Sperm are specialized for motility and contribute little more than their chromosomes to the zygote.

Ova are specialized for storing nutrients, and they contain the molecules and organelles needed to fuel, direct, and maintain the early development of the embryo. Nutrients are stored as **yolk** in the cytoplasm of the ovum. Consequently, mammalian ova are larger than body cells and in some species reach a diameter of 0.2 mm. In the frog, additional yolk increases the diameter of the ovum to 2 mm, and in the chicken it reaches about 3 cm.

MATERIALS

Per student:

- compound microscope, lens paper, bottle of lens cleaning solution (optional), lint-free cloth (optional), dropper bottle of immersion oil (optional)
- prepared slide with a whole mount of bull sperm
- glass microscope slide, a coverslip, and a dissecting needle
- one-piece plastic dropping pipet

Per student pair:

- unfertilized hen's egg
- several paper towels
- dissecting tray
- Syracuse dish
- 2 camel-hair brushes
- dissecting microscope

Per lab group (table or bench):

- a model of a frog ovum (optional)

Per lab section:

- live frog sperm in pond water
- live frog ova in pond water
- pond water

Per lab room:

- phase-contrast compound light microscope (optional)

PROCEDURE

A. Sperm

1. With your compound microscope, study a prepared slide of bull sperm and draw several in Fig. 19-1 as seen with the high-dry objective. Each sperm has three major segments: the *head*, *midpiece*, and *tail*. Label the major segments of one of the sperm in your diagram. The tail is composed primarily of a single *flagellum*.

2. *Skip this step if your compound microscope does not have an oil immersion objective.* Using the oil immersion objective, find the **acrosome** covering the *nucleus* in the head of a sperm and *mitochondria* in its midpiece. The acrosome contains enzymes that aid in the pene-

Figure 19-1 Drawing of bull sperm (_____×)

tration of the egg. Considering its cell size, why does a sperm have a lot of mitochondria?

_____

_____

3. Place a drop of frog sperm suspension on a glass microscope slide and make a wet mount. Observe the movement of their flagella using either the phase-contrast compound microscope (a microscope that increases the contrast of transparent specimens) or your compound microscope with the iris diaphragm partially closed to increase contrast. Describe what you see.

_____

_____

_____

_____

B. Ova

1. *Unfertilized chicken egg.*

a. Obtain an unfertilized chicken egg. Crack it open as you would in the kitchen and spill the contents into a hollow made from paper towels placed in a dissecting tray.

b. Only the "yolk" is the ovum. Look for the **blastodisc**, a small white spot just under the cell membrane. This area is free of yolk and contains the nucleus. This is where fertilization would have occurred if sperm had been present in the hen's oviduct.

c. The *albumin* (egg white) is secreted by the walls of the oviduct. Examine the *shell* and the two *shell membranes*. The shell membranes are fused except in the region of the air space at the blunt end of the egg.

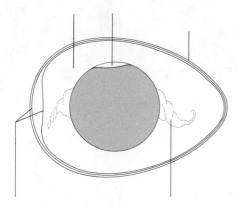

Labels: blastodisc, albumin, shell, shell membranes, chalazae

Figure 19-2   Unfertilized chicken egg

Note the two shock absorber-like *chalazae* (the singular is *chalaza*), which suspend the ovum between the ends of the egg. They are made of thickened albumin and may function to help rotate the ovum to keep the blastodisc always on top of the yolk.

d. Label Fig. 19-2.

2. *Frog egg.* Gently place a live frog egg in a Syracuse dish half filled with pond water and examine it with a dissecting microscope. Use two camel-hair brushes to transfer the egg. **Do not let the egg dry out.** The ovum is enclosed in a protective jelly membrane. Does the ovum float light or dark side up in the pond water?

Alternatively, study a model of a frog ovum. Ova from different species of animals vary in the amount and distribution of yolk. Frog ova have a moderate amount of yolk that is concentrated in the lower half of the ovum. This half of the ovum is called the **vegetal pole.** The nucleus is located in the upper, yolk-free half, the **animal pole.** Note that the animal pole is black. This is because it contains pigment granules. Label the animal and vegetal poles in Fig. 19-3. Why does a frog ovum have less yolk than a bird ovum?

Suggest one or more possible functions for the black pigment in the animal pole. (Hint: One function is the same as that for the pigment in your skin that increases when exposed to sunlight.)

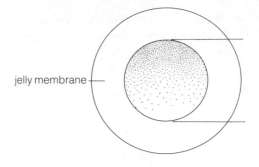

jelly membrane

Labels: animal pole, vegetal pole

Figure 19-3   Frog egg

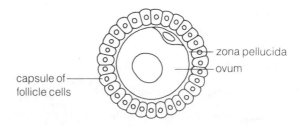

zona pellucida
ovum
capsule of follicle cells

Figure 19-4   Section through a human egg in the upper oviduct

3. *Human egg.* Figure 19-4 shows a human egg as it would appear in the upper oviduct. The egg is surrounded by a membrane, the *zona pellucida*, and a capsule of follicle cells (see Section II.B). Why do most mammalian ova have very little yolk?

II. GAMETOGENESIS

Gametogenesis is the production of the gametes, and is called **oogenesis** in the female and **spermatogenesis** in the male. The general scheme and terminology of gametogenesis are summarized in Table 19-1.

Table 19-1 Gametogenesis		
	Type of Cell	
Condition of Cell	Male	Female
mitotically active	spermatogonium	oogonium
before meiosis I	primary spermatocyte	primary oocyte
before meiosis II	secondary spermatocyte	secondary oocyte and first polar body
after meiosis II	spermatid	ovum and three polar bodies
after differentiation	sperm	—

Because you have already studied the events of meiosis that bring about the haploid number of chromosomes present in gametes, we will now examine the production of gametes in the gonads of mammals. The gonads are paired organs and are called **testes** (the singular is *testis*) in males and **ovaries** in females.

MATERIALS

Per student:

- prepared slide of a section of mammalian testis stained with iron hematoxylin
- prepared slide of a section of mammalian ovary with at least one mature follicle
- prepared slide of a section of mammalian ovary with at least one corpus luteum

PROCEDURE

A. Mammalian Spermatogenesis

Examine a prepared slide of the testis (Fig. 19-5). Most of the interior of a testis is filled with **seminiferous tubules**, which coil to and fro. Transverse and oblique sections will be present in your slide. Look for glandular **interstitial cells** between the seminiferous tubules. Interstitial cells secrete the male sex hormone **testosterone.** Now find a transverse section of a seminiferous tubule and increase the magnification of your compound light microscope until a portion of the tubule's wall fills the field of view.

The wall of the seminiferous tubule contains cells in various stages of spermatogenesis as well as **Sertoli cells** that function to nurture the developing sperm.

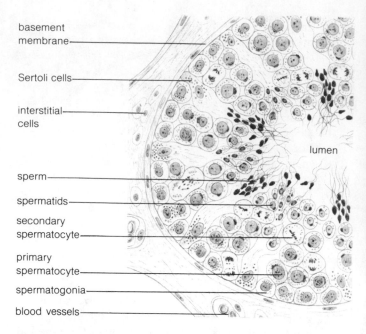

Figure 19-5 Transverse section of the testis. (After Creager, 1983.)

Spermatogenesis in most animal species is seasonal, its completion coinciding with mating. In humans, however, sperm production is continuous from puberty throughout a male's lifetime.

B. Mammalian Oogenesis

Examine a prepared slide of a section of an adult mammalian ovary (Fig. 19-6). After birth, in humans and most other mammals, oogonia are not present

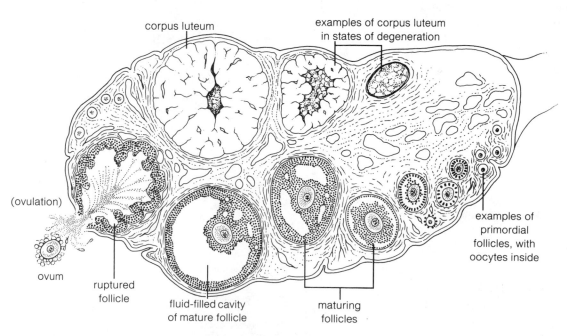

Figure 19-6 Idealized human ovary. (After Starr and Taggart, 1987.)

and meiosis is suspended in prophase of the first meiotic division. The primary oocytes are the largest cells in the section and are always surrounded by a **follicle** composed of smaller cells.

There is an excess supply of primary oocytes present at birth (about 2 million in a newborn girl). Each primary oocyte is initially surrounded by a *primordial follicle* whose thin walls are one cell thick. Most follicles degenerate, and in humans only about 300,000 primary oocytes remain at puberty. The majority of these oocytes will also degenerate. With each turn of the female cycle, several follicles begin to mature and one, rarely two or more, of each batch of oocytes finally bursts from the surface of the ovary and is swept into the oviduct. The release of oocytes from the ovary is called **ovulation.**

Follicles also function to secrete the female sex hormones, **estrogen** and **progesterone.**

Find in your section of a mammalian ovary the following stages of follicular development and draw them in Figs. 19-7, 19-8, and 19-9.

1. *Primordial follicle.* Look for primordial follicles. Both the follicles and their primary oocytes are small compared with maturing follicles and their primary oocytes. Primordial follicles tend to occur in groups located between maturing follicles or between maturing follicles and the ovarian wall.

2. *Mature follicle.* In maturing follicles, the size of the cells in the follicle wall and of the primary oocyte itself increases. Also, the follicle cells divide, causing the wall to become first two cells thick and then multilayered. As a follicle matures, a space appears between the follicle cells. This fluid-filled space increases in size until the primary oocyte and the follicle cells immediately around the primary oocyte are suspended in it. The mass of cells is connected to the rest of the wall by a narrow stalk of follicle cells. Just prior to ovulation, the follicle reaches its maximum size and bulges from the surface of the ovary.

3. *Corpus luteum.* Around the time of ovulation, the first meiotic division is completed, a polar body is split off, and the secondary oocyte enters but does not complete the second meiotic division. The oocyte is surrounded by the zona pellucida and a capsule of follicle cells (Fig. 19-4). A sperm must first penetrate this barrier before penetration of the egg membrane and fertilization can occur.

After ovulation, the follicle cells that remain in the ovary develop collectively into a large roundish structure called the **corpus luteum** (yellow body). The corpus luteum continues to secrete female sex hormones, especially progesterone.

If fertilization and implantation of the embryo in the uterus do not occur, the corpus luteum degenerates and is replaced by scar tissue.

Figure 19-7   Drawing of a primordial follicle (_____×)

Figure 19-8   Drawing of a mature follicle (_____×)

Figure 19-9   Drawing of a corpus luteum (_____×)

III. SPERM PENETRATION AND FERTILIZATION

Fertilization and subsequent development may be internal or external to the female's body. **External fertilization** with **external development** is common in invertebrates, fish, and amphibians. Why is external fertilization generally limited to aquatic animals?

_____

_____

_____

Land animals have **internal fertilization,** with either external development in the shell (true of most reptiles and birds) or **internal development** in the mother's *uterus* (true of most mammals). The uterus is the organ of the female reproductive system where most mammalian embryos develop until birth.

In many animals, meiosis is not complete at the time of sperm penetration. For example, in humans and most mammals sperm penetration triggers the completion of the second meiotic division and the splitting of the second polar body from the secondary oocyte to form the ovum proper.

MATERIALS

Per student pair:

- fertilization series of prepared slides with sections of the ovary of the horse roundworm (*Ascaris*) or sequential models of *Ascaris* fertilization stages
- Syracuse dish
- 2 camel-hair brushes

Per lab group (table or bench):

- model of a frog zygote (optional)

Per lab section:

- live frog zygotes in pond water
- pond water
- film or videotape of sperm penetration and fertilization in the frog (optional)

PROCEDURE

A. Sperm Penetration and Fertilization in the Horse Roundworm (*Ascaris*)

1. In this species, sperm penetration precedes the completion of the first meiotic division. At what stage of oogenesis are the eggs in this situation?

_____

2. Study Fig. 19-10 and find the equivalent stages of sperm penetration and fertilization in a series of prepared microscope slides, each bearing a section of the ovary of the horse roundworm or models of these stages. The slides and models are sequential in time. In this species the sperm are amoeboid and lack flagella. After sperm penetration (Fig. 19-10a), is the fertilized ovum haploid, diploid, or triploid?

_____

3. Note the change in shape of the sperm, the change in appearance of the cytoplasm immediately around the sperm, and the formation of the fertilization membrane as the first meiotic division takes place (Fig. 19-10b).

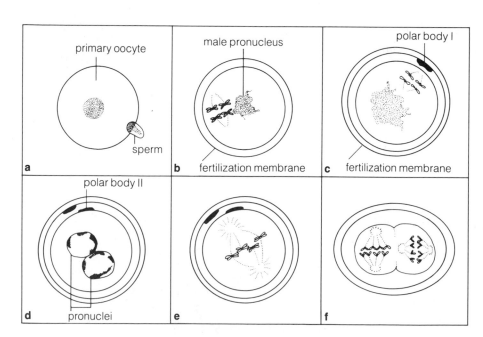

Figure 19-10 Fertilization in the horse roundworm. (After Mathews, 1972.)

4. The fertilization membrane then lifts from the surface of the secondary oocyte to form a space that prevents the penetration of other sperm. During this process, the first polar body is dragged away from the cell membrane (Fig. 19-10c). The second meiotic division takes place, and the second polar body now can be seen on the cell membrane.

5. Within the zygote, ovum and sperm *pronuclei* are formed (Fig. 19-10d). In this animal there is no fusion of the pronuclei; rather, the pronuclear membranes break down, the spindles of the pronuclei combine, fertilization occurs, and cleavage divisions commence (Fig. 19-10e and 19-10f).

B. Frog Zygote

At the beginning of the lab, or just before, your instructor induced ovulation in a female frog by injecting pituitary gland extract into the abdominopelvic cavity. Some of these ova were fertilized by sperm obtained from the shredded testes of a male frog.

Gently place a live frog zygote that has not yet begun to cleave in a Syracuse dish half filled with pond water and examine it with a dissecting microscope. Transfer the zygote using two camel-hair brushes. **Do not let the zygote dry out.** Does the zygote float animal or vegetal side up in the pond water?

_____

Is a fertilization membrane present or absent?

_____

Alternatively, study a model of a frog zygote, or watch a film or videotape on sperm penetration and fertilization in the frog. Is there a change in the pigmentation pattern? If so, describe what you observe.

PRE-LAB QUESTIONS

____ **1.** Most animals reproduce by (a) asexual means, (b) sexual means, (c) both of these means, (d) none of these means.

____ **2.** Most animals (a) are monoecious, (b) are dioecious, (c) have two sexes, (d) b and c.

____ **3.** Sperm (a) are male gametes, (b) are female gametes, (c) contain yolk, (d) are produced in the ovaries.

____ **4.** Ova (a) are male gametes, (b) are female gametes, (c) are specialized for motility, (d) a and c.

____ **5.** The production of gametes in the gonads is called (a) gametogenesis, (b) spermatogenesis, (c) oogenesis, (d) none of the above.

____ **6.** One primary oocyte will form (a) one ovum, (b) four ova, (c) up to three polar bodies, (d) a and c.

____ **7.** Which of the following could be found in a section of the testis? (a) secondary spermatocytes, (b) sperm, (c) Sertoli cells, (d) all of the above.

____ **8.** Oocytes are found in an ovary in (a) seminiferous tubules, (b) follicles, (c) corpora luteum, (d) none of the above.

____ **9.** Most mammals have (a) internal fertilization, (b) external fertilization, (c) internal development, (d) a and c.

____ **10.** Which process results in a zygote? (a) meiosis, (b) mitosis, (c) fertilization, (d) none of the above.

EXERCISE 19
ANIMAL DEVELOPMENT: GAMETOGENESIS AND FERTILIZATION

POST-LAB QUESTIONS

1. Define and characterize:

   a. gametes—

   b. gonads—

   c. gametogenesis—

2. Describe the similarities and differences between sperm and ova.

3. Why do you think four sperm cells are produced as a result of gametogenesis but only one ovum?

4. Why is meiosis a necessary part of gametogenesis?

5. Are oogonia present in adult human females?

   (yes or no) _____

6. In the frog, what is the relationship of the gray crescent to sperm penetration?

7. What substances are contained in the acrosome of a sperm? What is the function of these substances during sperm penetration?

8. As various newspaper articles, books, and movies suggest, the cloning of human beings—producing new individuals from activated somatic cells, perhaps followed by uterine implant—is a distinct possibility. Can you suggest any biological advantages or disadvantages to having the earth populated with clones of a few of the best examples of our species?

ANIMAL DEVELOPMENT: CLEAVAGE, GASTRULATION, AND LATE DEVELOPMENT

OBJECTIVES After completing this exercise you will be able to:

1. define blastomere, morula, blastula, blastocoel, blastoderm, blastocyst, inner cell mass, trophoblast, archenteron, blastopore, primitive streak, notochord, neural tube, somites, embryonic disk, implantation, amniotic cavity, placenta, umbilical cord, fetus;

2. list and define the five stages of development;

3. compare and contrast cleavage in the sea star, frog, chicken, and human;

4. explain differences in cleavage according to (a) the amount and distribution of yolk in the ovum and (b) the evolution of mammals;

5. describe gastrulation, the formation of the primary germ layers, and their derivatives;

6. list the four extra-embryonic membranes;

7. give the functions of the extra-embryonic membranes in birds and mammals.

INTRODUCTION Development has five stages: gametogenesis, fertilization, cleavage, gastrulation, and organogenesis. We have already studied the first two of these stages in the previous exercise, leaving the last three stages for this exercise.

(Starr, Chapter 32)

I. CLEAVAGE

Cleavage is a special type of cell division that occurs first in the zygote and then in the cells formed by successive cleavages, the **blastomeres.** Unlike normal cell division, there is no intervening period of cytoplasmic growth between cleavage stages. After a number of cleavages, the blastomeres form a solid cluster of cells called the **morula.** The formation of a hollow ball of cells, called the **blastula** in invertebrates and amphibians, marks the end of cleavage. Because there is no cytoplasmic growth, the size of the blastula is only slightly larger than that of the zygote.

In many organisms whose ova have little yolk (the sea star and human, for example), cleavage is *complete*

and nearly *equal,* resulting in separate blastomeres that are all about the same size. In amphibians like the frog and other animals with moderate amounts of yolk, cleavage is complete but *unequal.* There is so much yolk in the ova of many animals (most fish, reptiles, birds, and the two mammals that lay eggs—the platypus and spiny anteater) that complete cleavage is impossible. This type of cleavage is incomplete and is called *discoidal* (disklike).

MATERIALS

Per student:

- compound microscope, lens paper, bottle of lens cleaning solution (optional), lint-free cloth (optional), dropper bottle of immersion oil (optional)
- prepared slide of a whole mount of sea star development through gastrulation

Per student pair:

- 2 Syracuse dishes
- 2 blue camel-hair brushes
- 2 red camel-hair brushes
- dissecting microscope

Per student group (bench or table):

- models of early human development

Per lab section:

- frog embryos in pond water from eggs artificially inseminated one hour before lab

Per lab room:

- preserved specimens of two-, four-, and eight-cell cleavage stages, morulae (32-cell cleavage stage), and blastulae of frog in easily accessible screw-top containers
- source of distilled water
- models of early frog development

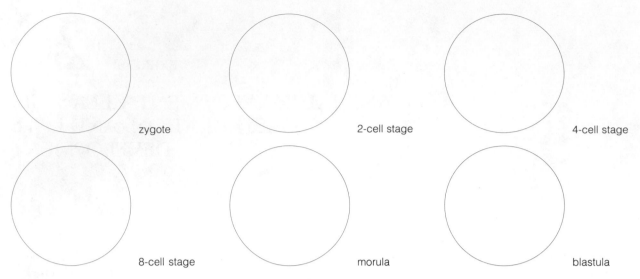

zygote   2-cell stage   4-cell stage

8-cell stage   morula   blastula

Labels: blastomeres, blastocoel

Figure 20-1 Drawings of early sea star developmental stages
(_____×)

PROCEDURE

A. Sea Star

With your compound microscope, observe a prepared slide with whole mounts of early sea star embryos. Find and draw in Fig. 20-1 a zygote; two-, four-, and eight-cell cleavage stages; a morula; and a blastula. Be sure to adjust the fine-focus knob to see the three-dimensional aspects of these stages. The blastula is a hollow ball of flagellated blastomeres surrounding a cavity called the **blastocoel.** Label the blastomeres and blastocoel in your diagram of a sea star blastula.

Are the first two cleavage planes parallel or perpendicular to each other?

What is the orientation of the third cleavage plane compared to the first and second cleavage planes?

B. Frog

1. If your instructor artificially inseminated frog eggs one hour before lab, you will be able to watch the first two cleavage divisions during this laboratory. At room temperature about an hour after addition of the sperm to the eggs, a region of less pigmented cytoplasm, the *gray crescent*, appears opposite the site of fertilization. The first cleavage division occurs about two hours after fertilization, the second a half hour later, and the third after another two hours.

Gently place several live frog zygotes in a Syracuse dish half filled with pond water and examine them with a dissecting microscope. Use two blue camel-hair brushes to transfer the egg. **Do not let the egg dry out.**

CAUTION *Preserved frog embryos are kept in a formalin preservative solution. Thoroughly wash any part of your body exposed to this solution with water. If the formalin solution is splashed into your eyes, wash them with the safety eyewash bottle for 15 minutes.*

2. If living embryos are unavailable, and for the eight-cell, morula, and blastula stages, examine preserved specimens under the dissecting microscope. Use two red camel-hair brushes to transfer each stage in turn to a Syracuse dish half filled with distilled water. When done, return the specimen to its container. The preserved embryos are in a formalin preservative solution, so **do not use the red brushes to manipulate the live embryos.**

As a supplement, study the models of the early stages of frog development.

3. In Fig. 20-2 draw the stages of frog development. Note that cleavage in the vegetal pole lags behind that of the animal pole. Why? (Hint: What is present in the vegetal pole that would hinder cleavage?)

C. Chicken

In the chicken, cleavage of the blastodisc forms a layer of cells called the **blastoderm,** which in time becomes separated from the yolk by a cavity, the blastocoel. Further development of the embryo will occur only in the blastoderm. Label Fig. 20-3.

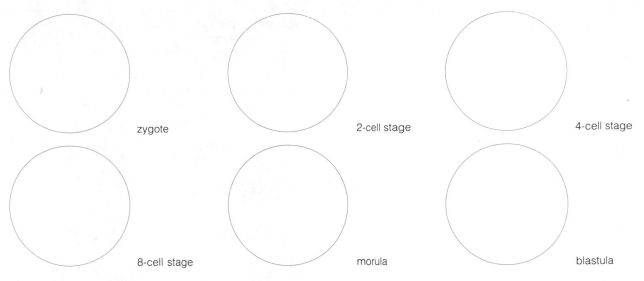

zygote

2-cell stage

4-cell stage

8-cell stage

morula

blastula

Figure 20-2 Drawings of early frog developmental stages
(_____×)

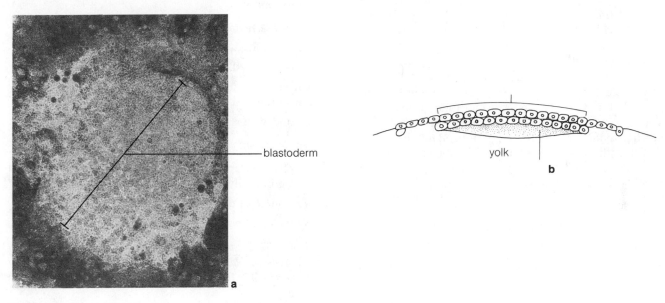

blastoderm

yolk

b

a

Labels: blastoderm, blastocoel

Figure 20-3 Late cleavage in the chicken. **(a)** surface view
(16×) (Photo courtesy Ripon Microslides, Inc.); **(b)** transverse
section.

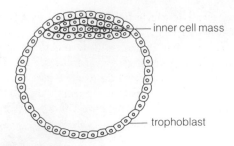
Figure 20-4 Section through a human blastocyst

Table 20-1 Tissue Derivatives of the Primary Germ Layers

Layer	Derivatives
ectoderm	nervous tissues; epidermis and its derivatives
mesoderm	muscle tissues, connective tissues, and epithelia of the excretory and reproductive systems
endoderm	epithelial lining of most of the digestive tract and respiratory tract, and associated glands (e.g., liver)

D. Human

Examine the models of early human development. Because the ovum has little yolk, cleavage is complete and nearly equal. At the end of cleavage a hollow ball of cells is formed. However, this **blastocyst** differs from a blastula in that a group of cells aggregate at one pole of the inner surface of the blastocoel (Fig. 20-4). These cells are called the **inner cell mass,** and, like the blastoderm of the chicken, further development of the embryo will proceed only here. The remaining cells that surround the blastocoel are called the **trophoblast.**

Current evolutionary thought is that modern reptiles, birds, and mammals evolved from earlier reptilian ancestors that had external development and an ovum rich in yolk. Internal development in mammals linked to nourishment of the embryo directly by the mother made large yolky eggs redundant. Excessive yolk would have been a biological liability; therefore, by natural selection, mammals have developed ova with little yolk. However, gastrulation and subsequent developmental events in reptiles, birds, and mammals are remarkably similar. This theme will be expanded in the discussion of extra-embryonic membranes later in this exercise.

II. GASTRULATION

Gastrulation is a time of growth and cell migration that produces, in most animals, three **primary germ layers** and the longitudinal axis of the body. In animals having at least the organ level of organization, the primary germ layers are called **ectoderm, mesoderm,** and **endoderm** and give rise to the four tissue types (Table 20-1).

MATERIALS

Per student:

- compound microscope, lens paper, bottle of lens cleaning solution (optional), lint-free cloth (optional), dropper bottle of immersion oil (optional)
- prepared slide of a whole mount of sea star development through gastrulation
- prepared slides with whole mounts of chicken embryos at 18 and 24 hours incubation

Per student pair:

- 2 Syracuse dishes
- 2 red camel-hair brushes
- dissecting microscope

Per student group (4):

- fertile hen's eggs incubated for 18 and 24 hours (optional)

Per lab section:

- frog embryos in pond water from eggs artificially inseminated 21 hours before lab
- frog embryos in pond water from eggs artificially inseminated 36 hours before lab
- pond water

Per lab room:

- preserved specimens of early and late gastrulae of frog in easily accessible screw-top containers
- a source of distilled water
- models of early and late gastrulae of frog

Note: See Scadding, S. R. (1985) How to culture chicken embryos in plastic wrap suspension. American Biology Teacher, *47: 107–108 for an excellent method for chicken embryos.*

A. Sea Star

The sea star has complete and nearly equal cleavage, resulting in a blastula with one layer of flagellated blastomeres that are slightly elongated at the *vegetal pole* (Fig. 20-5a).

Re-examine the prepared slide bearing whole mounts of the early stages of sea star development. Find an early gastrula (Fig. 20-5b). As gastrulation starts, the vegetal pole flattens and folds in like a pocket to create a new cavity, the **archenteron** ("ancient gut").

Find a late gastrula (Fig. 20-5c). The hole connecting the archenteron to the outside is called the **blastopore.** Gastrulation initially forms two layers of cells: (a) the ectoderm covering the outside of the gastrula and (b) the endoderm lining the archenteron. The mesoderm buds off from the inner tip of the archen-

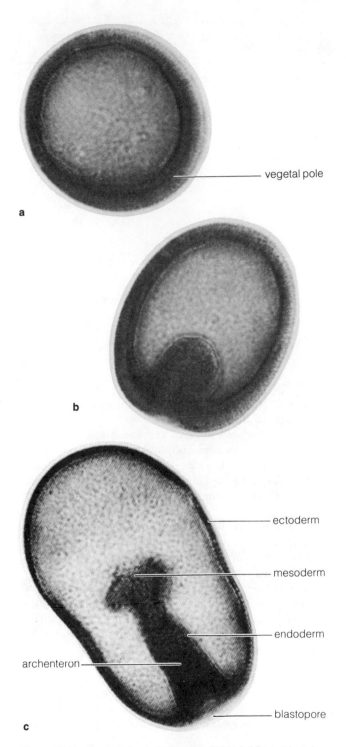

— vegetal pole

a

b

— ectoderm

— mesoderm

— endoderm

archenteron —

— blastopore

c

Figure 20-5 Gastrulation in the sea star (134×). **(a)** late blastula; **(b)** early gastrula; **(c)** late gastrula. (Photos by D. Morton.)

teron, and its cells migrate to form a third layer of cells between the ectoderm and endoderm.

The archenteron is the primitive gut, and the blastopore is situated in what will be the region of the anus in the adult sea star. This latter fact is important from an evolutionary viewpoint because it marks a major fork in the evolution of animals called *eucoelomates,* which have body cavities completely lined with

mesoderm. The phyla Echinodermata (e.g., sea star) and Chordata (e.g., frogs, humans) are called *deuterostomes* because the blastopore marks the region of the anus in the adult. Phyla Mollusca (e.g., snails, clams, octopuses), Annelida (e.g., earthworms), and Arthropoda (e.g., lobsters, insects) are called *protostomes* because the blastopore marks the region of the mouth in the adult.

B. Frog

Gastrulation in the frog is affected by the large amount of yolk in the vegetal hemisphere. Because the pigmented cells of the animal pole divide faster, they partially overgrow the yolk-laden cells of the vegetal pole.

Examine living or preserved specimens, as well as supplemental models of an early and a late gastrula as described in Section I.B for earlier developmental stages. Note that gastrulation does not occur simultaneously over the surface of the vegetal pole. It starts at a point just under what will be the anus of the adult frog and continues, forming a crescent (Fig. 20-6a) that will close to form a circle around a plug of yolk-

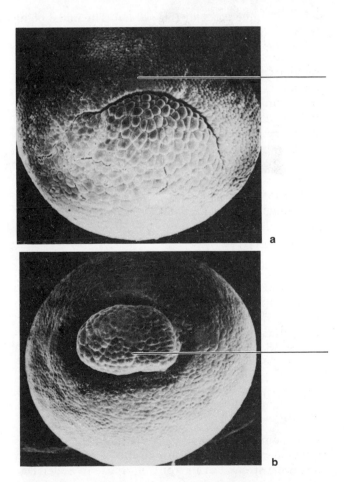

a

b

Labels: yolk plug, dorsal lip of the blastopore

Figure 20-6 Frog gastrulation. **(a)** early; **(b)** late. (Photos from R. Kessel and C. Shih, *Scanning Electron Microscopy in Biology,* Springer-Verlag, 1974.)

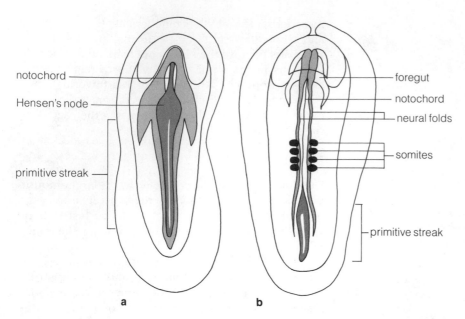

notochord
Hensen's node
primitive streak

foregut
notochord
neural folds
somites
primitive streak

a   b

Figure 20-7 Gastrulation in the chicken. **(a)** gastrulation (18 hours of incubation); **(b)** early embryo (24 hours of incubation).

laden cells, the *yolk plug* (Fig. 20-6b). The initial point of the folding in is referred to as the *dorsal lip of the blastopore*. Label Fig. 20-6.

C. Chicken

1. Obtain a prepared slide bearing a whole mount of a chicken gastrula (14 hours incubation) and one of an embryo incubated for 24 hours.

CAUTION *Do not use the high power objectives when examining whole mounts of thick material. You will break the coverslip.*

---

2. Observe the gastrula using only the low and medium power objectives. Like cleavage, gastrulation in the chicken is influenced by the large amount of yolk. The gastrula does not fold in like a pocket through a blastopore, but rather cells move or migrate into a groove on the blastoderm called the **primitive streak** (Fig. 20-7a). At one end of the primitive streak, the migrating cells pile up to form *Hensen's node.* This is thought to be equivalent to the dorsal lip of the blastopore.

3. Examine the embryo of 24 hours incubation. The three-layered embryo forms in the same axis as the primitive streak but in front of Hensen's node (Fig. 20-7b). The three layers from the top to the bottom are ectoderm, mesoderm, and endoderm. One of the

first recognizable structures in the embryo is the mesodermal **notochord.**

The notochord induces the formation of the embryo's nervous system from the ectoderm. *Neural folds* can be seen on either side of the notochord. With time the neural folds will fuse like a zipper, anterior to posterior, and in doing so form the **neural tube** (Fig. 20-11).

Likewise, the head of the embryo has lifted off the surface of the yolk and in doing so has formed the anterior portion of the digestive tract, the *foregut.*

Also prominent are two rows of **somites** on either side of the notochord. These are segmental condensations of mesoderm that will later develop into the skeleton, the skeletal muscles of the trunk, and the dermis of the skin. How many pairs of somites are present in your embryo?

---

4. If your instructor has fertile hen's eggs incubated for 14 and 24 hours, you will be instructed how to examine this material under the dissecting microscope.

D. Human

Human gastrulation follows much the same scenario as the chicken. However, before gastrulation a cavity forms between the inner cell mass and the trophoblast. This is the cavity of the amnion (Fig. 20-8). Also, cells from the inner cell mass grow downward, along the inner surface of the trophoblast, fuse, and form the yolk sac. Between the amniotic and yolk sac cavities is the embryonic disk. The **embryonic disk** in

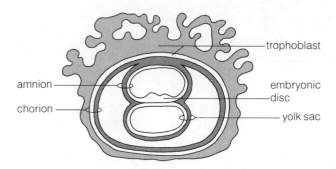

Figure 20-8 Transverse section through a human embryo at implantation

Labels: trophoblast, amnion, chorion, embryonic disc, yolk sac

mammals is the equivalent of the blastoderm in the chicken. Thus, gastrulation commences in the embryonic disk with the formation of a primitive streak.

Implantation of the embryo in the uterus occurs at the same time as the formation of the amnion and yolk sac, about eight days after fertilization of the ovum in the upper oviduct. Part of the trophoblast erodes away the maternal tissues so that the blastocyst can sink into the wall of the uterus. The remainder of the trophoblast forms the chorion. By this time, the ectoderm of the amnion and chorion as well as the endoderm of the yolk sac are coated with mesoderm derived from the inner cell mass.

III. EXTRA-EMBRYONIC MEMBRANES

The **amnion, yolk sac,** and **chorion** are three of the four **extra-embryonic membranes.** The fourth is the **allantois.**

MATERIALS

Per student group:

- fertile hen's eggs incubated for 5 days (optional)
- preserved pig fetus and placenta with injected vessels

PROCEDURE

A. Chicken

The amnion in the chicken forms from four folds of ectoderm and mesoderm: one in front of the developing embryo, one behind it, and one on each side. These in time fuse over the developing embryo's back. The outer wall of the fused folds becomes the chorion; the inner wall becomes the amnion. The chorion forms a sac that surrounds the developing embryo and the other extra-embryonic membranes. The amnion forms the fluid-filled **amniotic cavity,** in which the developing embryo is suspended.

The yolk sac is formed when endoderm, accompanied by mesoderm containing a rich network of blood vessels, spreads over the yolk. The yolk sac serves as the digestive organ for the developing embryo. The endodermal cells secrete enzymes that digest the yolk. The end products of digestion diffuse into the blood vessels and are carried into the developing embryo.

The floor of the hindgut folds out like a pocket to form the allantois, which is lined on the inside by endoderm and covered on the outside by mesoderm. The mesoderm forms a rich network of blood vessels. The cavity of the allantois functions as a dump for excretory wastes.

The yolk sac and allantois both function as embryonic respiratory organs.

If your instructor has fertile hen's eggs incubated for about five days, you will be instructed how to examine extra-embryonic membranes under the dissecting microscope.

The extra-embryonic membranes of the chicken are illustrated in Fig. 20-9.

B. Mammals

The formation of the extra-embryonic membranes in mammals is quite similar to that of the chicken. One difference is the earlier formation of the amnion and allantois. Because there is little yolk in the zygote of most mammals, the yolk sac is generally smaller. In Fig. 20-10 label the extra-embryonic membranes. The dashed lines are mesoderm, and the solid lines are ectoderm or endoderm.

After implantation, the trophoblast and the maternal tissues of the uterus start to form the **placenta.** When complete, the placenta brings the blood of the mother and the embryo very close to each other but does not allow them to mix. Diffusion across this thin barrier allows the placenta to function as the digestive, respiratory, and excretory organs of the developing embryo.

The mesoderm of the allantois forms the **umbilical cord,** and the *umbilical arteries* and *umbilical vein* contained therein. The mesoderm of the allantois also directs the formation of connecting blood vessels in the placenta. By the time the embryo's circulatory system is established, a circuit of vessels to and from the placenta is complete.

Once the organs and basic body shape of an embryo are established, the embryo is called a **fetus.** This transition occurs about one-third of the way through the time spent in the uterus. This time spent in the uterus is called the *gestation period.*

Examine the preserved pig fetus and its placenta. The umbilical arteries have been injected with red latex, the umbilical vein with blue latex, and the maternal vessels with yellow latex. Identify the fetus, umbilical cord, placenta, umbilical arteries, umbilical vein, and maternal vessels.

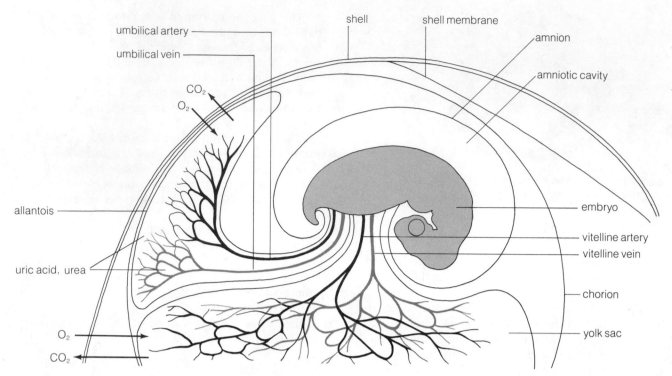

umbilical artery

umbilical vein

CO_2

O_2

shell

shell membrane

amnion

amniotic cavity

allantois

uric acid, urea

embryo

vitelline artery

vitelline vein

chorion

yolk sac

O_2

CO_2

Figure 20-9 Extra-embryonic membranes of the chicken.
(After B. M. Patten, National Sigma Xi Lecture, reprinted in
American Scientist, 39: 225–243, 1951. Used by permission.)

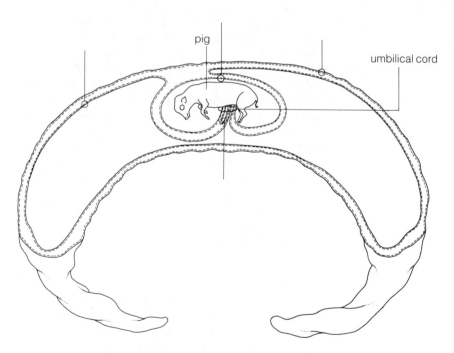

pig

umbilical cord

Labels: chorion, amnion, allantois

Figure 20-10 Extra-embryonic membranes of the pig

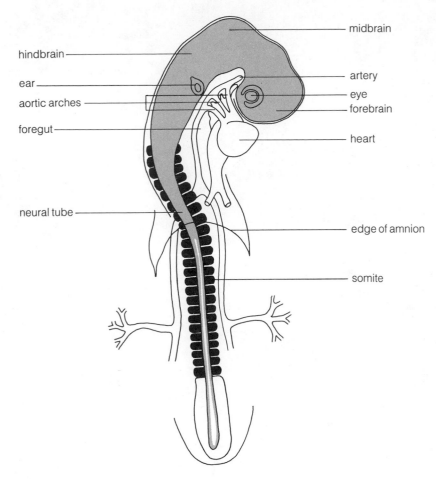

hindbrain

ear

aortic arches

foregut

neural tube

midbrain

artery

eye

forebrain

heart

edge of amnion

somite

Figure 20-11   Development of the chicken—48 hours of incubation

IV. ORGANOGENESIS

Organogenesis is the final stage of development. As the name suggests, it is during this period that the developing animal's organs and adult form are achieved.

MATERIALS

Per student:

- prepared slides with whole mounts of chicken embryos at 48 and 72 hours of incubation

Per student group:

- fertile hen's eggs incubated for 48 and 72 hours and longer (optional)

PROCEDURE

1. Obtain and examine under the low-power lens a slide of a whole mount of a chicken embryo incubated for 48 hours (Fig. 20-11). At this time, development of the front end of the neural tube has produced the *forebrain, midbrain,* and *hindbrain.* The head of the embryo has turned to the right, and growth of the brain has caused it to bend over itself. The *primitive eye* can now be seen forming on both sides of the forebrain. The *primitive ear* has just formed on both sides of the hindbrain.

The first portion of the circulatory system to develop is the heart. The *primitive heart* can be seen bulging to the embryo's right. Three pairs of *aortic arches* are present as well as other blood vessels.

Has the number of somites increased compared with those of the chicken embryo incubated for 24 hours?

(yes or no) _____

Note that the front half of the embryo is enclosed by the folds that form the amnion and chorion. Only the amnion, especially its lower edge, can be seen.

2. Now examine a chicken embryo incubated for 72 hours (Fig. 20-12). By this stage the embryo is lying on its right side, and the rate of development is accelerat-

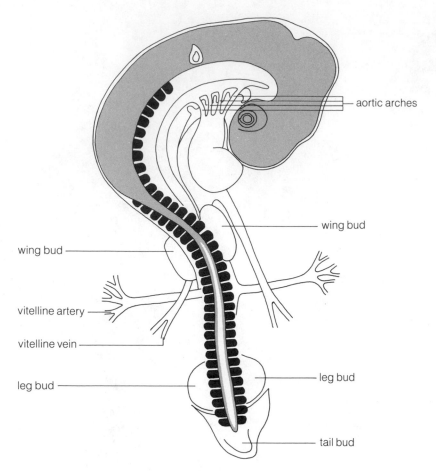

aortic arches

wing bud

wing bud

vitelline artery

vitelline vein

leg bud

leg bud

tail bud

Figure 20-12 Development of the chicken—72 hours of incubation

ing. New features include another pair of *aortic arches*, the *wing buds, hindlimb buds*, and a *tail bud*. The heart has established contact with the vitelline *arteries* and *vein*, which connect the embryo with its source of nourishment, the yolk in the yolk sac.

3. If your instructor has fertile hen's eggs incubated for 48 hours, 72 hours, and longer, you will be instructed how to examine this material under the dissecting microscope.

PRE-LAB QUESTIONS

_____ **1.** Choose the correct chronological order for the five stages of development listed below: (1) fertilization, (2) cleavage, (3) organogenesis, (4) gametogenesis, (5) gastrulation. (a) 1,2,3,4,5; (b) 1,4,2,5,3; (c) 2,1,4,5,3; (d) 4,1,2,5,3.

_____ **2.** The three primary germ layers are formed during (a) fertilization, (b) cleavage, (c) organogenesis, (d) gastrulation.

_____ **3.** Muscle tissues, connective tissues, and the epithelia lining the excretory and reproductive systems arise from (a) mesoderm, (b) endoderm, (c) ectoderm, (d) two of the above.

_____ **4.** Cleavage is complete and unequal in the (a) sea star, (b) frog, (c) chicken, (d) human.

_____ **5.** The cavity within a blastula is called the (a) archenteron, (b) blastocoel, (c) blastocyst, (d) blastopore.

_____ **6.** The entrance into the cavity formed during gastrulation in the sea star and frog is called the (a) archenteron, (b) blastocoel, (c) blastocyst, (d) blastopore.

_____ **7.** The blastocyst consists of the (a) inner cell mass, (b) trophoblast, (c) placenta, (d) a and b.

_____ **8.** In which of the following species do all of the cells derived from the zygote contribute to the body of the embryo? (a) frog, (b) chicken, (c) human, (d) b and c.

_____ **9.** How many extra-embryonic membranes are there? (a) one, (b) two, (c) three, (d) four.

_____ **10.** The majority of mammals derive most of their nourishment during development from (a) the yolk sac, (b) the placenta, (c) the fluid in the amnionic cavity, (d) none of the above.

EXERCISE 20
ANIMAL DEVELOPMENT: CLEAVAGE, GASTRULATION,
AND LATE DEVELOPMENT

POST-LAB QUESTIONS

1. Define and characterize:

   a. cleavage—

   b. gastrulation—

   c. organogenesis—

2. How do the amount and distribution of yolk in animal zygotes affect cleavage?

3. In what ways is gastrulation in humans more like that of the chicken than that of the sea star?

4. List the primary germ layers formed by gastrulation. What tissues will they form in the adult?

5. List the extra-embryonic membranes of a chicken embryo and give their functions.

6. Describe the formation of the placenta in mammals. What are its functions?

EVOLUTIONARY AGENTS

OBJECTIVES After completing this exercise you will be able to:

1. define evolutionary agent, natural selection, fitness, directional selection, stabilizing selection, migration, gene flow, divergence, speciation, mutation, genetic drift, bottleneck effect, founder's effect;

2. determine the gene frequencies for a model population;

3. calculate expected ratios of phenotypes based on Hardy-Weinberg proportions;

4. describe the effects of nonrandom mating, natural selection, migration, genetic drift, and mutation on a model population;

5. describe the effects of different selection pressures on identical model populations;

6. identify the level at which selection operates in a population;

7. describe the impact of the founder's effect on the genetic structure of populations.

INTRODUCTION The Hardy-Weinberg Principle says that heredity itself cannot cause changes in the frequencies of alternate forms of the same gene (alleles). If certain conditions are met, then the proportions of genotypes that make up a population of organisms should remain constant generation after generation according to Hardy-Weinberg equilibrium:

$$p^2 + 2pq + q^2 = 1.0 \text{ (for two alleles)}$$

If p is the frequency of one allele, "A," and q is the frequency of the other allele, "a," then

$$p + q = 1.0$$

If two alleles for coat color exist in a population of mice and the allele for white coats is present 70% of the time, then the alternate allele (black) must be present 30% of the time.

The Hardy-Weinberg Principle can be used to determine the proportions of phenotypes present in suc-

ceeding generations. In our example, since $p = 0.7$, we would expect 49% (p^2) of the mice in our population to be homozygous for white coats. 42% ($2pq$) would have one copy of each allele and would appear gray if the alleles are codominants (i.e., both have equal expression in the phenotype). What percentage of our population is homozygous for black coats?

_____ %

In nature, however, the frequencies of genes in populations are not static (i.e., not unchanging). Natural populations never meet all of the assumptions for Hardy-Weinberg equilibrium. _Evolution is a process resulting in changes in the genetic make-up of populations through time_; therefore, factors that disrupt Hardy-Weinberg equilibrium are referred to as evolutionary agents. In random mating populations, natural selection, migration (gene flow), genetic drift, and mutation can all result in a shift in gene frequencies predicted by the Hardy-Weinberg formula. Nonrandom mating can also result in such changes. This exercise will demonstrate the effect of these agents on the genetic structure of a simplified model population.

MATERIALS

Per student group (4):

- plastic dishpan (12" × 7" × 2")
- 50 large (10-mm diameter) white beads (Beads can be obtained from Carolina Biological Supply. Contact Carolina directly if your catalogue does not contain this information.)
- 50 large red beads
- 50 large pink beads
- 1 large gray bead
- 4,000 small (8-mm diameter) white beads
- 4,000 small red beads (optional)
- pair of long forceps
- coarse sieve (9.5 mm)
- small bowl

PROCEDURE

(Starr, pp. 177–187)

A. The General Model

The populations you will be working with are composed of colored beads. White beads in our model represent individuals that are homozygous for the white allele (**R**). Red beads are homozygous for the red allele (**r**) and pink beads are heterozygotes (**Rr**). These beads exist in ponds that are represented by plastic dishpans filled with smaller beads. The smaller beads can be strained to retrieve all the "individuals" that make up the model population. When the "individuals" are recovered, the frequencies of the color alleles can be determined using the Hardy-Weinberg formula. The alleles in our population are codominant. Thus, each white bead contains two white alleles; each pink bead, one white and one red; and each red bead, two red alleles. The total number of color alleles in a population of 20 "individuals" is 40. If such a population contains five white beads and ten pink beads, the frequency of the white allele is:

$$p = \frac{(2 \times 5) + 10}{40} = 0.5$$

Because $p + q = 1.0$, the frequency of the red allele (q) must also be 0.5 if there are only two color alleles in this population.

B. Natural Selection

Natural selection disturbs Hardy-Weinberg equilibrium by discriminating between individuals with respect to their ability to produce young. Those individuals that survive and reproduce will perpetuate more of their genes in the population. These individuals are said to exhibit greater **fitness** than those who leave no offspring or fewer offspring. We will model the effect of natural selection by mimicking predation on our population.

Students should begin working in groups of four.

1. Place 10 large white beads, 10 large red beads, and 20 large pink beads into a dishpan filled with small white beads (to a depth of at least 5 cm).

2. One student is the predator. After the beads are mixed, the predator searches the "pond" and removes as many prey items (large beads) as possible in 30 seconds. In order to more closely model the handling time required by "real" predators, you must search for and remove beads with a pair of long forceps.

3. Because some of the large beads are cryptically colored (they blend into the environment), the proportions of beads taken may not reflect the original proportions. Sift the "pond," count the number of large white, pink, and red beads, and calculate the frequency of the white and red alleles remaining in the

population. Record the differences in Table 21-1. For example, if five large white, eight red, and eight pink beads remain, the frequency of the white allele is:

$$p = \frac{(2 \times 5) + 8}{42} = 0.43$$

Table 21-1   Gene Frequencies with One Generation of Selection

Initial Population	1st Generation After Selection
$p = 0.5$	$p =$ _____
$q = 0.5$	$q =$ _____

4. Using the new values for allele frequencies, calculate the number of white, red, and pink individuals that would be "reproduced" in the next generation (i.e., if p now equals 0.43, then $p^2 = 0.18$ and a population with 50 individuals would have $0.18 \times 50 = 9.0$ white beads). Assume 50 individuals in all successive generations.

5. Repeat steps 2–4 for three more generations. A different student should be the "predator" in each new generation. Between generations, calculate the new allele frequencies, record the frequency of the red allele in each generation in Table 21-2, and plot the change in gene frequencies in Figure 21-1.

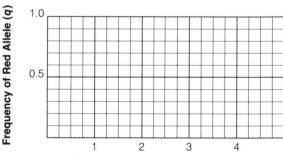

Figure 21-1   Effects of predation on gene frequencies

If you had started with small red beads (this you may do if time permits) as a background, how would the gene frequencies change?

_____

_____

_____

Selection that favors one extreme over the other and causes gene frequencies to change in a predictable direction is known as **directional selection.** When selection favors an intermediate phenotype rather than one at the extremes, it is known as **stabilizing selection.**

It is important to realize that selection operates on the entire phenotype so that the overall fitness of an organism is based on the result of interactions of thousands of genes. The model presented here is very simple. Occasionally simple genetic differences like the one you have modeled are critical to the survival of different phenotypes. For an example, read a description of natural selection in the peppered moth discovered by H. B. D. Kettlewell (e.g., *Biology: Concepts and Applications,* Starr, p. 184).

If two identical populations were in different environments (such as in our red and white "ponds"), how would the frequency of the color genes in each "pond" compare after a large number of generations?

_____

_____

_____

As two populations become genetically different through time (**divergence**), individuals from these populations may lose the ability to interbreed. If this happens, two species form from one ancestral species. This process is called **speciation.**

C. Migration

Migration causes a change in gene frequencies in a population due to new organisms entering the population or to organisms leaving. This may be a powerful force in evolution. To demonstrate its effect:

1. Establish an initial population as in Section B.1.

2. Begin selection as before, except in this exercise add five new red beads to each generation before the new gene frequencies are determined. These beads represent migrants from a population where the red allele confers greater fitness. For each generation, record in Table 21-2 the gene frequencies obtained with selection and migration.

Table 21-2 Frequency of Red Allele Due to Selection and Migration

Generation	Selection Alone (Section B)	Selection and Migration
1	$q =$ _____	$q =$ _____
2	$q =$ _____	$q =$ _____
3	$q =$ _____	$q =$ _____
4	$q =$ _____	$q =$ _____

How does migration influence the effectiveness of selection in this example?

_____

_____

_____

How would migration have influenced the change in gene frequencies if white instead of red individuals had entered the population?

_____

_____

When migrating individuals interbreed with members of the population into which they have migrated, it is referred to as **gene flow.** Some level of gene flow is necessary to keep local populations of the same species from becoming more and more different from each other. Things that serve as barriers to gene flow may accelerate the production of new species. Migration may also introduce new genes into a population and produce new genetic combinations. Imagine the result of a black allele being introduced into our model population and the new heterozygotes (gray and dark red) it would produce.

D. Mutation

Another way to introduce new genetic information into a population is through **mutation.** This usually represents an actual change in the information encoded by the DNA of an organism. As such, many mutations are harmful to the organism and will be eliminated by natural selection. Nevertheless, mutations do provide the raw material for evolution by introducing new genetic information. The following exercise may be done outside the laboratory as a "thought" experiment.

1. Establish an initial population as in Section B.1, but do not place these beads in a "pond." Place them, instead, in a small bowl without small beads.

2. One member of the group should choose, without looking, 20 of these large beads at random.

3. Calculate the gene frequencies of the "individuals" selected:

$p =$ _____

$q =$ _____

4. Now replace one of the white beads with a gray bead. This represents a mutation in one of the parents of the next generation. The gray "individual" has new genetic information for color production. (Of course, the only mutations important in evolution are those that accumulate in gametes. That is the way changes in information can be passed on to future generations.)

5. Recalculate the new gene frequencies with the frequency of the new color allele equal to "r." (*Hint:* The gray bead must be a heterozygote unless both parents had the same mutation at the same time, a very unlikely event.)

$p = $ _____

$q = $ _____

$r = $ _____

If the next generation contains 50 individuals, how many of each phenotype would you expect? (*Hint:* Three alleles are present $[p + q + r = 1.0]$, so the equilibrium formula must be expanded $[p^2 + 2pq + q^2 + 2pr + 2qr + r^2 = 1.0]$ or the proportion of white + pink + red + gray + dark red + black phenotypes expanded.)

white _____

pink _____

red _____

gray _____

dark red _____

black _____

Imagine a population made up of individuals in these proportions. What effect will natural selection have on these phenotypes in a white "pond"?

_____

_____

How could conditions change to favor the selection of the rare black allele?

_____

_____

E. Genetic Drift

Chance is another factor that affects the kind of gametes in a population that are involved in fertilization. As a result, shifts in gene frequencies can occur between generations just because of the random aspects of fertilization. This phenomenon is known as **genetic drift.** In this portion of the experiment, we'll simulate genetic drift.

1. Establish an initial population as in Section D.1.

2. One student in the group should, without looking, place his or her hand in the bowl and choose 10 large beads at random.

Table 21-3   Frequencies Produced by Genetic Drift

Expected Frequencies	Actual Frequencies ($n = 10$)	Actual Frequencies ($n = 30$)
$p = 0.5$	$p = $ _____	$p = $ _____
$q = 0.5$	$q = $ _____	$q = $ _____

3. Record the gene frequencies in Table 21-3 that would result from the 10 "individuals" you have chosen.

If the "individuals" you selected were the only individuals to reproduce this generation, what would be the effect on gene frequencies compared to those initially present in the population?

_____

_____

4. Now replace the 10 beads you removed in step 2. Select beads at random again, but this time select 30 large beads.

5. Calculate the gene frequencies from the 30 large beads and record these frequencies in Table 21-3.

How do the frequencies generated by the beads in step 4 compare to those selected in step 2?

_____

_____

Generally, the larger the breeding population, the smaller the sampling effect that we call genetic drift. In small populations, genetic drift can cause fluctuations in gene frequencies that are great enough to eliminate an allele from a population such that p becomes 0.0 and the other allele becomes fixed ($q = 1.0$). Genetic variation in such a population is reduced. Populations that become very small may lose much of their genetic variation. This is known as a **bottleneck effect.**

Another way in which chance affects gene frequencies in a population is when new populations are established by migrants from old populations.

6. To model this effect, choose at random six individuals from an initial population to represent the migrants.

7. Move these "individuals" to a new unoccupied "pond." (It is not necessary to actually set up a new pond for this demonstration—use your imagination.)

Table 21-4 Gene Frequencies in a Founder Population

Initial Population	Founder Population
$p = 0.5$	$p = $ _____
$q = 0.5$	$q = $ _____

8. Now calculate the gene frequencies in the new pond and record them in Table 21-4. How do they compare with the frequencies that were characteristic of the pond from which these migrants came?

_____

_____

_____

The genetic make-up in future generations in the new population will more closely resemble the six migrants than the population from which the migrants came. This effect is known as the **founder's effect**. The founder's effect may not be an entirely random process because organisms that migrate from a population may be genetically different from the rest of the population to begin with. For example, if wing length in a population of insects is variable, one might expect insects with longer wings to be better at founding new populations because they may be carried farther by winds.

F. Nonrandom Mating

Hardy-Weinberg equilibrium is also disturbed if individuals in a population do not choose mates randomly. Some members of a population may show a strong preference for mates with similar genetic make-ups. To model this effect, conduct the following exercise. This exercise may also be done as a "thought" experiment if necessary.

1. Establish an initial population as in Section D.1.

2. Assume that all red "individuals" will mate with only other red individuals and white "individuals" select only other white members as mates. The pink heterozygotes will also mate only with each other. Arbitrarily assign sex to every bead so there are equal numbers of males and females in each color group.

3. If each pair of beads produces four offspring, how many of each phenotype will be present in the next generation? Remember that the pink "pairs" will produce one red, one white, and two pink individuals on the average:

red _____ white _____ pink _____

4. Calculate the genotype frequencies in this generation and compare these with the frequencies in the initial generation.

Table 21-5 Genotype Frequency Changes Due to Nonrandom Mating

Initial Generation	Next Generation
$p^2 = 0.25$	$p^2 = $ _____
$2pq = 0.5$	$2pq = $ _____
$q^2 = 0.25$	$q^2 = $ _____

What will happen to the frequency of the heterozygote in subsequent generations?

_____

_____

Now that you have modeled the five major factors that disrupt Hardy-Weinberg equilibrium in natural populations, attempt to answer the post-lab questions concerning the effect of these agents on the genetic make-up of such populations. Remember, the models you have used are very simple, while the phenotypes of real organisms are the result of interactions of thousands of genes. Also remember that most real populations are very complex mixtures of phenotypes, and the factors you have examined operate on these phenotypes as a whole—they do not affect some genes in an individual without affecting others.

PRE-LAB QUESTIONS

_____ **1.** If all conditions of Hardy-Weinberg equilibrium are met, (a) gene frequencies move closer to 0.5 each generation, (b) gene frequencies change in the direction predicted by natural selection, (c) gene frequencies stay the same, (d) all gene frequencies increase.

_____ **2.** If a population is in Hardy-Weinberg equilibrium and $p = 0.6$, (a) $q = 0.5$, (b) $q^2 = 0.4$, (c) $q = 0.16$, (d) $q^2 = 0.16$.

_____ **3.** Natural selection operates directly on (a) the genotype, (b) individual alleles, (c) the phenotype, (d) color only.

_____ **4.** The process that discriminates between phenotypes with respect to their ability to produce offspring is known as (a) natural selection, (b) migration, (c) genetic drift, (d) cytokinesis.

_____ **5.** Two populations that have no gene flow between them are likely to (a) become more different with time, (b) become more alike with time, (c) become more alike if the directional selection pressures are different, (d) stay the same unless mutations occur.

_____ **6.** A process that results in individuals of two populations losing the ability to interbreed is referred to as (a) stabilizing selection, (b) fusion, (c) speciation, (d) differential migration.

_____ **7.** Two ways in which new alleles can become incorporated in a population are (a) mutation and drift, (b) selection and drift, (c) selection and mutation, (d) mutation and migration.

_____ **8.** If a new allele appears in a population, the Hardy-Weinberg formula (a) cannot be used because no equilibrium exists, (b) can be used but only for two alleles at a time, (c) can be used by lumping all but two phenotypes in one class, (d) can be expanded by adding more terms.

_____ **9.** A shift from expected gene frequencies due to random effects is known as (a) natural selection, (b) genetic drift, (c) fission, (d) migration.

_____ **10.** Genetic drift is a process that has a greater effect on populations that (a) are large, (b) are small, (c) are not affected by mutation, (d) do not go through bottlenecks.

EXERCISE 21
EVOLUTIONARY AGENTS

POST-LAB QUESTIONS

1. What two evolutionary agents are most responsible for decreases in genetic variation in a population?

2. How can selection cause two populations to become different with time?

3. What effect would increasing gene flow between two populations have on their genetic make-up?

4. New genetic information can be introduced into a population through what mechanisms?

5. Nonrandom mating can exert what kind of effect on a population?

6. Describe how the effects of directional selection may be offset by migration.

7. What is the fate of most new mutations?

8. If a population has three color alleles and the frequencies are $p = 0.5$, $q = 0.3$, and $r = 0.2$, how many phenotypes are possible?

9. In question 8, if the alleles are represented by yellow (p), red (q), and blue (r), what are the phenotypes and their proportions?

10. In humans, birth weight is an example of a character affected by stabilizing selection. What does this mean about the long-term average birth weight of human babies?

MONERANS AND PROTISTANS

OBJECTIVES After completing this exercise you will be able to:

1. define decomposer, producer, consumer, antibiotic, symbiont, symbiosis, parasitism, commensalism, mutualism, nitrogen fixation, pellicle, diatomaceous earth, red tide, pseudopodium, phagocytosis, vector, plasmodium;

2. describe characteristics distinguishing monerans from protistans;

3. identify and classify the organisms studied in this exercise;

4. identify structures in the organisms studied;

5. distinguish Gram-positive and Gram-negative bacteria, indicating their susceptibility to certain antibiotics;

6. suggest measures that might be used to control malaria.

INTRODUCTION The monerans (kingdom Monera) and protistans (kingdom Protista) are among the simplest of living organisms. Both kingdoms consist of unicellular organisms, but that's where the similarity ends. The members of the kingdom Monera are *prokaryotic* organisms, meaning that their DNA is free in the cytoplasm, unbound by a membrane. They lack organelles. By contrast, the kingdom Protista consists of unicellular *eukaryotic* organisms: The genetic material contained within the nucleus and many of their cellular components is compartmentalized into membrane-bound organelles. Prokaryotic and eukaryotic organization was introduced in Exercise 3.

Monerans are such organisms as bacteria and cyanobacteria. Most bacteria are heterotrophic, dependent upon an outside source for nutrition, while cyanobacteria are autotrophic (photosynthetic), able to produce their own carbohydrates. The protistans are also a diverse assemblage of organisms, both green (photosynthetic) and nongreen (heterotrophic).

Some bacteria cause plant and animal diseases. But most are **decomposers,** breaking down and recycling the waste products of life. Some are nitrogen-fixers, capturing the gaseous nitrogen in the at-

mosphere and making it available to plants via a symbiotic association with their roots.

Both monerans and protistans are at the base of the food chain. Many are autotrophic **producers,** capturing the energy of the sun. They are eaten by the heterotrophic **primary consumers;** these in turn are eaten by heterotrophic **secondary consumers** and so on. From an ecological standpoint, these simple organisms are among the most important organisms on our planet. Ecologically, they're much more important than we are.

I. KINGDOM MONERA (Starr, pp. 225–229)

MATERIALS

Per student:

- nutrient agar culture plate
- sterile cotton swab
- china marker
- bacteria type slide
- microscope slide
- coverslip
- dissecting needle
- compound microscope

Per group:

- distilled water (dH$_2$O) in dropping bottle

Per lab room:

- Gram-stained bacteria (three demonstration slides)
- *Oscillatoria*—living culture; disposable pipet
- *Azolla*—living plants

PROCEDURE

A. Bacteria (Heterotrophic Monerans)

Obtain a petri dish containing sterile nutrient agar. Open the dish and expose it to various environments

Figure 22-1 Drawings of the three bacterial shapes
(_____×)

of the classroom by first running a sterile cotton swab over the surface of the object you wish to sample, then over the surface of the agar.

_____

CAUTION *Be careful that you do not break the agar surface.*

_____

Replace the cover of the dish. Using a china marker, label your culture with your name and item sampled. Tape the lid securely to the bottom half of the dish and place the culture in a desk drawer to incubate until the next class period. At that time, examine your culture for bacterial colonies, noting the color and texture of the bacterial growth.

_____

CAUTION *Leave the lid on as you examine the cultures to prevent the spread of any potentially pathogenic (disease-causing) organisms.*

_____

Describe what you see.

Source: _____ _____

Bacteria come in three shapes: *coccus* (the plural is *cocci*; spherical), *bacillus* (the plural is *bacilli*; rods), and *spirillum* (the plural is *spirilla*; spirals). Study a bacteria type slide illustrating these three shapes.

Table 22-1 Gram Stain Reaction of Various Bacteria	
Bacterial Species	Gram Reaction (+ or −)

You'll need to use the highest magnification available on your compound microscope. In Fig. 22-1, draw the bacteria you are observing.

In addition to being differentiated on the basis of their shape, bacteria can be separated according to how they react to a staining procedure, called **Gram stain,** in honor of a nineteenth-century microbiologist, Hans Gram. **Gram-positive** bacteria are purple after being stained by the Gram stain procedure, while **Gram-negative** bacteria appear pink. The Gram stain reaction is important to bacteriologists because it is one of the first steps in identifying an unknown bacterium. Furthermore, the Gram stain reaction indicates a bacterium's susceptibility or resistance to certain **antibiotics,** substances that inhibit the growth of bacteria.

Examine the *demonstration slides* illustrating Gram-stained bacteria. Gram-positive bacteria are susceptible to penicillin, while Gram-negative bacteria are not. In Table 22-1, list the species of bacteria that you have examined and their staining characteristics.

B. Cyanobacteria (Blue-green Algae)

The cyanobacteria (sometimes called blue-green algae) are distinguished from the heterotrophic bacteria by being photosynthetic.

From the culture provided, obtain filaments of

Figure 22-2 Drawing of *Oscillatoria* (_____×)

Oscillatoria. Make a wet mount slide and examine with your compound microscope, starting with the medium power objective and finally with the highest magnification available (oil immersion, if possible). Note the individual cells are joined and so form the filament.

Do all the *Oscillatoria* cells look alike, or is there differentiation of certain cells within the filament?

Oscillatoria is widespread, often forming a black ooze on the surface of flower pots or other surfaces that are usually wet. The color is a consequence of the photosynthetic accessory pigments that for the most part mask the chlorophyll. Draw a portion of the filament in Fig. 22-2.

Some cyanobacteria live as **symbionts** within other organisms. Literally, symbiosis means "living together." There are three types of symbiosis. In a parasitic symbiosis (**parasitism**), one organism lives at the expense of the other; that is, the parasite benefits while the host is harmed. A commensalistic symbiosis (**commensalism**) occurs when effects are positive for one species and neutral for the other. In a mutualistic symbiosis (**mutualism**) both organisms benefit from living together.

Place a leaf of the tiny water fern *Azolla* on a clean glass slide. Use a dissecting needle to crush the leaf into very small pieces. Now add a drop of water and a coverslip.

Scan your preparation with the medium power objective of your compound microscope, looking for long chains (composed of numerous beadlike cells) of the filamentous cyanobacterium *Anabaena*. Switch to higher magnification when you find *Anabaena*. Within the filament, locate the **heterocysts,** cells that are a bit larger than the other cells. Heterocysts are believed to be able to convert nitrogen in the air (or water) to a form that the cyanobacterium can use for cellular metabolism. This process is called **nitrogen fixation.** Presumably, the nitrogen fixed by *Anabaena* is harvested by the water fern, which in turn uses it for its own metabolic needs.

Which type of symbiosis is the association between *Anabaena* and the water fern?

Examine Fig. 22-3, an electron micrograph of *Anabaena.* The single large cell with the electron-dense regions at either end is the heterocyst. In the other cells, note the numerous wavy **thylakoids,** membranes on and in which the photosynthetic pigments are found. Identify the large electron-dense **storage granules** within the cytoplasm and the cell wall.

Is a nucleus present within the cells of *Anabaena?* Explain.

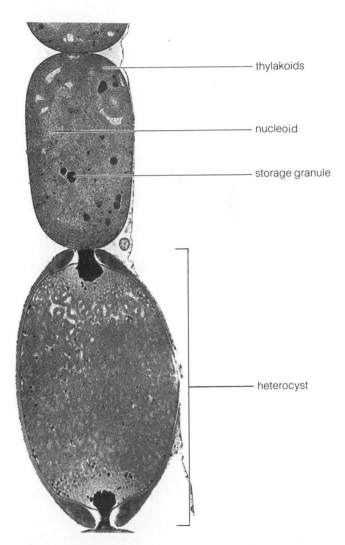

Figure 22-3 Electron micrograph of *Anabaena.* (Photo courtesy R. D. Warmbrodt.)

II. KINGDOM PROTISTA (Starr, pp. 229–237)

MATERIALS

Per student:

- prepared slide of a dinoflagellate (e.g., *Gymnodinium, Ceratium,* or *Peridinium*)
- prepared slide of *Trypanosoma* in blood smear
- prepared slide of *Trichonympha*
- plate culture of slime mold (*Physarum*)
- depression slide
- microscope slide
- coverslip
- dissecting needle
- compound microscope

Per group (table):

- methylcellulose in dropping bottles (2)
- diatomaceous earth
- distilled water (dH$_2$O) in dropping bottles (2)
- carmine, in screw-cap bottle (2)
- tissue paper
- acetocarmine stain in dropping bottles (2)
- box of toothpicks

Per lab room:

- *Euglena*—living culture; disposable pipet
- diatom—living culture; disposable pipet
- dinoflagellate—living culture; disposable pipet (optional)
- *Amoeba*—living culture on demonstration at dissecting microscope; disposable pipet
- demonstration slide of *Plasmodium vivax,* sporozoites
- demonstration slide of *P. vivax,* merozoites
- demonstration slide of *P. vivax,* immature gametocytes
- *Paramecium caudatum*—living culture; disposable pipet
- Congo red—yeast mixture; disposable pipet
- demonstration of slime mold (*Physarum*) sporangia

PROCEDURE

A. Phylum Euglenophyta: Euglenids

Euglenids are motile, unicellular, photosynthetic protistans. From the culture provided, prepare a wet mount slide of *Euglena* and observe with the medium power objective of your compound microscope. Notice the motion of these green cells as they swim through the medium. If they're swimming too rap-

Figure 22-4   Scanning electron micrograph of *Euglena* (1,700×). (Photo from C. Shih and R. G. Kessel, *Living Images,* Science Books International, 1982, page 8. Reprinted with permission of the present publisher, Jones and Bartlett Publishers, Inc.)

idly, prepare another slide, but add a drop of methylcellulose to the cell suspension before adding a coverslip. Switch to the high-dry objective for more detailed observation.

Within the **cytoplasm,** identify the **chloroplasts** and, if possible, the centrally located **nucleus.** By closing the microscope's diaphragm to increase the contrast (see Exercise 1), you may be able to locate the **flagellum** at one end of the cell. Search for the orange **eyespot,** a photoreceptive organelle located within the cytoplasm at the base of the flagellum.

Besides seeing the swimming motion caused by the flagellum, you may observe a contractionlike motion of the entire cell (*euglenoid movement*). *Euglena* is able to perform this contortion because it lacks a rigid cell wall. Instead, flexible helical interlocking proteinaceous strips within the cell membrane delimit the cytoplasm. These strips plus the cell membrane form the **pellicle.** Euglenoid movement provides a means of locomotion for mud-dwelling organisms. Figure 22-4 is a scanning electron micrograph (SEM) showing the helical strips of the pellicle and the flagellum.

In Fig. 22-5, make a series of sketches illustrating the different shapes *Euglena* takes on during euglenoid movement.

Now label the diagram of *Euglena* provided in Fig. 22-6.

B. Phylum Chrysophyta: Diatoms

Often considered as algae (see Exercise 23), the diatoms are called the organisms that live in glass houses because their cell walls are composed largely of opaline *silica* (SiO$_2$ · nH$_2$O). Diatoms are important as primary producers in the food chain of aquatic environments, and their cell walls are used for a wide variety of industrial purposes, ranging from the polishing agent in toothpaste to a reflective roadway paint additive. Massive deposits of cell walls of long-dead diatoms make up **diatomaceous earth.**

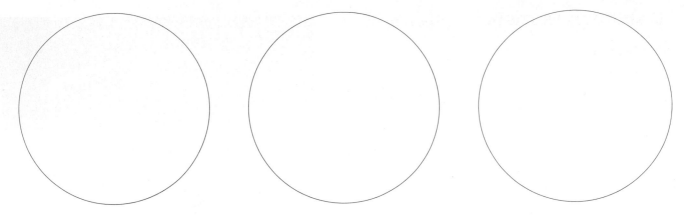

Figure 22-5 Different shapes possible in living *Euglena* exhibiting euglenoid movement

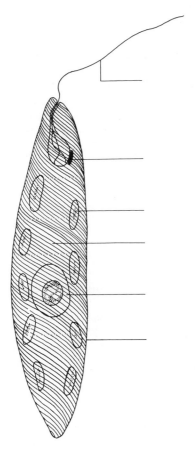

Labels: cytoplasm, chloroplast, nucleus, flagellum, eyespot, pellicle

Figure 22-6 *Euglena*

Examine microscopically a bit of diatomaceous earth by preparing a wet mount slide. Then prepare a wet mount of living diatoms. Use the high-dry objective to note the pigmentation within the cytoplasm and the numerous perforations in the cell walls

Figure 22-7 Drawing of diatoms (_____×) .

(shells). In Fig. 22-7 make a sketch of several of the diatoms you are observing.

C. Phylum Pyrrophyta: Dinoflagellates

Dinoflagellates are commonly called "whirling whips" because of the spinning motion they exhibit. On occasion, populations of certain dinoflagellates may increase dramatically, causing the seas to turn red or brown. These are the **red tides,** which may devastate fish populations because neurotoxins produced by the dinoflagellates poison fish that feed on them.

With the high-dry objective of your compound microscope, examine a prepared slide or living representative. Dinoflagellates are encased in stiff cellulosic plates. The junction of these plates forms two grooves in which flagella are located. Find the plates and grooves. If you are examining living specimens, chloroplasts may be visible beneath the cellulose

Figure 22-8 Drawing of a dinoflagellate (_____×)

plates. Draw the dinoflagellate specimen you are observing (Fig. 22-8).

D. Phylum Sarcomastigophora: Protozoans

Protozoans are heterotrophic, generally motile, single-celled organisms. Some cause human disease. Amoebic dysentery and giardiasis are examples of illnesses caused by drinking water contaminated with the causal protozoans.

What type of symbiosis is exemplified by these disease-causing organisms?

1. Subphylum Mastigophora: Flagellate protozoans
Examine a prepared slide of human blood that contains the parasitic flagellate *Trypanosoma* (Fig. 22-9), the cause of African sleeping sickness. This flagellate is transmitted from host to host by the bloodsucking tsetse fly. Note the **flagellum** arising from one end of the cell.

Another example of a flagellated protozoan is the termite-inhabiting *Trichonympha* (Fig. 22-10). Prepare a wet mount of these organisms. Examine the gut of the termite with the high-dry objective to find *Trichonympha*. Note the large number of **flagella** covering the upper portion of the cell, the more or less centrally located **nucleus,** and wood fragments in the cytoplasm. Label Fig. 22-10.

The association of the termite and *Trichonympha* is an example of **obligate mutualism,** neither organism being capable of surviving without the other. Termites lack the enzymes to metabolize cellulose, a major component of wood. Wood particles ingested by termites are engulfed by *Trichonympha,* whose enzymes break the cellulose into soluble carbohydrates that are released for use by the termite.

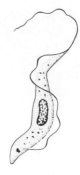

Figure 22-9 *Trypanosoma* (From Stanier, Ingraham, Wheelis, Painter, *The Microbial World*, 5th ed., © 1986. Reprinted by permission of Prentice-Hall, Inc.)

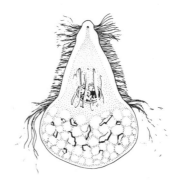

Labels: flagella, nucleus, wood fragments

Figure 22-10 *Trichonympha* (From Stanier, Ingraham, Wheelis, Painter, *The Microbial World*, 5th ed., © 1986. Reprinted by permission of Prentice-Hall, Inc.)

2. Subphylum Sarcodina: Amoeboid protozoans
The best-known amoeboid protozoans are the amoebas, organisms that are continually changing shape through formation of projections called **pseudopodia** (the singular is pseudopodium, "false foot").

Observe the *Amoeba*-containing culture on the stage of a demonstration dissecting microscope. (One species of amoeba has the scientific name *Amoeba*.) The microscope has been focused on the bottom of the culture dish, where the amoebas are located. Look for gray, irregularly shaped masses moving among the food particles in the culture. Using a clean pipet, remove an amoeba and place it, along with some of the culture medium, in a depression slide. Examine with your compound microscope using the medium power objective. You will need to adjust the diaphragm to increase the contrast (see Exercise 1) because *Amoeba* is nearly transparent.

Locate the pseudopodia. At the periphery of the cell, identify the **ectoplasm,** a thin, clear layer that surrounds the inner, granular **endoplasm.** Watch the organism as it changes shape. Which region of the endoplasm appears to stream, the outer or the inner?

This region, called the *plasmasol*, consists of a fluid matrix that can undergo phase changes with the semi-solid *plasmagel*, the outer layer of the endoplasm. Pseudopodium formation occurs as the plasmasol flows into new environmental frontiers and then changes to plasmagel. (This phenomenon was discussed in Exercise 3.)

Numerous granules will be found within the endoplasm. Some of these are organelles; others are food particles. Within the endoplasm, try to locate the **nucleus,** a densely granular, spherical structure around which the cytoplasm is streaming. You may be able to see clear, spherical **contractile vacuoles,** which regulate water balance within the cell. Watch for a minute or two to observe the action of contractile vacuoles. Label Fig. 22-11.

Amoeba feeds by a process called **phagocytosis,** engulfing its food. Pseudopodia form around food particles, and then the pseudopodia fuse, creating a **food vacuole** within the cytoplasm. Enzymes are then emptied into the food vacuole, where the food particle is digested into a soluble form that can pass through the vacuolar membrane.

You can stimulate feeding behavior by drawing carmine under the coverslips. Place a drop of distilled water (dH$_2$O) against one edge of the coverslip, pick up some carmine crystals by dipping a dissecting needle into the bottle, and deposit them into the water droplet. Draw the suspension beneath the coverslip by holding a piece of absorbent tissue against the coverslip on the side *opposite* the carmine suspension. Observe the *Amoeba* again—you may catch it in the act.

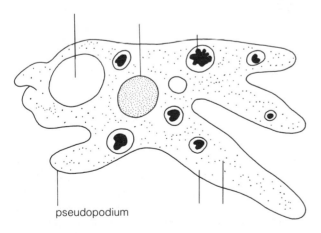

pseudopodium

Labels: ectoplasm, endoplasm, nucleus, contractile vacuole, food vacuole

Figure 22-11 *Amoeba*

E. Phylum Apicomplexa: Sporozoans

All sporozoans are parasites, infecting a wide range of animals, including humans. *Plasmodium vivax* causes one type of malaria in humans. We will study its life cycle (Fig. 22-12) with demonstration slides.

P. vivax is transmitted to humans through the bite of an infected female *Anopheles* mosquito. The mosquito serves as a **vector,** a means of transmitting the organism from one host to another. Male mosquitos cannot serve as vectors, because they lack the mouth parts for piercing skin and sucking blood. If the mosquito is carrying the pathogen, **sporozoites** enter the host's bloodstream with the saliva of the mosquito.

Examine the demonstration slide of sporozoites.

The sporozoites travel through the bloodstream to the liver, where they penetrate certain cells, grow, and multiply. When released from the liver cells, the parasite is in the form of a **merozoite** and infects the red blood cells.

Examine the demonstration slide illustrating merozoites in red blood cells.

Within the red blood cells, merozoites divide, increasing the merozoite population. At intervals of 48 or 72 hours, the infected red blood cells break down, releasing the merozoites. At this time, the infected individual exhibits disease symptoms, including fever, chills, and shaking caused by the release of merozoites and metabolic wastes from the red blood cells. Some of these merozoites return to the liver cells, where they repeat the cycle and are responsible for recurrent episodes of malaria.

Merozoites within the bloodstream may develop into **gametocytes.** For development of a gametocyte to be completed, the gametocyte must enter the gut of the mosquito. This occurs when a mosquito feeds upon an infected (diseased) human.

Observe the demonstration slide of an **immature gametocyte** in a red blood cell.

Within the gut of the mosquito, the gametocyte matures into gametes. When gametes fuse, they form a zygote that matures into an **oocyst.** Within each oocyst, sporozoites form, completing the life cycle of *Plasmodium vivax*. These sporozoites migrate to the mosquito's salivary glands to be injected into a new host.

F. Phylum Ciliophora: Ciliated Protozoans

Most members of the subphylum Ciliata are covered with numerous short locomotory structures called **cilia.** One of the largest ciliates is the predatory *Paramecium caudatum.*

From the culture provided, prepare a wet mount of *Paramecium* on a clean microscope slide. Observe their rapid motion with the medium power objective of your microscope.

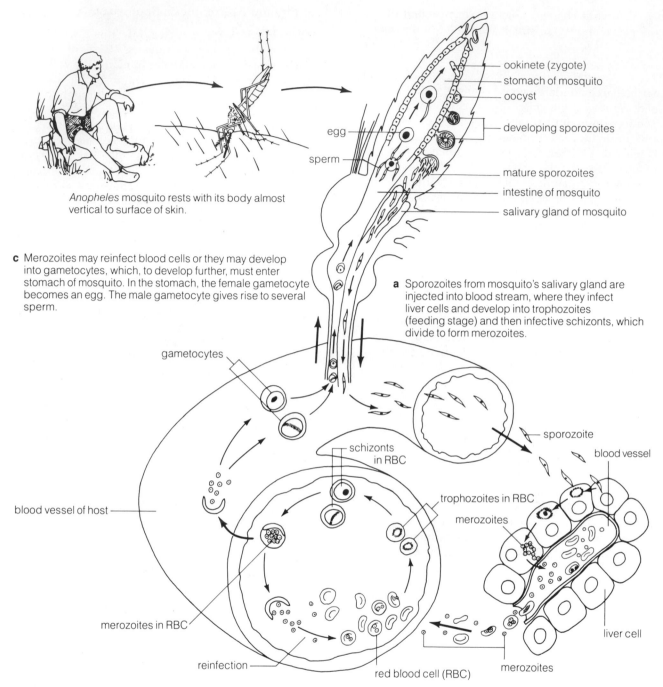

Anopheles mosquito rests with its body almost vertical to surface of skin.

ookinete (zygote)
stomach of mosquito
oocyst
developing sporozoites
egg
sperm
mature sporozoites
intestine of mosquito
salivary gland of mosquito

c Merozoites may reinfect blood cells or they may develop into gametocytes, which, to develop further, must enter stomach of mosquito. In the stomach, the female gametocyte becomes an egg. The male gametocyte gives rise to several sperm.

a Sporozoites from mosquito's salivary gland are injected into blood stream, where they infect liver cells and develop into trophozoites (feeding stage) and then infective schizonts, which divide to form merozoites.

gametocytes

sporozoite
blood vessel
schizonts in RBC
trophozoites in RBC
merozoites
blood vessel of host
merozoites in RBC
reinfection
red blood cell (RBC)
merozoites
liver cell

b Merozoites released into bloodstream infect red blood cells where they undergo developmental stages similar to those in liver.

Figure 22-12 Life cycle of *Plasmodium vivax*, causal agent of malaria. (After P. Abramoff and R. G. Thomson, *Laboratory Outlines in Biology-III*. Copyright © 1962, 1963 Peter Abramoff and Robert G. Thomson. Copyright © 1964, 1966, 1972, 1982 W. H. Freeman and Company. Used by permission.)

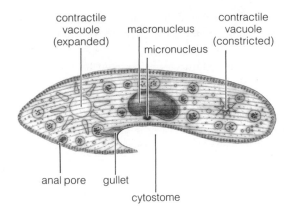

Figure 22-13 *Paramecium caudatum* (From Starr and Taggart, 1987.)

Paramecium possesses two nuclei. One is a larger, centrally located **macronucleus,** which is associated with nonreproductive functions. The other is a smaller **micronucleus** adjacent to the macronucleus. The micronucleus regulates reproductive functions. Add a drop of acetocarmine stain, drawing it beneath the coverslip by touching a piece of absorbent tissue to the opposite side of the coverslip. The acetocarmine should make the nuclei more visible.

On another slide, add a drop of methylcellulose and a drop of Congo red yeast mixture. Stir the mixture with a toothpick, then add a drop of *Paramecium* culture. Place a coverslip on your wet mount and observe with the medium power objective. Label the drawing of *Paramecium* (Fig. 22-13) with the structures described below.

Paramecium is covered with numerous **cilia,** which serve two functions: locomotion and directing food into the opening of the digestive tract. This opening is at the base of a depression called the **oral groove.** Locate the oral groove. The actual entrance (cytostome) is probably not obvious, but the tubule leading from the oral groove, the **gullet,** should be identified.

Observe the yeast cells as they are eaten by *Paramecium.* The yeast will be taken into **food vacuoles,** where they are digested. Congo red is a pH indicator. As digestion occurs within the food vacuoles, the indicator will turn blue because of the increased acidity.

At either end of the organism, find the **contractile vacuoles,** which regulate water content within the organism. Watch them in operation.

G. Phylum Gymnomycota: Slime Molds

The slime molds have both plantlike and animal-like characteristics. Because they engulf their food and lack a cell wall in their vegetative (nonreproductive) state, they are placed in the kingdom Protista. How-ever, when they reproduce, they produce spores with a rigid cell wall.

PROCEDURE

Physarum: A Plasmodial Slime Mold
The vegetative (nonreproductive) body of the plasmodial slime molds consists of a naked multinucleate mass of protoplasm known as a **plasmodium.**

Obtain a petri dish culture of *Physarum* and remove the cover. After examining it with your unaided eye, place the culture dish on the stage of your compound microscope and examine it with the low power objective.

Watch the cytoplasm. The motion that you see within the plasmodium is cytoplasmic streaming (Exercise 3). Is the cytoplasmic streaming unidirectional, or does the flow reverse?

As you noticed, the *Physarum* culture was stored in the dark. That's because light (along with other factors) stimulates the plasmodium to switch to the reproductive phase. Examine the spore-containing sporangia of *Physarum* that are on demonstration. Spores released from the sporangia germinate, producing a new plasmodium.

Because the plasmodium is multinucleate (whereas the spores are uninucleate), what event must occur following spore germination?

Examine Fig. 22-14, an illustration of the plasmodium of *Physarum.*

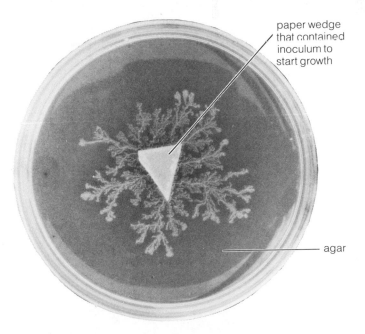

Figure 22-14 Culture dish containing the plasmodial slime mold, *Physarum.*
(Photo by J. W. Perry.)

PRE-LAB QUESTIONS

____ **1.** Members of the kingdom Monera *lack* (a) a nucleus, (b) organelles, (c) chloroplasts, (d) all of the above.

____ **2.** Unicellular eukaryotic organisms are placed in the kingdom (a) Monera, (b) Protista, (c) Animalia, (d) Plantae.

____ **3.** Which of the following organisms are phototrophic? (a) all monerans, (b) all protistans, (c) *Oscillatoria*, (d) *Trypanosoma*.

____ **4.** Which organisms are characterized as decomposers? (a) bacteria, (b) diatoms, (c) amoebas, (d) dinoflagellates.

____ **5.** Organisms capable of nitrogen fixation (a) include some bacteria, (b) include some cyanobacteria, (c) may live as symbionts with other organisms, (d) all of the above.

____ **6.** A spherical bacterium would be called a (a) bacillus, (b) coccus, (c) spirillum, (d) none of the above.

____ **7.** Gram stain would be used to distinguish between different (a) bacteria, (b) protistans, (c) dinoflagellates, (d) all of the above.

____ **8.** Which of the organisms (or parts of the organism) listed below might you find in toothpaste? (a) cyanobacteria, (b) diatoms, (c) amoebas, (d) *Trypanosoma vivax*.

____ **9.** Red tides are caused by (a) dinoflagellates, (b) bacteria, (c) diatoms, (d) monerans.

____ **10.** Those organisms that are covered by numerous, tiny locomotory structures belong to the Phylum (a) Gymnomycota, (b) Sarcomastigophora, (c) Apicomplexa, (d) Ciliophora.

EXERCISE 22
MONERANS AND PROTISTANS

POST-LAB QUESTIONS

1. Both monerans and protistans are unicellular organisms. What major characteristic distinguishes the organisms within the two kingdoms?

2. Do you think predation might be considered a type of symbiosis? If so, what kind? Explain your answer.

3. From an energetics standpoint, is it more efficient to eat a primary producer or a primary consumer? Explain your answer.

4. Of what significance are thylakoids in photosynthetic organisms?

5. The protistan *Euglena* is often studied in plant-related courses because it is photosynthetic. What characteristic of the pellicle makes *Euglena* different from true plants?

6. List several commercial uses for diatoms.

7. What characteristic separates the euglenids, diatoms, and dinoflagellates from protozoans?

8. How does phagocytosis differ from endocytosis?

9. Based upon your knowledge of the life history of *Plasmodium vivax*, suggest two methods for controlling malaria. Explain why each method would work.

   a.

   b.

10. Why would an organism such as *Paramecium* need contractile vacuoles?

FUNGI AND ALGAE

OBJECTIVES After completing this exercise you will be able to:

1. define parasite, saprophyte, mycologist, gametangium, oogonium, antheridium, hypha, mycelium, multinucleate, monoecious, dioecious, sporangium, rhizoid, zygospore, ascus, conidium, ascospore, ascocarp, basidium, basidiospore, basidiocarp, pathogen, phytoplankton, pyrenoid, phycobilin, agar, fucoxanthin, algin, kelp, colony, fragmentation, isogamous, thallus;

2. recognize representatives of the major divisions of fungi;

3. distinguish structures that are used to place various representatives of the fungi in their proper divisions;

4. list reasons why fungi are important;

5. recognize selected members of the red, brown, and green algae;

6. identify the pigments that are characteristic of each algal division;

7. distinguish between the structures associated with asexual and sexual reproduction described in this exercise.

INTRODUCTION The fungi and the algae are sometimes called *thallophytes*. Thallophytes are organisms that lack a highly differentiated body. Unfortunately, there are organisms that are neither fungi nor algae that fit this description as well.

Fungi are *heterotrophic* organisms; that is, they are incapable of producing their own food material. They secrete enzymes that digest their food source, which is then taken into the body, primarily by diffusion. By contrast, algae are photosynthetic. Despite their dissimilarity the fungi and algae are considered in this single exercise.

I. FUNGI: KINGDOM FUNGI (Starr, pp. 239–244)

As you walk into the woods following a warm late summer rain you are likely to be met by a vast assemblage of colorful fungi. Some of them are growing on dead or diseased trees, some on the surface of the soil, some in pools of water. Some are edible, some deadly poisonous. Some are **parasites,** organisms that obtain their nutrients from the organic material of another living organism. Others are **saprophytes,** growing on nonliving organic matter.

Fungi (kingdom Fungi), along with the bacteria, are essential components of the ecosystem as decomposers. These organisms recycle the products of life, making the products of death available so that life may continue. Without them we would be hopelessly lost in our own refuse. Fungi and fungal metabolism are responsible for some of the food products that enrich our lives—the mushrooms of the field, the blue cheese of the dairy case, even the citric acid used in making soft drinks.

The following five major divisions of the kingdom Fungi are those traditionally recognized by **mycologists,** individuals who study fungi:

Division	Common Name
Oomycota	egg fungi
Zygomycota	zygospore-forming fungi
Ascomycota	sac fungi
Basidiomycota	club fungi
Deuteromycota	imperfect fungi

The basis for taxonomic separation among the divisions is the mode of sexual reproduction that each exhibits.

A. Division Oomycota: The Egg Fungi

Some of the most notorious and historically important plant destroyers known to man are egg fungi. Included in this group is *Phytophthora infestans,* the fungus that causes the disease known as late blight of potato. This disease spread through the potato fields of Ireland between 1845 and 1847, resulting in the deaths of 1 million Irish working-class citizens who had come to depend on potatoes as their primary food source. Another 2 million emigrated, many to the United States.

The egg fungi are so named because during sexual reproduction all members produce large nonmotile female gametes, the eggs. These eggs are contained in sex organs (**gametangia;** the singular is *gametangium*) called **oogonia.** By contrast the male gametes are nothing more than *sperm nuclei* contained in **antheridia,** the male gametangia.

Some members of the Oomycota are referred to as the water molds because they are commonly found in aquatic environments.

MATERIALS

Per student:

- culture of *Saprolegnia* or *Achlya*
- glass microscope slide
- compound microscope

PROCEDURE

Saprolegnia or *Achlya:* A Water Mold
Examine the water cultures of either *Saprolegnia* or *Achlya.* These fungi are chiefly saprophytes and are growing on a sterilized hemp seed that provides a carbohydrate source. Label Fig. 23-1 as you study this organism.

Notice the numerous filamentous hyphae that radiate from the hemp seed. A **hypha** (the plural is *hyphae*) is the basic unit of the fungal body. Collectively, all the hyphae constitute the **mycelium** (the plural is *mycelia;* Fig. 23-1a).

The mycelium of the egg fungi is **multinucleate,** meaning that there are many nuclei in each cell, not just one. Each nucleus is diploid (2N). Moreover, cross walls separating the mycelium into distinct cells are infrequent, forming only when reproductive organs are formed.

Place a glass slide on the stage of your compound microscope. This slide will serve as a platform for the culture dish that contains the fungal culture, allowing you to use the mechanical stage of the microscope to move the culture (if the microscope is so equipped).

Remove the lid from the culture dish, carefully place the culture dish on the platform, and examine with the low power objective. Look first at the edge of the mycelium, that is, at the tips of the youngest hyphae. Find the tips that appear more dense (darker) than the rest of the hyphae. These dense tips are cells specialized for asexual reproduction. They are the **zoosporangia** (the singular is *zoosporangium;* Fig. 23-1b), which produce biflagellate **zoospores** (Fig. 23-1c). Note the cross wall separating the zoosporangium from the rest of the hypha.

Zoospores are released from the zoosporangia, swim about for a period of time, lose their flagella, and encyst (Fig. 23-1d); that is, they form a thick wall around the cytoplasm. When the encysted zoospore germinates, it will produce a second type of zoospore (Fig. 23-1e) that also will eventually encyst (Fig. 23-1f). When the second cyst stage germinates, it produces a hypha (Fig. 23-1g) that will proliferate into a new mycelium (Fig. 23-1a). As you see, this is asexual reproduction; no sex organs were involved in the formation of zoospores.

Sexual reproduction in the egg fungi occurs in the older portion of the mycelium. Scan the colony to find the spherical female gametangia, the **oogonia** (the singular is *oogonium;* Fig. 23-1h). Meiosis has taken place within the oogonium to form haploid eggs. Switch to the medium power objective and study a single oogonium in greater detail. Depending upon the stage of development, you will find either *eggs* (Fig. 23-1i) or *zygotes* (fertilized eggs) (Fig. 23-1j) within the oogonium.

The male gametangium, the *antheridium* (the plural is *antheridia*), is a short fingerlike hypha that attaches itself to the wall of the oogonium (Fig. 23-1i), much as you would wrap your finger around a baseball. Find an antheridium. Because the nuclei of the antheridium have undergone meiosis, each nucleus is haploid.

Fertilization takes place when tiny fertilization tubes penetrate the wall of the oogonium (Fig. 23-1i). Rather than forming special male gametes, the haploid nuclei within the antheridium flow through the fertilization tubes to fuse with the egg nuclei.

After fertilization, each zygote forms a thick wall (Fig. 23-1j). Following a maturation period that may last several months, the zygote germinates by forming a germ tube (Fig. 23-1k) that grows into a new mycelium (Fig. 23-1a).

Search your culture to see if you can locate any thick-walled zygotes within oogonia.

Notice that both male and female gametangia were produced on the same mycelium. The term describing this condition is **monoecious** (from the Greek words for "one house"). Organisms producing one type of sex organ on one body and the other type of sex organ on another body are **dioecious** (from the Greek words for "two" and "house"). Are humans monoecious or dioecious?

B. Division Zygomycota: Zygospore-Forming Fungi

All members of the Zygomycota produce a thick-walled zygote, a **zygospore.** Most fungi in this division are saprophytes, including the common black bread mold *Rhizopus.* Before the introduction of chemical preservatives into bread, *Rhizopus* was an almost certain invader, especially if the humidity was high.

Rhizopus is dioecious. Consequently, growth of two different mycelia in close proximity is necessary before sexual reproduction will occur. (The difference in the mycelia is genetic rather than structural. Be-

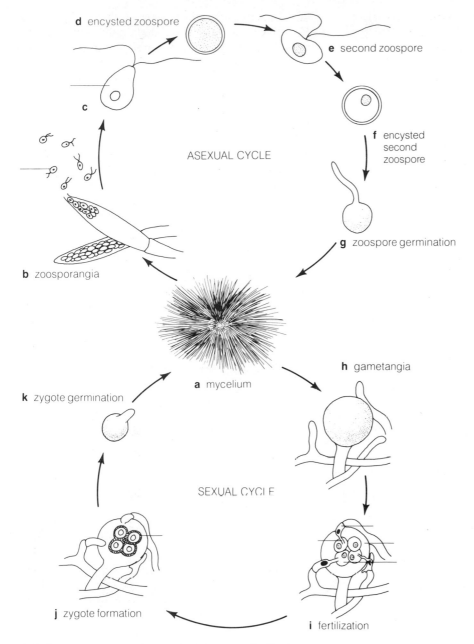

d encysted zoospore

e second zoospore

f encysted second zoospore

c

ASEXUAL CYCLE

g zoospore germination

b zoosporangia

a mycelium

h gametangia

k zygote germination

SEXUAL CYCLE

j zygote formation

i fertilization

Labels: zoospore, oogonium, antheridium, egg, fertilization tube, thick-walled zygote

Figure 23-1 Life cycle of an egg fungus such as *Saprolegnia* or *Achlya*. (Modified from artwork by Carolina Biological Supply Company.)

cause they are impossible to distinguish, the mycelia are referred to as + and − strains, as indicated in Fig. 23-2.)

MATERIALS

Per student:

- culture of *Rhizopus*
- prepared slide of *Rhizopus*
- dissecting needle

- glass microscope slide
- coverslip
- compound microscope

Per student pair:

- distilled water (dH$_2$O) in dropping bottle

Per lab room:

- demonstration culture of *Rhizopus* zygospores, on dissecting microscope

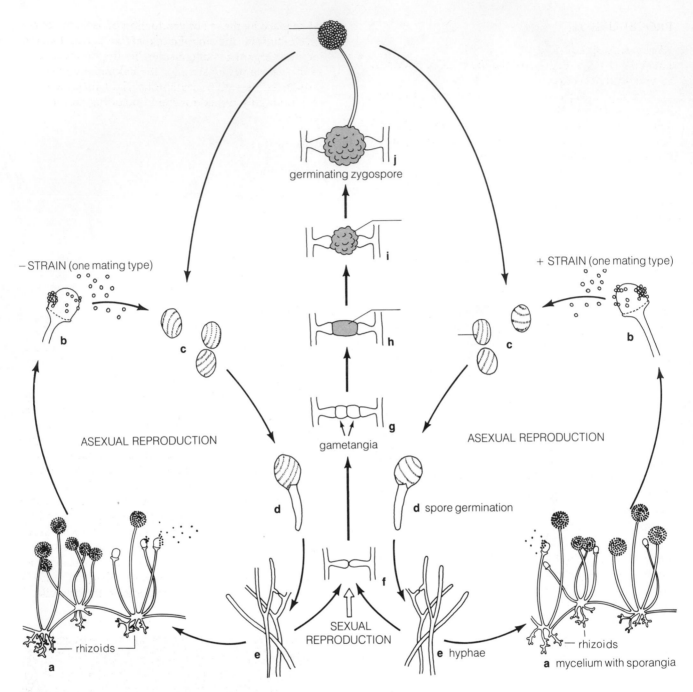

– STRAIN (one mating type)

+ STRAIN (one mating type)

germinating zygospore **j**

i

h

gametangia **g**

ASEXUAL REPRODUCTION

ASEXUAL REPRODUCTION

b   **c**

c   **b**

d

d spore germination

f

SEXUAL REPRODUCTION

— rhizoids —

e

e hyphae

— rhizoids —

a

a mycelium with sporangia

Labels: sporangium, spore, zygote, zygospore

Figure 23-2   Life cycle of *Rhizopus,* black bread mold. (Shaded structures are 2N.) (Modified from artwork by Carolina Biological Supply Company.)

PROCEDURE

Rhizopus: A Bread Mold
Label Fig. 23-2, a diagram of the life cycle of *Rhizopus,* as you study this organism.

Examine a petri dish culture containing the mycelium, which consists of many hyphae (Fig. 23-2a). The numerous black "dots" are the **sporangia** (the singular is *sporangium;* Figs. 23-2a, 2b). Sporangia are containers of **spores** (Figs. 23-2b, 2c, 2d) by which *Rhizopus* reproduces asexually.

Using a dissecting needle, remove a small portion of the culture to prepare a wet mount. Examine your preparation with the high-dry objective of your compound microscope. It's likely that when you added the coverslip, you crushed the sporangia, liberating the spores. Are there many or few spores within a single sporangium?

The hypha that elevates the sporangium off the surface is the **sporangiophore** (Fig. 23-2a). Identify the **rhizoids** (Fig. 23-2a) at the base of the sporangiophore. Rhizoids serve to anchor the mycelium to the substrate.

Now observe the demonstration culture illustrating sexual reproduction in *Rhizopus*. The black line running down the center of the culture plate consists of numerous **zygospores,** the products of sexual reproduction.

Sexual reproduction in *Rhizopus* (Fig. 23-2f through 23-2j) occurs when two sexually compatible mycelia are in close proximity.

As the hyphae from each mating type grow close, chemical messengers signal them to produce protuberances (Fig. 23-2f). When the protuberances make contact, **gametangia** (Fig. 23-2g) are produced at their tips. The wall between the two gametangia then dissolves, and the cytoplasms of the gametangia mix. Eventually the haploid nuclei from each gametangium fuse. The resulting cell is the *zygote* (Fig. 23-2h). A thick, bumpy wall forms about this diploid cell, and it is now referred to as the **zygospore** (Fig. 23-2i).

On prepared slides, find the stages of sexual reproduction in *Rhizopus,* including gametangia, zygotes, and zygospores. Use the medium power and high-dry objectives of your compound microscope in making your observations.

Meiosis occurs within the zygospore, which then germinates to produce a sporangiophore and a sporangium (Fig. 23-2j). Some of the spores give rise to mycelia of one mating type, others to the other mating type (Fig. 23-2e).

C. Division Ascomycota: Sac Fungi

Members of the Ascomycota produce spores in a sac, the **ascus** (the plural is *asci*), which develops as a result of sexual reproduction. Asexual reproduction takes place by means of production of asexual spores called **conidia** (the singular is *conidium*). The division includes organisms of considerable importance, such as the yeasts responsible for the baking and brewing industries, as well as numerous plant pathogens. A few are highly prized for food, including morels and truffles.

MATERIALS

Per student:

- prepared slide of *Peziza*
- compound microscope

Per student group (table):

- 2 large preserved specimens of *Peziza* or another cup fungus

PROCEDURE

Peziza: A Cup Fungus
The cup fungi are found commonly on soil during cool early spring and fall weather. Observe the preserved specimen of a cup fungus (Fig. 23-3).

Actually, the structure we identify as a cup fungus is the "fruiting body," produced as a result of sexual reproduction by the fungus. Most of the organism is present within the soil as an extensive *mycelium*. Specifically, the fruiting body is called an **ascocarp.**

Obtain a prepared slide of the ascocarp of *Peziza* or a related cup fungus. Examine the slide with the medium power and high-dry objectives of your compound microscope. Identify the elongate fingerlike **asci,** which contain dark-colored, spherical **ascospores.** In Fig. 23-4, label the section of an ascocarp.

The structures that are involved in ascocarp formation are difficult to demonstrate; the sexual cycle is complex and will not be considered in this exercise.

Figure 23-3 Cup fungus (0.5×). (Photo by J. W. Perry.)

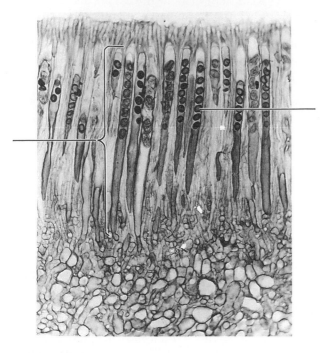

Labels: ascus, ascospore

Figure 23-4 Cross section of an ascocarp from a cup fungus (800×). (Photo courtesy Triarch, Inc.)

Labels: stalk, cap, gills

Figure 23-5 Mushroom basidiocarp (0.5×). (Photo by J. W. Perry.)

D. Division Basidiomycota: Club Fungi

Members of this group of fungi are probably what the average person thinks of as fungi, because the division contains those organisms called mushrooms. Actually, the "mushroom" is only a portion of the fungus—it's the "fruiting body" containing the sexually produced haploid **basidiospores**. These basidiospores are produced by a club-shaped **basidium** for which the group is named. Much (if not most) of the fungal mycelium grows out of sight, within the substrate upon which the basidiocarp is found.

MATERIALS

Per student:

- commercial mushroom
- prepared slide of mushroom pileus, cross section (*Coprinus*)
- compound microscope

Per lab room:

- demonstration specimens of various club fungi

PROCEDURE

1. Gill Fungi: The Mushrooms

Obtain a fresh fruiting body, more properly called a **basidiocarp** (Fig. 23-5). Identify the **stalk** and **cap.** Look at the bottom surface of the cap, noting the numerous gills. It is on the surface of these gills that the haploid basidiospores are borne. Remember that all the structures you are looking at are composed of aggregations of fungal hyphae. Label Fig. 23-5.

Study a prepared slide of a cross section of the cap of a mushroom, labeling Fig. 23-6 as you proceed. Observe the slide first with the low power objective of your compound microscope. In the center of the **cap** identify the **stalk.** The **gills** radiate from the edge of the cap to the stalk much as spokes of a bicycle wheel radiate to the hub.

Switch to the high-dry objective to study a single gill (Fig. 23-7). Note that the component hyphae produce club-shaped structures at the edge. These are the **basidia** (the singular is *basidium*). Each basidium produces four haploid **basidiospores** (all four may not be in the same plane of section).

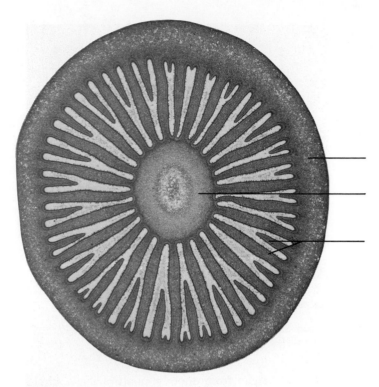

Labels: cap, stalk, gills

Figure 23-6 Cross section of mushroom cap (10×). (Photo courtesy Ripon Microslides, Inc.)

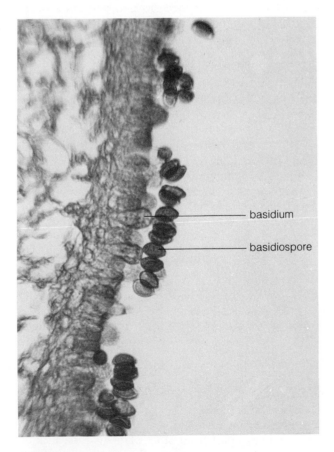

basidium

basidiospore

Figure 23-7 High magnification of a mushroom gill (500×). (Photo courtesy Ripon Microslides, Inc.)

Each basidiospore is attached to the basidium by a tiny hornlike projection. As the basidiospore matures, it is shot off the projection due to buildup of turgor pressure within the basidium.

Other than the basidiocarp just examined we will not study the details of the life cycle of the club fungi.

2. Other Club Fungi

Examine the representatives of fruiting bodies of other members of the club fungi that have been put on demonstration. These include puffballs and shelf fungi. The basidiospores of puffballs are contained within a spherical basidiocarp that develops a pore at the apex. Basidiospores are released when the puffball is crushed or hit by driving rain.

The presence of a basidiocarp of the familiar shelf fungi is an indication that an extensive network of fungal hyphae is growing within a tree, digesting the cells of the wood. As a forester assesses a woodlot to determine the amount of usable wood it might yield, one of the things noted is the presence of shelf fungi, indicating low-value (diseased) trees.

One of the most common shelf fungi is frequently called a "polypore" because of the numerous holes or pores on the lower surface of the fruiting body. These pores are lined with basidia bearing basidiospores. Observe the undersurface of the basidiocarp that is on demonstration at a dissecting microscope and note the pores.

E. Division Deuteromycota: Fungi Imperfecti

The Deuteromycota consists of fungi for which no known sexual stage is known and are hence called "imperfect." Reproduction takes place primarily by production of asexual **conidia.**

These fungi are among the most economically important, producing antibiotics (for example, one species of *Penicillium* produces penicillin). Others produce citric acid used in the soft drink industry and are used in the manufacture of cheese. Some are important as **pathogens** (disease-causing organisms) of both plants and animals.

MATERIALS

Per student:

- dissecting needle
- glass microscope slide
- coverslip
- prepared slide of *Penicillium* conidia (optional)
- compound microscope

Per student pair:

- distilled water (dH$_2$O) in dropping bottle

Per lab room:

- demonstration of *Penicillium*-covered foodstuff or plate cultures

PROCEDURE

Penicillium
Examine demonstration specimens of moldy oranges or other foodstuffs. The blue color is attributable to a pigment in the numerous conidia produced by this fungus, *Penicillium.*

With a dissecting needle, scrape some of the conidia from the surface of the moldy specimen (or from a culture plate containing *Penicillium*) and prepare a wet mount. Observe your preparation using the high-dry objective. (Prepared slides may also be available.)

Identify the **conidiophore** (Fig. 23-8), and the numerous tiny, spherical **conidia.** *Penicillium* comes from the Latin word *penicillus*, meaning "a brush." (Appropriate, isn't it?) Label Fig. 23-8.

II. ALGAE: KINGDOM PLANTAE

(Starr, pp. 245–247, 251)

Algae—pond scum, frog spittle, seaweed, the stuff that clogs your aquarium if it's not cleaned routinely, the debris on an ocean beach after a storm at sea; the nuisance organisms of a lake. These are the images that probably pop into your mind when you first think about the organisms called algae. But let's con-

Labels: conidiophore, conidia

Figure 23-8 *Penicillium* (300×). (Photo courtesy Ripon Microslides, Inc.)

sider the algae from another point of view. **Phytoplankton,** the weakly swimming or floating algae, are at the base of the aquatic food chain. They are among the smallest of the photosynthetic organisms producing their own food, themselves serving as food for animal life. As photosynthesizers, algae return vast amounts of oxygen to the water and in turn, to the atmosphere.

The term *algae* was once a bonafide taxonomic category. More recently, however, the organisms once grouped under this term have been placed in separate kingdoms. The organisms we will examine in this exercise are placed in the kingdom Plantae. (Other "algae" were considered in the exercise concerning the Monera and Protista kingdoms.) Today, the term *algae* (the singular is *alga*) is still used commonly without taxonomic implication to include all photosynthetic aquatic microbes as well as larger nonvascular aquatic plants.

The classification system we will use is as follows:

KINGDOM PLANTAE	
Division	**Common Name**
Rhodophyta	red algae
Phaeophyta	brown algae
Chlorophyta	green algae

This portion of the exercise will acquaint you with these three divisions of algae. These divisions are composed of mostly unrelated organisms that share few features. All possess chlorophyll *a* as the primary photosynthetic pigment (but so do most other photosynthetic organisms). Perhaps the one feature that sets the algae apart from other plants is their lack of multicellular sex organs (and there are even exceptions to this "rule").

Three characteristics are typically used to separate the divisions: (1) the photosynthetic pigments other than chlorophyll *a;* (2) the type of stored food materials; and (3) the characteristics of locomotory structures, when present.

A. Division Rhodophyta: Red Algae

Characteristics:

- Photosynthetic pigments: chlorophylls *a* and *d* phycobilins
- Stored food: floridean starch
- Motile cells: none

MATERIALS

Per lab room:

- demonstration slide of *Porphyridium*
- demonstration specimen and slide of *Porphyra* (nori)

PROCEDURE Although commonly called red algae, members of the Rhodophyta vary in color from red to green to purple to greenish-black. The color depends upon the quantity of their accessory pigments, the **phycobilins,** which are blue and red. These accessory pigments allow capture of light energy across the entire visible spectrum. This energy is passed on to chlorophyll for photosynthesis. One phycobilin, the red phycoerythrin, allows some red algae to live at great depths, where red wavelengths, those of primary importance for green and brown algae, fail to penetrate.

Which wavelengths (colors) would be absorbed by a red pigment?

Most abundant in warm marine waters, red algae are the source of **agar,** a substance extracted from their cell walls. Agar is the solidifying agent in media on which some microorganisms are cultured.

Representatives of the red algae are on demonstration.

1. *Porphyridium:* A Unicellular Red Alga. Unicellular red algae are exemplified by *Porphyridium.* Examine these cells at the demonstration microscope, noting the reddish chloroplast. In Fig. 23-9, sketch a cell of *Porphyridium.*

2. *Porphyra:* A Multicellular Membranous Form. At the other extreme of the morphological spectrum is *Porphyra,* a membranous form. Examine a portion of this organism and the wet mount specimen on demonstration. *Porphyra* is used extensively as a food substance in Asia, where it is commonly sold under the name "nori." In Japan, nori production is valued at $20 million annually. In Fig. 23-10, make a sketch of the nori demonstrations, describing both its macroscopic and microscopic appearance.

Figure 23-9   Drawing of *Porphyridium* (Rhodophyta; _____ ×)

Figure 23-10   Drawing of *Porphyra* (nori) (Rhodophyta; _____ ×)

B. Division Phaeophyta: Brown Algae

Characteristics:

- Photosynthetic pigments: chlorophylls *a* and *c* xanthophylls, including fucoxanthin
- Stored food: laminarin
- Motile cells: zoospores and gametes

MATERIALS

Per lab room:

- demonstration specimen of *Laminaria*
- demonstration specimen of *Macrocystis*

PROCEDURE   The vast majority of the brown algae are found in cold, marine environments. All members

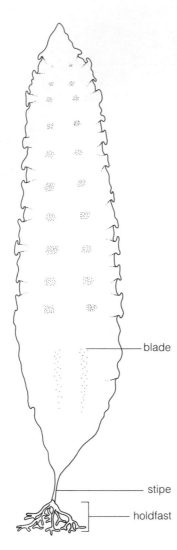

blade

stipe

holdfast

Figure 23-11 *Laminaria*

are multicellular and most are macroscopic. Their color is due to the accessory pigment fucoxanthin, which is so abundant that it masks the green chlorophylls. Some species are used as food, while others are harvested for fertilizers. Of primary economic importance is **algin,** a cell wall component of brown algae that is used to make ice cream smooth, cosmetics soft, and paint uniform in consistency, among other uses.

Kelps: *Laminaria* and *Macrocystis*
Kelps are large (up to 100 meters long), complex brown algae. Examine specimens of *Laminaria* (Fig. 23-11) and *Macrocystis* (Fig. 23-12). On each, identify the rootlike **holdfast** that anchors the alga to the substrate; the **stipe,** a stemlike structure; and the leaflike **blades.**

C. Division Chlorophyta: Green Algae

Characteristics:

- Photosynthetic pigments: chlorophylls *a* and *b* carotenoids: xanthophyll, carotene
- Stored food: starch
- Motile cells: zoospores and gametes

MATERIALS

Per student:

- glass microscope slide
- coverslip
- depression slide
- small culture dish
- prepared slide of *Oedogonium*
- prepared slide of *Spirogyra*
- compound microscope

Per student pair:

- diluted India ink in dropping bottle
- tissue paper
- I₂KI in dropping bottle
- dissecting microscope

Per lab room:

- living culture of *Chlamydomonas*, disposable pipet
- living culture of *Volvox*
- demonstration slide of *Volvox* zygotes
- living culture of *Oedogonium*
- living culture of *Spirogyra*
- demonstration specimen of *Ulva*

PROCEDURE The Chlorophyta is a diverse assemblage of green organisms, ranging from motile and nonmotile unicellular forms to colonial, filamentous, membranous, and multinucleate forms. Not only do they include the most species, they are also important phylogenetically because ancestral green algae are believed to have given rise to the land plants. We will examine the green algae from a morphological standpoint, starting with unicellular forms.

1. *Chlamydomonas:* A Motile Unicell
Prepare a wet mount of *Chlamydomonas* cells from the culture provided. Examine first with the low power objective of your compound microscope. Notice the numerous small cells swimming across the field of view. It will be difficult to study the fast-swimming cells, so kill the cells by adding a drop of I₂KI to the edge of the coverslip; draw the I₂KI under the coverslip by touching a folded tissue to the opposite edge. To observe the cells, switch to the highest power objective available.

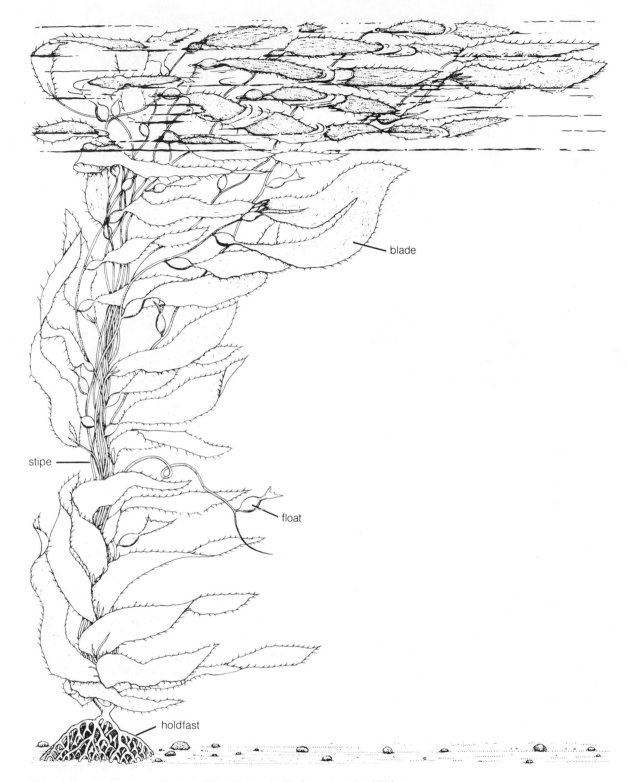

blade

stipe

float

holdfast

Figure 23-12 *Macrocystis* (0.01×). (After M. Neushul in Scagel *et al.*, 1982.)

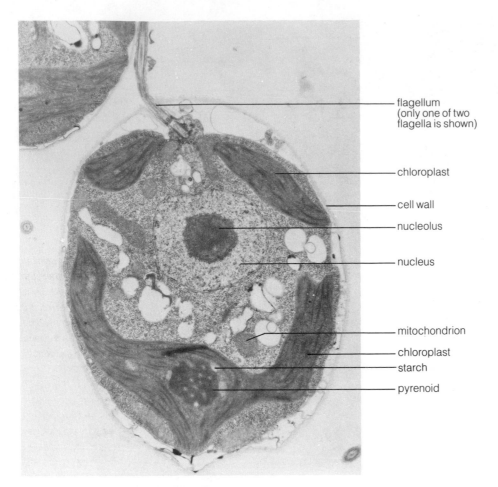

flagellum
(only one of two
flagella is shown)

chloroplast

cell wall

nucleolus

nucleus

mitochondrion

chloroplast

starch

pyrenoid

Figure 23-13 Transmission electron micrograph of *Chlamydomonas* (9,750×). (Photo courtesy H. Hoops.)

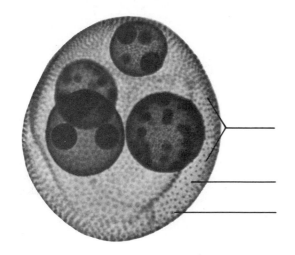

Labels: gelatinous matrix, cell, autocolony (daughter colony)

Figure 23-14 *Volvox*, with autocolonies (110×). (Photo courtesy Ripon Microslides, Inc.)

Most of the *Chlamydomonas* cytoplasm is filled with a very large, green cup-shaped *chloroplast*.

Chlamydomonas stores its excess photosynthate as *starch grains*, which will appear dark blue or black when stained with I_2KI. Locate the starch grains. The green algae have a specialized center for starch synthesis, the **pyrenoid.** The starch grains you are observing are clustered around the pyrenoid. If the orientation of the cell is just right, you may be able to detect an orange **stigma** (eyespot) that serves as a light receptor. Finally, find the two **flagella** at the anterior end of the cell.

Examine Fig. 23-13, a transmission electron micrograph of *Chlamydomonas*. Notice the magnification. The electron microscope makes much more obvious the structures you could barely see with your light microscope.

2. *Volvox:* A Motile Colony

From the culture provided, place a drop of *Volvox*-containing culture solution on a depression slide. (A prepared slide may be substituted if living specimens are not available.) Observe the large motile **colonies** (Fig. 23-14), first with the microscope's low and medium power objectives. The colony is a hollow cluster of mostly identical, *Chlamydomonas*-like cells that are held together by a **gelatinous matrix.** Identify the gelatinous matrix, which appears as the transparent region between individual cells.

Each cell possesses flagella. As the flagella beat, the entire colony rolls through the water. (The scientific name, *Volvox*, comes from the Latin word *volvere*, which means to roll.)

Asexual reproduction takes place by **autocolony (daughter colony) formation.** Certain cells within the colony divide and then round up into a sphere, the **autocolony.** Find the autocolonies within your specimen.

Sexual reproduction in *Volvox* is by **oogamy:** Relatively large, nonmotile *egg cells* are produced within gametangia called **oogonia,** while numerous, small, motile sperm are produced in other cells at the margin of the colony. When sperm fertilize the eggs, **zygotes** are produced.

Observe the demonstration slide of a spiny-walled zygote of *Volvox*.

Label Fig. 23-14.

3. *Oedogonium:* A Nonmotile Filament
Oedogonium (Fig. 23-15) is a filamentous, zoospore-producing green alga that is found commonly attached to sides of aquaria and slow-moving freshwater streams.

Prepare a wet mount slide from the living culture provided and observe with the medium power and high-dry objectives of your compound microscope. Within each cell, locate the single, netlike *chloroplast* with its many pyrenoids. The cell that attaches the filament to the substrate is specialized as a **holdfast.** Scan your specimen to determine if any holdfasts are present at the ends of the filaments. Each cell in the filament contains a single nucleus and a large, central vacuole (both of which are difficult to distinguish).

Asexual reproduction takes place by **zoospores.** If present, zoosporangia (the cells producing zoospores) will be found within the filament.

Sexual reproduction takes place by **oogamy.** What is meant by "oogamy"?

Obtain a prepared slide of a monoecious species of *Oedogonium,* which bears both male and female gametangia on the same filament. Look for the large, spherical *oogonia*.

Oogonia are

_____ (male or female)

sex organs.

Within each oogonium locate a single, large *egg cell*. Now find the male gametangia, the *antheridia,* which appear as short, boxlike cells.

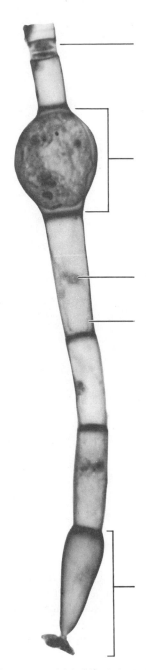

Labels: chloroplast, pyrenoid, holdfast, oogonium, antheridium

Figure 23-15 *Oedogonium*, with gametangia (100×). (Photo courtesy Triarch, Inc.)

Antheridia are

_____ (male or female)

gametangia.

Each antheridium produces two *sperm*.

With fertilization, a *zygote* is formed. The zygote develops a thick, heavy wall. Some oogonia may contain zygotes. When the zygote germinates, the diploid nucleus undergoes meiosis, producing motile spores. When the spore settles down, mitosis and cell division occur, producing the haploid filament you have examined.

Label Fig. 23-15, a photomicrograph of *Oedogonium*.

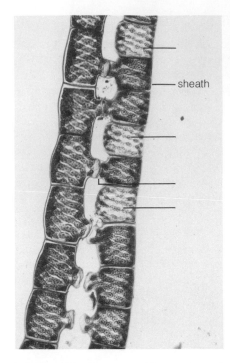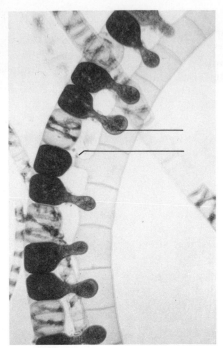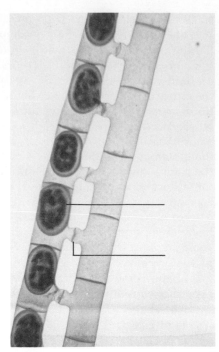

— sheath

Labels: chloroplast, pyrenoid, vacuole, conjugation tube, isogamete, zygote (some labels used more than once)

Figure 23-16 *Spirogyra*, vegetative and sexual (400×). (Photo courtesy Ripon Microslides, Inc.)

4. *Spirogyra:* A Nonmotile Filament

Another filamentous green alga common to freshwater ponds is *Spirogyra* (Fig. 23-16). This alga is often called "pond scum" because it forms a bright green, frothy mass on and just below the surface of the water.

Observe the *Spirogyra* in the large culture dish. Pick up some of the mass, noting the slimy sensation. This is due to the watery sheath surrounding each filament.

Place a few filaments on a slide and add a drop of diluted India ink before adding a coverslip. Observe the filaments with the medium power and high-dry objectives of your compound microscope. (Label Fig. 23-16 as you proceed.) The **sheath** will appear as a bright area off the edge of the cell wall. Note the spiral-shaped *chloroplast* with the numerous **pyrenoids.** (Prepared slides may be used if living filaments are not available.) Each cell contains a large central **vacuole** and a single **nucleus.** Locate the nucleus. Remember, the chloroplast is located within the cytoplasm, as is the nucleus. The nucleus, however, is suspended in strands of cytoplasm, much as a spider might be found in the center of a web.

Asexual reproduction occurs by means of **fragmentation;** that is, a small portion of the filament simply breaks off and continues to grow. Zoospores are not formed.

Sexual reproduction is **isogamous,** meaning that there is no visible differentiation into male and female filaments or gametes.

Obtain a prepared slide illustrating sexual reproduction in *Spirogyra* (Fig. 23-16).

Find two filaments that are joined by cytoplasmic bridges known as **conjugation tubes.** The entire cytoplasmic contents serve as **isogametes** (that is, gametes of similar size) in *Spirogyra,* with one isogamete moving through the conjugation tube into the other cell, where it fuses with the other gamete. Find stages illustrating conjugation and label Fig. 23-16.

Eventually, the two nuclei of each gamete fuse to form a *zygote,* which develops a thick wall. This thick-walled zygote serves as an overwintering structure. In the spring, the zygote nucleus undergoes meiosis. Three of the four nuclei die, leaving one functional, haploid nucleus. Germination of this haploid cell results in the formation of a haploid filament.

5. *Ulva*: A Membranous Form

A final representative of the green algae illustrates the fourth morphological form in the group, those having a membranous (tissuelike) body. Examine living or preserved specimens of *Ulva*, commonly known as "sea lettuce." The broad, leaflike body is called a **thallus,** a general term describing a vegetative body with relatively little cell differentiation. The thallus originates from a single cell that undergoes cell division in three planes, but only one division occurs in one of the planes, giving rise to a two-cell-thick body.

Make a sketch of *Ulva* in Fig. 23-17.

Figure 23-17 Drawing of *Ulva* (Chlorophyta; _____ ×)

PRE-LAB QUESTIONS

____ **1.** An organism that grows specifically on nonliving organic material is called a (an) (a) autotroph, (b) heterotroph, (c) parasite, (d) saprophyte.

____ **2.** Taxonomic separation into fungal divisions is based upon (a) sexual reproduction, or lack thereof; (b) whether or not the fungus is a parasite or saprophyte; (c) the production of certain metabolites, like citric acid; (d) the edibility of the fungus.

____ **3.** Fungi is to fungus as _____ is to _____. (a) mycelium, mycelia; (b) hypha, hyphae; (c) mycelia, mycelium; (d) zoospore, zoospores.

____ **4.** A fungus that is dioecious (a) requires two different, sexually compatible mycelia for sexual reproduction; (b) produces both sex organs on the same mycelium; (c) reproduces only by asexual means; (d) all of the above.

____ **5.** Which of the following is *not* true of the Deuteromycota? (a) they reproduce sexually by means of conidia; (b) they form an ascocarp; (c) sex organs are present in the form of oogonia and antheridia; (d) all of the above.

____ **6.** Small, green, floating or weakly swimming organisms are specifically called (a) algae, (b) plankton, (c) phytoplankton, (d) zooplankton.

____ **7.** Which of the following is *not* characteristic of most algae? (a) gametes are usually motile and come in contact by swimming; (b) well-developed water conducting tissue is present; (c) they lack a protective covering that reduces water loss in land plants; (d) none of the above.

____ **8.** Agar is derived from (a) red algae, (b) brown algae, (c) green algae, (d) all of the above.

____ **9.** Which set of characteristics is found in the green algae? (a) chlorophyll *a* and *d*, phycobilins, floridean starch; (b) chlorophyll *a* and *b*, fucoxanthin, laminarin; (c) chlorophyll *a* and *c*, fucoxanthin, laminarin; (d) chlorophyll *a* and *b*, carotenoids, starch.

____ **10.** Which is the correct plural form of the word for the organisms studied in this exercise? (a) alga, (b) algae, (c) algas, (d) algaes.

EXERCISE 23
FUNGI AND ALGAE

POST-LAB QUESTIONS

1. Define: parasite

   Give an example of a fungal parasite.

2. Distinguish between a hypha and a mycelium.

3. Fungi are often studied in botany courses but they might also be studied in a zoology course. Make a case for studying them in an animal course.

4. Distinguish between sexual and asexual reproduction.

   Sexual reproduction is "cost-intensive," requiring a large expenditure of energy for the production of one to a few offspring. On the other hand, asexual reproduction as occurs by conidia results in numerous offspring with minimal energy production. What advantage is there to sexual reproduction that warrants this large expenditure of energy?

5. In the blanks provided, give the correct singular or plural form of the word provided.

Singular	Plural
a. hypha	_____
b. _____	mycelia
c. zygospore	_____
d. _____	asci
e. basidium	_____
f. _____	conidia

6. What color are the marker lights at the edge of an airport taxiway?

_____

Are the wavelengths of this color long or short, relative to the other visible wavelengths?

_____

Which wavelengths penetrate deepest into water, long or short?

_____

Make a statement regarding why phycobilin pigments are present in deep-growing red algae.

What benefit is there to the color of airport taxiway lights for a pilot attempting to taxi during foggy weather?

7. What are the principal photosynthetic pigments in the green algae?

The most highly evolved land plants, the flowering plants, also contain the same principal photosynthetic pigments as do the green algae. Phylogenetically, of what importance might this be?

8. How is an algal holdfast cell similar to a root?

How is it different?

9. Why do you suppose the Swedish automobile manufacturer, Volvo, chose this company name? (Hint: Go to the library and find out what the Latin word *volvere* means.)

10. List three positive reasons why algae are important to life.

   a.

   b.

   c.

MOSSES AND FERNS

OBJECTIVES After completing this exercise you will be able to:

1. define alternation of generations, dioecious, antheridium, spermatogenous tissue, archegonium, hygroscopic, sporogenous tissue, protonema;

2. produce a cycle diagram illustrating alternation of generations;

3. recognize mosses as nonvascular land plants and ferns as seedless vascular plants;

4. identify the sporophytes, gametophytes, and associated structures of mosses and ferns;

5. describe the function of the sporophyte and the gametophyte;

6. postulate why most mosses and ferns are restricted to environments where free water is often available;

7. speculate why mosses are generally small land plants while ferns reach much greater proportions.

INTRODUCTION Step into almost any moist, temperate forest, look down, and what do you see? More than likely, covering the bases of tree trunks, on decaying logs, and on rocks you'll find mosses (division Bryophyta). It's equally likely that you'll find an abundance of ferns (division Pterophyta).

Mosses and ferns (kingdom Plantae) represent land plants that are dependent upon water for fertilization. Their sperm are flagellated and swim to the egg cell. This, among other features, suggests that these plants were among the first to invade the land mass.

It's generally agreed that land plants arose from the green algae. Evidence for this includes identical food reserves (starch), the same photosynthetic pigments (chlorophylls *a* and *b*, and the carotenoids), and similarities in structure of their flagella. Some biologists have gone so far as to suggest that the land plants are nothing more than highly evolved green algae.

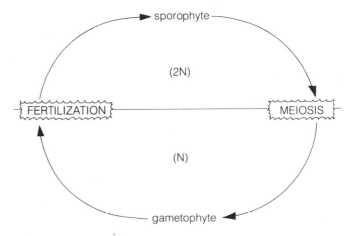

Figure 24-1 Summary of alternation of generations

One major mystery is the origin of a feature common to *all* land plants, **alternation of generations.** In alternation of generations, two distinct phases exist: A diploid **sporophyte** alternates with a haploid **gametophyte,** as summarized in Fig. 24-1.

At first, alternation of generations is difficult to envision. As animals, we find this concept foreign. But think of it as the existence of two body forms of the same organism. The primary reproductive function of one body form, the gametophyte, is to produce gametes (eggs and/or sperm) by *mitosis*. The primary reproductive function of the other, the sporophyte, is to produce spores by *meiosis*.

During the course of evolution, two major lines of divergence took place in the plant kingdom. The plants in one line had as the dominant phase the gametophytic generation, meaning that the sporophyte never was free-living but was permanently attached to and dependent upon the gametophyte for nutrition. Today these plants are represented by the bryophytes, mosses and their relatives. It seems that this line is an example of *dead-end evolution*, with no other group of plants present today arising from it.

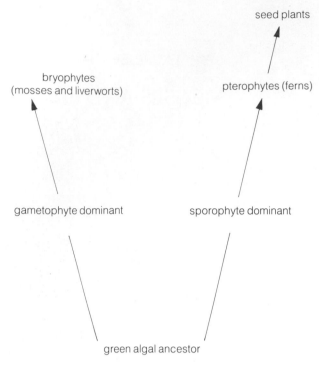

seed plants

bryophytes
(mosses and liverworts)

pterophytes (ferns)

gametophyte dominant

sporophyte dominant

green algal ancestor

Figure 24-2 Evolution of land plants

In the other line, the sporophyte led an independent existence, the gametophyte being quite small and inconspicuous. Ferns (pterophytes) are the least complex members of this line of evolution. We believe the fern line continued to evolve, ultimately resulting in the seed plants. Figure 24-2 summarizes these two evolutionary lines.

This exercise will acquaint you with the mosses and ferns that are part of today's flora.

I. DIVISION BRYOPHYTA: MOSSES

(Starr, pp. 248–249, 251)

MATERIALS

Per student:

- *Polytrichum,* male and female gametophytes, the latter with attached sporophytes
- prepared slide of moss antheridial head, l.s.
- prepared slide of moss archegonial head, l.s.
- prepared slide of moss sporangium (capsule), l.s.
- glass microscope slide
- coverslip
- compound microscope
- dissecting microscope
- dissecting needle

Per student pair:

- distilled water (dH$_2$O) in dropping bottle

Per student group (table):

- moss protenemata growing on culture medium

PROCEDURE   As you study the stages of the life history of the mosses, label Fig. 24-3, a diagrammatic representation of their life cycle.

Obtain a living or preserved specimen of a moss that grows in your area. The hairy-cap moss, *Polytrichum,* is a good choice because one of the 10 species in the genus is certain to be found wherever you live in North America.

Lacking vascular tissues, the mosses do not have true roots, stems, or leaves, although they do have structures that are rootlike, stemlike, and leaflike, and which function for the same purposes as do the true organs.

Identify the rootlike **rhizoids** at the base of the plant. What function do you think rhizoids serve?

Notice that the leaflike organs are arranged more or less radially about the stemlike *axis.* You are examining the **gametophyte generation** of the moss. In terms of the life cycle of the organism, what function does the gametophyte serve?

Polytrichum is **dioecious,** meaning that there are separate male and female plants. The male gametophyte can usually be distinguished by the flattened rosette of leaflike structures at its tip. Examine a male gametophyte (Fig. 24-3a), noting this feature. Embedded within the rosette are the male sex organs, **antheridia.**

With the low and medium power objectives of your compound microscope, examine a prepared slide with the antheridia of a moss (Fig. 24-3c). Identify the *antheridia.* Study a single antheridium, using the high-dry objective. Locate the jacket layer surrounding the **spermatogenous tissue.** With maturity, the spermatogenous tissue gives rise to numerous biflagellate *sperm.* Scattered among the antheridia, find the numerous sterile *paraphyses.* These do not have a reproductive role (and hence are called sterile) but instead function to hold capillary water, preventing the sex organs from drying out.

Examine a female gametophyte of *Polytrichum* (Fig. 24-3b). Before the development of the sporophyte, the female gametophyte can usually be distinguished by the absence of the rosette at its tip. Nonetheless, the apex of the female gametophyte contains the female sex organs, **archegonia.**

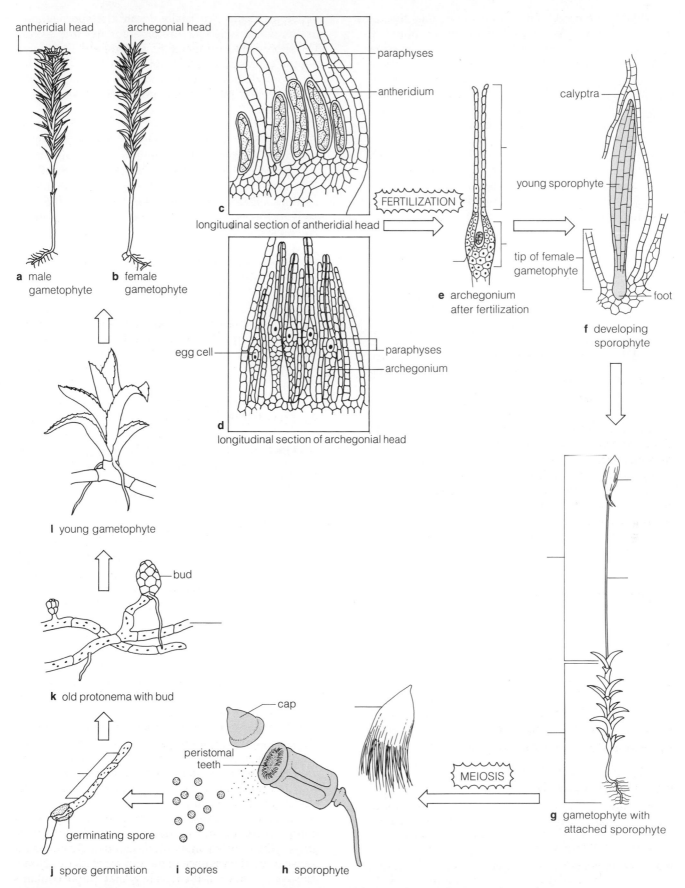

antheridial head   archegonial head

paraphyses

antheridium

c longitudinal section of antheridial head

a male gametophyte

b female gametophyte

FERTILIZATION

calyptra

young sporophyte

tip of female gametophyte

foot

e archegonium after fertilization

f developing sporophyte

egg cell

paraphyses

archegonium

d longitudinal section of archegonial head

l young gametophyte

bud

k old protonema with bud

cap

peristomal teeth

germinating spore

j spore germination

i spores

h sporophyte

MEIOSIS

g gametophyte with attached sporophyte

Labels: spermatogenous tissue, neck, venter, embryo sporophyte, mature sporophyte, stalk, sporangium, calyptra, protonema

Figure 24-3   Life cycle of a representative moss. (Shaded structures are 2N.) (Modified from Carolina Biological Supply Company diagrams.)

Obtain a prepared slide of the archegonial tip of a moss (Fig. 24-3d). Start with the low power objective of your compound microscope to gain an impression of the overall organization. Find the sterile *paraphyses* and the *archegonia*. Switch to the medium power objective to study a single archegonium, identifying the long **neck** and the slightly swollen **venter** (Fig. 24-3d). Within the venter, locate the **egg cell.**

Remember, you are looking at a *section* of a three-dimensional object. The archegonium is very much like a long-necked vase, except that it's solid. The venter is analogous to the base of the vase, while the egg cell is like a marble suspended in the middle of the base.

The central core of the archegonial neck contains cells that break down when the egg is mature, liberating a fluid that is rich in sucrose and that attracts sperm that are swimming in dew or rainwater. (Note that sperm are capable of swimming only short distances and so must be present close by.) Fertilization of the egg produces the diploid **zygote** (Fig. 24-3e), the first cell of the **sporophyte generation.** Numerous mitotic divisions produce an **embryo** (embryo sporophyte, Fig. 24-3f), which differentiates into the **mature sporophyte** (Fig. 24-3g) that protrudes from the tip of the gametophyte. What is the function of the sporophyte?

Now examine a female gametophyte that has an attached *mature sporophyte* (Fig. 24-3g). Is the sporophyte green?

What would you conclude about its ability to produce its own food?

Grasp the *stalk* of the sporophyte and detach it from the gametophyte. The base of the stalk absorbs water and nutrients from the gametophyte. At the tip of the sporophyte locate the *sporangium* covered by a papery hood. The cover is a remnant of the tissue that surrounded the archegonium and is called the **calyptra** (Figs. 24-3f, 3g, 3h). The calyptra of *Polytrichum* is covered with tiny hairs, hence the common name, "hairy-cap moss."

Remove the calyptra to expose the sporangium (Fig. 24-3h). Notice the small cap at the top of the sporangium. Remove the cap and observe the interior of the sporangium with a dissecting microscope. Find the *peristomal teeth* that point inward from the margin of the opening. The peristomal teeth are **hygroscopic,** meaning that they absorb water and aid in dispersing the **spores** (Fig. 24-3i) from the sporangium.

Study a prepared slide of a longitudinal section of a sporangium. At the top you will find sections of the peristomal teeth. Internally, locate the **sporogenous tissue,** which when mature differentiates into *spores*. The sporogenous tissue of the sporangium is diploid, but the spores are haploid and are the first cells of the gametophyte generation. What type of nuclear division must take place for the sporogenous tissue to become spores?

When the spores are shed from the sporangium, they are carried by wind and water currents to new sites. If conditions are favorable, the spore germinates to produce a filamentous **protonema** (Fig. 24-3j; the plural is *protonemata*) that looks much like a filamentous green alga.

Use a dissecting needle to remove a protonema from the culture provided and make a wet mount. Examine with the medium power and high-dry objectives of your compound microscope. Note the cellular composition and numerous green *chloroplasts*. If the protonema is of sufficient age, you should find *buds* (Fig. 24-3k) that grow into the leafy gametophyte (Figs. 24-3l, 3a, 3b).

II. DIVISION PTEROPHYTA: FERNS

(Starr, pp. 250–251)

MATERIALS

Per student:

- fern sporophytes, fresh, preserved, or herbarium specimens
- prepared slide of fern rhizome, cross section
- fern gametophytes, living, preserved, or whole mount prepared slides
- fern gametophyte with young sporophyte, living or preserved
- compound microscope
- dissecting microscope

Per lab room:

- squares of fern sori, in moist chamber (*Polypodium aurem* recommended)
- demonstration slide of fern archegonium, median l.s.
- other fern sporophytes, as available

PROCEDURE Obtain a fresh or herbarium specimen of a typical fern. As you examine the structures described below, label Fig. 24-5, a diagram representing the life cycle of a fern.

The sporophyte of many ferns (Fig. 24-5a) consists of *true* roots, stems, and leaves; that is, these possess vascular tissue. Identify the horizontal stem, the **rhizome** (which produces true **roots**), and upright leaves. The leaves of ferns are called **fronds** and are often compound.

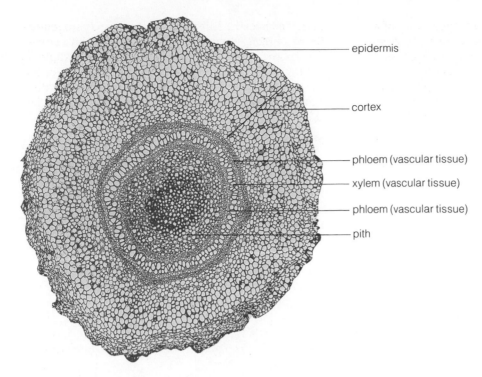

epidermis

cortex

phloem (vascular tissue)

xylem (vascular tissue)

phloem (vascular tissue)

pith

Figure 24-4 Cross section of a fern rhizome. (Photo courtesy Triarch, Inc.)

Ferns, unlike mosses, are vascular plants, their sporophytes containing xylem and phloem. With the low power objective of your compound microscope, examine a prepared slide of a cross section of a fern stem (rhizome). Using Fig. 24-4 as a reference, find the *epidermis, cortex,* and *vascular tissue.* Within the vascular tissue, distinguish between the *phloem* (of which there are outer and inner layers) and the thick-walled *xylem* sandwiched between the phloem layers.

Now examine the undersurface of the frond. Locate the dotlike **sori** (the singular is **sorus;** Fig. 24-5a). Each sorus is a cluster of **sporangia.** Using a dissecting microscope, study an individual sorus.

Each sporangium (Fig. 24-5b) contains *spores* (Fig. 24-5c). Although the sporangium is part of the diploid (sporophytic) generation, spores are the first cells of the haploid (gametophytic) generation. What process occurred within the sporangium to produce the haploid spores?

_____

Obtain a single, sorus-containing square of the hare's foot fern (*Polypodium aureum*), place it sorus-side up on a glass slide (DON'T ADD A COVERSLIP), and examine with the low power objective of your compound microscope. Note the row of brown, thick-walled cells running over the top of the sporangium, the *annulus.* The annulus is hygroscopic. Changes in moisture content within the cells of the annulus cause the sporangium to crack open. Watch what happens as the sporangium dries out.

As the water evaporates from the cells of the annulus, a tension develops that pulls the sporangium apart. Separation of the halves of the sporangium begins at the thin-walled *lip cells* (Fig. 24-5b). As the water continues to evaporate, the annulus pulls back the top half of the sporangium, exposing the spores. The sporangium continues to be pulled back until the tension on the water molecules within the annulus exceeds the strength of the hydrogen bonds holding the water molecules together. When this happens, the top half of the sporangium flies forward, throwing the spores out (Figs. 24-5b, 5c). The fern sporangium is a biological catapult!

If spores land in a suitable environment, one that is generally moist and shaded, they germinate (Figs. 24-5d, 5e, 5f), eventually growing into the heart-shaped adult gametophyte. Using your dissecting microscope, examine a living, preserved, or prepared slide whole mount of the gametophyte (Fig. 24-5g). What color is the gametophyte?

_____

What does the color indicate relative to the ability of the gametophyte to produce its own carbohydrates?

_____

Examine the undersurface of the gametophyte. Find the *rhizoids,* which serve to anchor the gametophyte and perhaps absorb water.

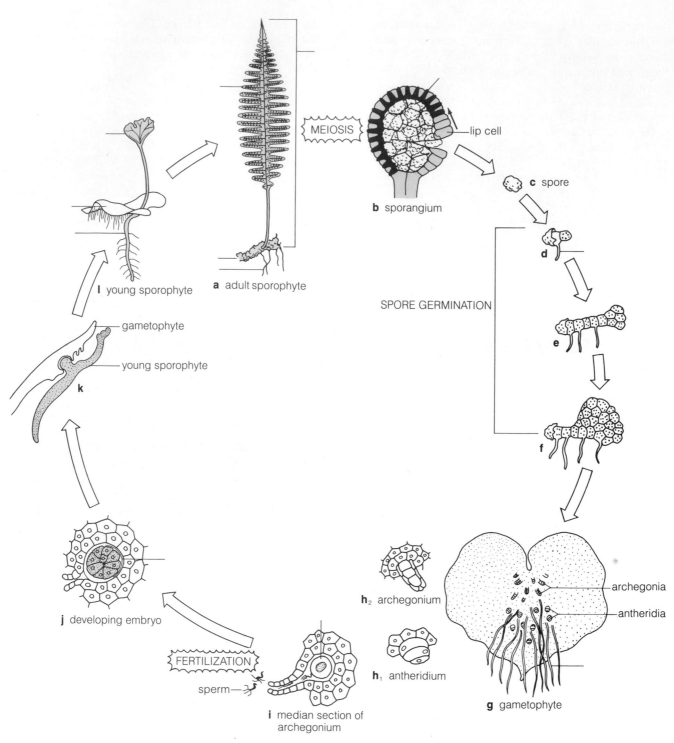

MEIOSIS

b sporangium

lip cell

c spore

d

SPORE GERMINATION

e

f

a adult sporophyte

l young sporophyte

gametophyte

young sporophyte

k

j developing embryo

FERTILIZATION

sperm

i median section of archegonium

h₂ archegonium

h₁ antheridium

g gametophyte

archegonia

antheridia

Labels: rhizome, root, frond, sorus, annulus, rhizoid, egg, neck, embryo, gametophyte, young sporophyte

Figure 24-5 Life cycle of a typical fern. (Shaded structures are 2N.)

Locate the *gametangia* (sex organs) clustered among the rhizoids (Fig. 24-5g). There are two types of gametangia: **antheridia** (Fig. 24-5h$_1$), which produce the flagellated **sperm;** and **archegonia** (Fig. 24-5h$_2$), which produce **egg cells.**

Study the demonstration slide of an archegonium (Fig. 24-5i). Identify the *egg* within the swollen basal portion of the archegonium. Note that the *neck* of the archegonium protrudes from the surface of the gametophyte.

The archegonia secrete chemicals that attract the flagellated sperm, which swim in a water film down a canal within the neck of the archegonium. One sperm fuses with the egg to produce the first cell of the sporophyte generation, the **zygote.** With subsequent cell divisions, the zygote develops into an embryo (embryo sporophyte; Fig. 24-5j). As the embryo grows, it pushes out of the gametophyte and develops into a young sporophyte (Fig. 24-5k).

Obtain a specimen of a young sporophyte that is attached to the gametophyte (Fig. 24-5l). Identify the **gametophyte,** *primary leaf,* and *primary root* of the young sporophyte. As the sporophyte continues to develop, the gametophyte withers away.

Examine any other specimens of ferns that may be on demonstration, noting the incredible diversity in form. Look for sori on each specimen.

PRE-LAB QUESTIONS

____ **1.** Land plants are believed to have evolved from (a) mosses, (b) ferns, (c) green algae, (d) fungi.

____ **2.** In the bryophytes, the sporophyte is (a) the dominant generation, (b) dependent upon the gametophyte generation, (c) able to produce all of its own nutritional requirements, (d) a and c above.

____ **3.** Mosses and ferns utilize which of the following pigments for photosynthesis? (a) chlorophylls *a* and *b,* (b) carotenes, (c) xanthophylls, (d) all of the above.

____ **4.** Which of the following *best* describes the concept of alternation of generations? (a) one generation of plants is skipped every other year; (b) there are two phases, a sporophyte and a gametophyte; (c) the parental generation alternates with a juvenile generation; (d) a green sporophyte phase produces food for a nongreen gametophyte.

____ **5.** In ferns (a) xylem and phloem are present in the sporophyte, (b) the sporophyte is the dominant generation, (c) the leaf is called a frond, (d) all of the above.

____ **6.** An organ that is hygroscopic is (a) sensitive to changes in moisture, (b) exemplified by the peristomal teeth in the sporophyte of mosses, (c) exemplified by the annulus on the sporangium of ferns, (d) all of the above.

____ **7.** The capsule of a sporophyte (a) is the same as the sporangium, (b) is the container in which meiosis takes place, (c) contains spores, (d) all of the above.

____ **8.** Sperm find their way to the archegonium (a) by swimming, (b) due to a chemical gradient diffusing from the archegonium, (c) as a result of sucrose being released during the breakdown of the neck canal cells of the archegonium, (d) all of the above.

____ **9.** A protonema (a) is part of the sporophyte generation of a moss, (b) is the product of spore germination of a moss, (c) looks very much like a filamentous brown alga, (d) produces the sporophyte when a bud grows from it.

____ **10.** The spores of a fern are (a) produced by mitosis within the sporangium, (b) diploid cells, (c) the first cells of the gametophyte generation, (d) a and b above.

EXERCISE 24
MOSSES AND FERNS

POST-LAB QUESTIONS

1. Explain why water must be present for the bryophytes to complete the sexual portion of their life cycle.

2. Why do you suppose the bryophytes never attained tree-sized proportions?

3. In what structure does meiosis occur in the bryophytes? How does this compare with the location of meiosis in the majority of the algae?

4. The ancestors of the bryophytes are believed to be green algae. Cite four distinct lines of evidence to support this belief.

   a.

   b.

   c.

   d.

5. Why are the bryophytes characterized as lacking true roots, stems, and leaves?

6. How do you suppose this old tale originated: "If you are lost in the woods, note the side of the tree that mosses grow on because this is the north side"?

7. Within a single sporangium, assuming the absence of crossing over, what is the maximum number of genotypes present among the spores? Explain.

8. Complete this diagram of a "generic" alternation of generations.

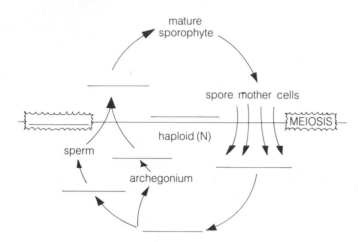

9. What tissue absent from the mosses and present in ferns has allowed the ferns to become much larger in size?

10. Describe in your own words the difference between a sporophyte and a gametophyte.

SEED PLANTS I: GYMNOSPERMS

OBJECTIVES After completing this exercise you will be able to:

1. define gymnosperm, pulp, heterosporous, homosporous, pollination, fertilization, dioecious, monoecious;

2. describe the characteristics that distinguish seed plants from other vascular plants;

3. produce a cycle diagram illustrating heterosporous alternation of generations;

4. list uses for conifers;

5. recognize and describe the life cycle of a pine;

6. distinguish between a male and a female pine cone;

7. describe the method by which pollination occurs in pines;

8. describe the process of fertilization in pines;

9. recognize members of the divisions Cycadophyta, Ginkgophyta, and Gnetophyta.

INTRODUCTION The division Coniferophyta (kingdom Plantae) represents seed-bearing plants known as conifers. They are a part of a more general assemblage of plants commonly known as **gymnosperms.** *Gymnosperm* translates literally as "naked seed," referring to the production of seeds on the *surface* of reproductive structures. This contrasts with the situation in the angiosperms (see Exercise 26), whose seeds are contained within a fruit.

During the evolution of vascular plants, the development of the seed was one of the most striking events to occur. Seeds have remarkable survival value and seem to be one of the reasons for the dominance of seed plants today.

Let's examine the characteristics of seeds and seed plants.

1. All seed plants produce **pollen grains.** Pollen grains serve as carriers for sperm. This characteristic is one factor accounting for the widespread distribution of seed plants. As a consequence of pollen production, the sperm of seed plants do not need free water to swim to the egg. Thus, seed plants are capable of reproducing in harsh climates where non-seed plants are much less successful.

2. All seeds have some type of **stored food** that the embryo uses as it emerges from the seed during germination.

3. All seeds have a **seed coat,** a protective covering enclosing the embryo and its stored food.

A seed coat and stored food are particularly important for survival. An embryo within a seed is protected from an inhospitable environment. Consider, for example, that a seed may be produced during a severe drought. Water is necessary for growth of the embryo. If none is available, the seed may remain dormant until growing conditions are favorable. When germination occurs, a ready food source is present to get things underway, providing nutrients until the developing plant can produce its own carbohydrates by photosynthesis (see Exercise 6).

4. As was the case with the mosses and ferns, seed plants exhibit **alternation of generations.**

5. All seed plants are **heterosporous;** that is, they produce *two* types of spores. Mosses and most ferns are **homosporous,** producing only *one* spore type.

Examine Fig. 25-1, a diagram representing heterosporous alternation of generations. Contrast it with the diagram you completed in Exercise 24, Post-lab Question 8.

I. DIVISION CONIFEROPHYTA: CONIFERS
(Starr, pp. 250–253)

Pine (*Pinus*) is one type of conifer, other examples being hemlock, firs, and spruces (see Exercise 2). The conifers are among the most important plants economically, because their wood is used in building construction. Millions of hectares (1 hectare = 2.47 acres) are devoted to growing conifers for this purpose, not to mention the numerous plantations that grow Christmas trees. In many areas conifers are

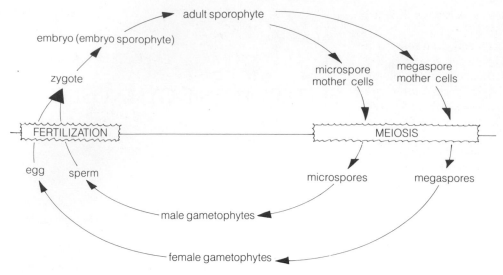

Figure 25-1 Heterosporous alternation of generations

used for **pulp,** the moistened cellulose used to manufacture paper. In this exercise, we'll examine reproduction in pine.

MATERIALS

Per student:

- cluster of male cones
- prepared slide of male strobilus, l.s., with microspores
- prepared slide of female strobilus, l.s., with megaspore mother cell
- prepared slide of pine seed, l.s.
- compound microscope
- dissecting microscope
- single-edged razor blade

Per student group (table):

- young sporophyte, living or herbarium specimen

Per lab room:

- demonstration slide of female strobilus
- demonstration slide of fertilization
- pine seeds, soaking in water
- pine seedlings, 12 weeks old
- pine seedlings, 36 weeks old

PROCEDURE As you do the exercise, label Fig. 25-3, representing the life cycle of a pine tree. To refresh your memory, look at a specimen of a small pine tree. This is the adult **sporophyte** (Fig. 25-3a). Remember, all pines are conifers, but not all conifers are pines (*Pinus*). Give the *scientific name* (genus) of a conifer that is not a pine (see Exercise 2).

The structure of a pine tree is just like that of the plant you examined in Exercise 12. Identify the stem and leaves. You probably know the main stem of a woody plant as the "trunk." The leaves of conifers are often called "needles" because most are shaped like needles.

A. Male Reproductive Structures and Events

Obtain a cluster of **male cones** (Fig. 25-3b). The function of male cones is to produce **pollen;** consequently, they are typically produced at the ends of branches, where the wind currents can catch the pollen as it is being shed. Note all the tiny scalelike structures that make up the male cones. These are *microsporophylls.*

Translated literally, a sporophyll would be a "spore-bearing leaf." The prefix *micro* refers to "small." But rather than having the literal interpretation "small, spore-bearing leaf," a microsporophyll is one that will produce *male* spores, called *microspores.* These develop into winged, immature *male gametophytes* called **pollen grains.** (Why they are immature is a logical question. The male gametophyte is not mature until it produces sperm.)

Remove a single microsporophyll and examine its lower surface with a dissecting microscope (Fig. 25-3c). Identify the two **microsporangia,** also called **pollen sacs.**

Study a prepared slide of a longitudinal section of a male cone (also called a male *strobilus*), first with a dissecting microscope to gain an impression of the cone's overall organization and then with the medium power objective of your compound microscope (Fig. 25-3d). Identify the *cone axis* bearing numerous **microsporophylls.** Switch to the high-dry objective to observe more closely a single microsporophyll. Note that it contains a cavity; this is the

microsporangium (also called a *pollen sac*), which contains numerous *pollen grains.*

As the male cone grows, several events occur in the microsporophylls that lead to the production of pollen grains. Microspore mother cells within the microsporangia undergo meiosis to form *microspores.* Cell division within the microspore wall and subsequent differentiation result in the formation of the pollen grain.

Examine a single pollen grain with the high-dry objective (Fig. 25-3e). Identify the earlike *wings* on either side of the body. The body consists of four cells, the two most obvious of which are the **tube cell** and smaller **generative cell.** (The nucleus of the tube cell is almost as large as the entire generative cell.) **Pollination,** the transfer of pollen from the male cone to the female cone, is accomplished by wind, occurring in the spring of the year.

B. Female Reproductive Structures and Events

Development and maturation of the female cone takes two to three years, the exact time depending upon the species. Female cones are typically produced on higher branches of the tree. Because the individual tree's pollen is generally shed downward, this arrangement favors crossing between different individuals.

Obtain a young female cone (Fig. 25-3f), noting the arrangement of the cone scales. Unlike the male cone, the female cone is a complex structure, each scale consisting of an **ovuliferous scale** fused atop a *sterile bract.* Remove a single scale-bract complex (Fig. 25-3g). On the top surface of the complex find the two **ovules,** the structures that eventually will develop into the **seeds.**

Examine a prepared slide of a longitudinal section of a female cone (Fig. 25-3h) first with the dissecting microscope. Note the spiral arrangement of the scales on the cone axis. Distinguish the smaller *sterile bract* from the *ovuliferous scale.*

Now examine the slide with the low power objective of your compound microscope. Look for a section through an ovuliferous scale containing a very large cell; this is the **megaspore mother cell** (Fig. 25-3i). The tissue surrounding the megaspore mother cell is the **megasporangium.** Protruding inward toward the cone axis are flaps of tissue surrounding the megasporangium, the **integument.** Find the integuments and the opening between them, the **micropyle.**

Think for a moment about the three-dimensional nature of the ovule: it's much like a short vase lying on its side on the ovuliferous scale. The "neck" of the vase is the integument, the opening the micropyle. The integument extends around the base of the vase. If you poured liquid rubber inside the base of a vase, suspended a marble in the middle, and allowed the rubber to harden, you'd have a model of the megasporangium and the megaspore mother cell.

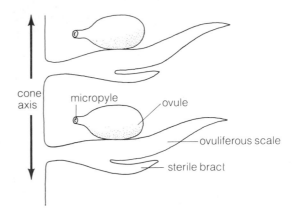

Figure 25-2 Ovule and ovuliferous scale/sterile bract complex

Figure 25-2 gives you an idea of the three-dimensional structure.

The megaspore mother cell undergoes meiosis to produce four haploid **megaspores** (Fig. 25-3j), but only one survives, the other three degenerating. The functional megaspore repeatedly divides mitotically to produce the multicellular **female gametophyte** (Fig. 25-3k). At the same time, the female cone is continually increasing in size to accommodate the developing female gametophytes. (Remember, there are numerous ovuliferous scale/sterile bract complexes on each cone.)

The female gametophyte of pine is produced

_____ (within or outside of) the megasporangium.

Archegonia eventually develop within the female gametophyte. Examine the demonstration slide showing archegonia (Fig. 25-3l). Identify the single large **egg cell** that fills the entirety of the archegonium. (The nucleus of the egg cell may be visible as well. The other generally spherical structures are protein bodies within the egg.)

Fertilization occurs when the **pollen tube,** an outgrowth of the pollen grain's tube cell, penetrates the megasporangium and enters the archegonium. The generative cell of the pollen grain has divided to produce two sperm, one of which fuses with the egg. (The second sperm nucleus degenerates.) Examine the demonstration slide illustrating fertilization in *Pinus.* Identify the *zygote,* the product of fusion of egg and sperm (Fig. 25-3l).

After fertilization, numerous mitotic divisions of the zygote take place, eventually producing an **embryo** (*embryo sporophyte*). Fertilization also triggers changes in the integument, causing it to harden and become the seed coat.

With the low power objective of your compound microscope, study a prepared slide of a longitudinal section through a pine seed (Fig. 25-3m). Starting

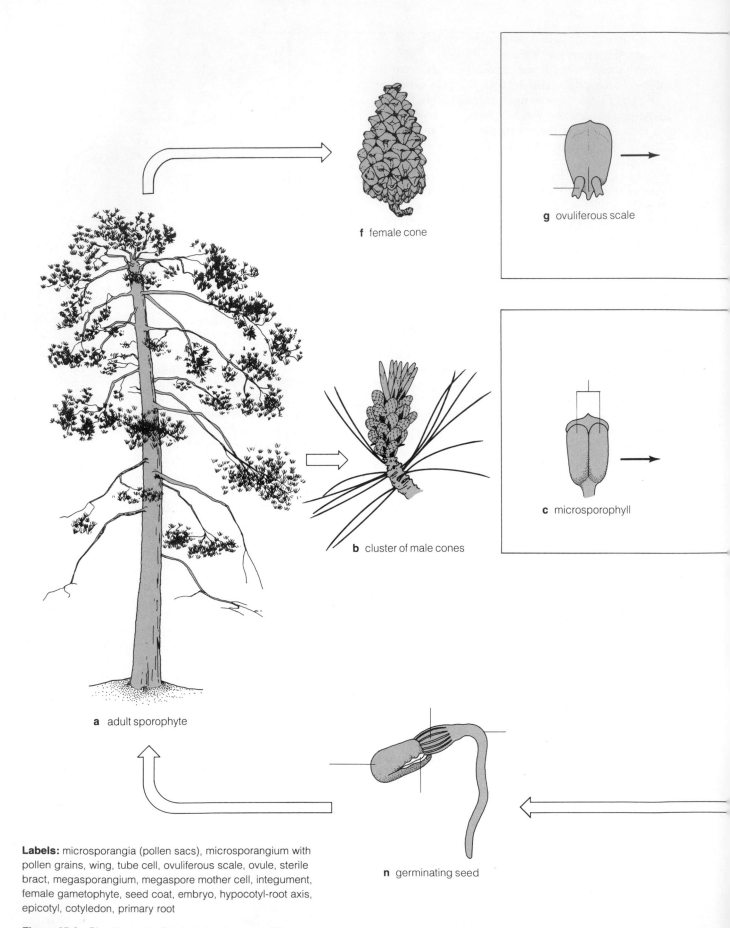

f female cone

g ovuliferous scale

c microsporophyll

b cluster of male cones

a adult sporophyte

n germinating seed

Labels: microsporangia (pollen sacs), microsporangium with pollen grains, wing, tube cell, ovuliferous scale, ovule, sterile bract, megasporangium, megaspore mother cell, integument, female gametophyte, seed coat, embryo, hypocotyl-root axis, epicotyl, cotyledon, primary root

Figure 25-3 Pine life cycle. Shaded structures are 2N.

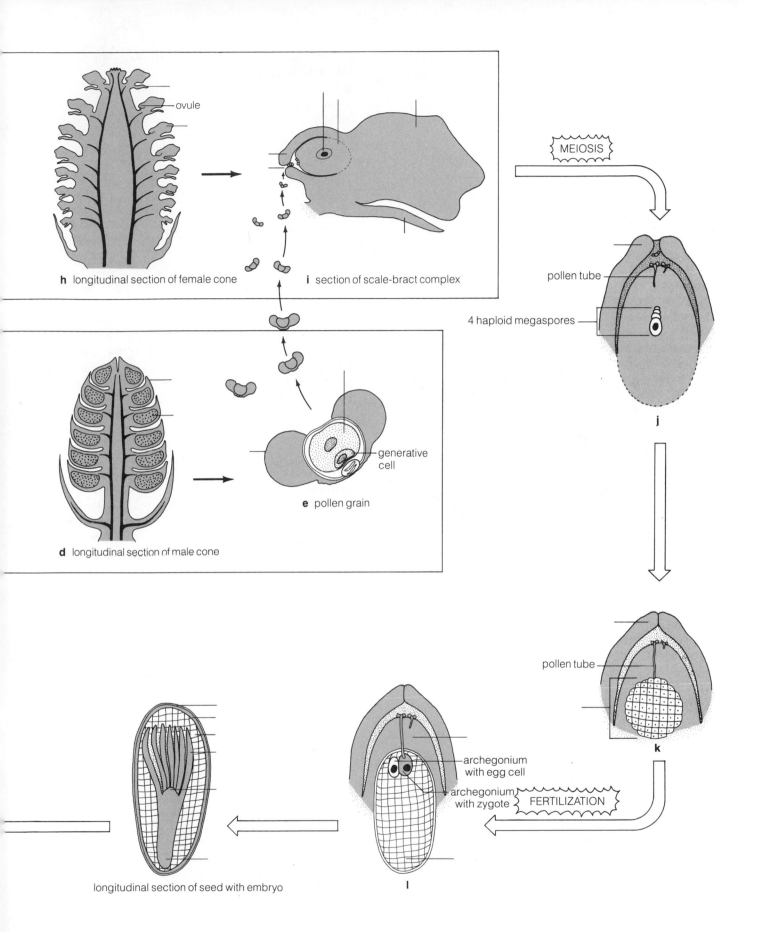

h longitudinal section of female cone

ovule

i section of scale-bract complex

MEIOSIS

pollen tube

4 haploid megaspores

j

d longitudinal section of male cone

generative cell

e pollen grain

pollen tube

k

FERTILIZATION

archegonium with egg cell

archegonium with zygote

l

longitudinal section of seed with embryo

from the outside, identify the **seed coat, megasporangium** (a very thin, papery remnant), the female gametophyte, and **embryo** (embryo sporophyte). The embryo consists of a **hypocotyl-root axis** and numerous **cotyledons,** in the center of which is the **epicotyl.** Identify these parts. The female gametophyte will serve as a food source for the embryo sporophyte when germination takes place.

Obtain a pine seed that has been soaked in water to soften the seed coat. Remove it and make a free-hand longitudinal section with a sharp razor blade. Again, identify the papery remnant of the *megasporangium*, the white *female gametophyte*, and *embryo* (embryo sporophyte). How many cotyledons are present?

Examine the culture of pine seeds that were planted in sand several weeks ago. Note the germinating seeds (Fig. 25-3n), finding the *hypocotyl-root axis, cotyledons, female gametophyte*, and *seed coat*. The cotyledons serve two functions. One is to absorb the nutrients stored in the female gametophyte during germination. As the cotyledons are exposed to light, they turn green. What then is the second function of the cotyledons?

Finally, examine somewhat older sporophyte seedlings. Notice that eventually the cotyledons wither away as the epicotyl produces new leaves.

II. OTHER GYMNOSPERMS (Starr, p. 251)

Although the conifers are the most important and widely distributed gymnosperms, there are three other divisions. Let's take a look at representative sporophytes of each of these.

MATERIALS

Per lab room:

- demonstration specimens of: *Zamia* or *Cycas, Ginkgo, Ephedra*

PROCEDURE

A. Division Cycadophyta

Examine the demonstration specimen of *Zamia* or *Cycas*. Both have the common name cycad. Do these plants resemble any of the conifers you know?

Zamia and *Cycas* are limited to the subtropical regions of North America. They're often planted as ornamentals in Florida. During the age of the dinosaurs (200 million years ago), cycads were extremely numerous.

All of the cycads are **dioecious;** that is, there are distinct male and female plants. *Zamia* produces male and female cones, while *Cycas* has only male cones, the female structures being much more leaflike. (Pine is **monoecious,** meaning that both male *and* female structures are produced on the *same* plant.)

The female cones of some other genera become extremely large, weighing as much as 30 kilograms.

Notice the leaves of the cycads. They resemble more closely the leaves of the ferns than those of the conifers.

B. Division Ginkgophyta

Examine the demonstration specimen of *Ginkgo biloba*, the maidenhair tree. *Ginkgo* is the only living representative of the division. It is sometimes called a "living fossil" because it has changed little in the last 80 million years. At one time, in fact, it was believed to be extinct; the Western world knew it from the fossil record before trees were discovered in remote China.

Note the fan-shaped leaves. *Ginkgo* is a highly prized ornamental tree that is now commonly planted in our urban areas. The tree has a reputation for being resistant to most insect pests and atmospheric pollution.

Like the cycads, *Ginkgo* is dioecious. Male trees are preferred as ornamentals, because the female trees produce seeds whose seed coat has a disagreeable odor.

C. Division Gnetophyta

Examine the demonstration specimen of *Ephedra*. Most species are found in desert or arid regions of the world. In the desert southwest of the United States, *Ephedra* is a common shrub known as "Mormon tea," because its stems were once harvested and used to make a tea in Utah. (Utah has the largest Mormon population in the United States.)

Many scientists now believe that the Gnetophyta is very closely related to the flowering plants. Look for the reproductive structures on the specimens before you. They look very much like flowers. Moreover, the structure of their water-conducting tissue (xylem) is more like that of the flowering plants than is true of the other gymnosperms.

PRE-LAB QUESTIONS

_____ **1.** Which of the following statements about conifers is *not* true? (a) conifers are gymnosperms, (b) all conifers belong to the genus *Pinus*, (c) all conifers have naked seeds, (d) conifers are heterosporous.

_____ **2.** Seed plants (a) have alternation of generations, (b) are heterosporous, (c) develop a seed coat, (d) all of the above.

_____ **3.** A pine tree is (a) a sporophyte, (b) a gametophyte, (c) diploid, (d) a and c above.

_____ **4.** The male pine cone (a) produces pollen, (b) contains a female gametophyte, (c) bears a megasporangium containing a megaspore mother cell, (d) gives rise to a seed.

_____ **5.** The male gametophyte of a pine tree (a) is produced within a pollen grain, (b) produces sperm, (c) is diploid, (d) a and b above.

_____ **6.** Which of the following are produced _directly_ by meiosis in pine? (a) sperm cells, (b) pollen grains, (c) microspores, (d) microspore mother cells.

_____ **7.** An ovule (a) is the structure that develops into a seed, (b) contains the microsporophyll, (c) is produced on the surface of a male cone, (d) all of the above.

_____ **8.** The process by which pollen is transferred to the ovule is called (a) transmigration, (b) fertilization, (c) pollination, (d) all of the above.

_____ **9.** Which of the following is true of the female gametophyte of pine? (a) it's a product of repeated cell divisions of the functional megaspore; (b) it's haploid; (c) it serves as the stored food to be used by the embryo sporophyte upon germination; (d) all of the above.

_____ **10.** The seed coat of a pine seed (a) is derived from the integuments, (b) was produced by the micropyle, (c) surrounds the male gametophyte, (d) is divided into the hypocotyl-root axis and epicotyl.

NAME _____   SECTION NUMBER _____

EXERCISE 25
SEED PLANTS I: GYMNOSPERMS

POST-LAB QUESTIONS

1. Think about the structures you've seen in the seed plants you've examined in this exercise. What survival advantage does a seed have that has allowed the seed plants to be the most successful of all plants?

2. List four uses for conifers.

   a.

   b.

   c.

   d.

3. Below is a diagrammatic representation of a seed. Give the ploidy level (N or 2N) of each part listed.

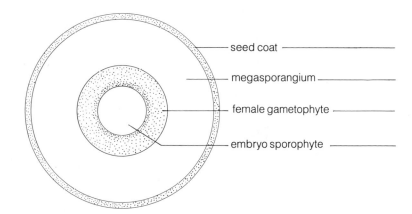

seed coat _____

megasporangium _____

female gametophyte _____

embryo sporophyte _____

4. It is sometimes said that a seed is a new sporophyte generation within the gametophyte generation, both of which are contained in the old sporophyte generation. Explain.

5. Distinguish between pollination and fertilization.

6. How does the complexity of the gametophyte generation of the Coniferophyta compare with that of the Pterophyta? Assuming that the Coniferophyta represents greater evolutionary development than the Pterophyta, what can be said about the changes that have taken place in the gametophytic generation over the course of evolution?

7. Are antheridia present in conifers? Archegonia?

8. What environmental factor necessary for fertilization in mosses and ferns is not required in conifers?

9. To which previously studied group of plants do you think the cycads might be most closely related? Why?

10. Are all gymnosperms evergreen?

Are all conifers evergreen?

SEED PLANTS II: ANGIOSPERMS

OBJECTIVES After completing this exercise you will be able to:

1. define angiosperm, fruit, pollination, double fertilization, endosperm, seed, germination, annual, biennial, perennial;

2. describe the significance of the flower, fruit, and seed for the success of the angiosperms;

3. identify the structures of the flower;

4. recognize the structures and events that take place in angiosperm reproduction;

5. describe the origin and function of fruit and seed;

6. identify the characteristics distinguishing angiosperms from gymnosperms.

INTRODUCTION The **angiosperms,** seed plants that produce flowers, are placed in the division Anthophyta. The word "angiosperm" literally means "vessel seed," referring to the seeds being borne in a fruit. There are more flowering plants in the world today than any other group of plants. Assuming that numbers indicate success, it must be said that flowering plants are the most successful plants to evolve.

Our lives and diets revolve around flowering plants. The most important characteristic that distinguishes the Anthophyta from other seed plants is the presence of flower parts that mature into a **fruit,** a container that protects the seeds, allowing them to be dispersed without coming into contact with the rigors of the external environment for some time. In many instances, the fruit also contributes to the dispersal of the seed. For example, some fruits stick to fur (or clothing) of animals, and are brushed off some distance from the plant that produced them. Others are eaten by animals. The undigested seeds may pass out of the digestive tract, falling into an environment often far removed from the seeds' source.

Fruits enrich our lives, and include such things as apples, oranges, tomatoes, beans, peas, corn, wheat, walnuts, pecans . . . the list goes on and on. Moreover, even when we are not eating fruits, we're eating flowering plant parts. Cauliflower, broccoli, potatoes, celery, and carrots all are parts of flowering plants.

The number of different kinds of flowers is so large it's difficult to pick a single example to be representative of the entire division. Nonetheless, there is enough similarity among flowers that, once you've learned the structure of one representative, you'll be able to recognize the parts of most.

Flower parts are believed to have originated as leaves modified during the course of evolution to increase the probability for fertilization. For instance, some flower parts are colorful, attracting animals that serve to transfer the sperm-producing pollen to the receptive female parts.

Figure 26-5 represents the life cycle of a typical flowering plant. Label Fig. 26-5 as you study the specimens in this exercise.

I. EXTERNAL STRUCTURE OF THE FLOWER
(Starr, pp. 254–255)

MATERIALS

Per student:

- flower for dissection (gladiolus or hybrid lily, for example)
- single-edged razor blade
- dissecting microscope

PROCEDURE Obtain a flower provided for dissection. The flower parts are arranged in whorls atop a swollen stem tip, the **receptacle.** The outermost whorl is frequently green (although not always) and is the **calyx.** Individual components of the calyx are called **sepals.** The calyx surrounds the rest of the flower in the bud stage. Identify these parts, labeling them on Fig. 26-5a.

Moving inward, locate the next whorl of the flower, the usually colorful **corolla** made up of **petals.** It is usually the petals that we appreciate for their color.

Labels: ovules

Figure 26-1 Drawing of cross sections and longitudinal sections of an ovary

Remember, however, that attractive flowers evolved because their color put them at a selective advantage for perpetuating their species.

Both the calyx and corolla are sterile; that is, they do not contain any of the sexual reproductive parts.

The next whorl of flower parts consists of the male, pollen-producing parts, the **stamens,** also called **microsporophylls** ("microspore-bearing leaves"; Fig. 26-5b). Examine a single stamen in greater detail. Each stamen consists of a stalklike **filament** and an **anther.** The anther consists of four **microsporangia,** also called **pollen sacs.**

Next locate the female portion of the flower, the **pistil** (Figs. 26-5a, 5c). A pistil consists of one or more **megasporophylls** ("megaspore-bearing leaves"), also called **carpels.** If the pistil consists of more than one megasporophyll, they are usually fused together, making it difficult to distinguish the individual components.

Identify the different parts of the pistil (Fig. 26-5c): at the top, the **stigma,** which serves as a receptive region on which pollen is deposited; a necklike **style;** and a more-or-less globose **ovary.** The illustration in Fig. 26-5c is oversimplified; very few flowers have only one ovule per ovary. Note that the only members of the plant kingdom to have ovaries are angiosperms.

With a sharp razor blade, make a section of the ovary. (Some students should cut the pistil longitudinally; others should cut the ovary crosswise. Then compare the different sections.) Examine the sections with a dissecting microscope, finding the numerous small **ovules** within the ovary. In Fig. 26-1, make and label two sketches: one of the cross section and the other of a longitudinal section of the ovary. After fertilization, the ovules will develop into *seeds,* and the ovary will enlarge and mature into the *fruit.* Notice

that the ovules are completely enclosed within the ovary.

As you may recall from Exercise 12, there are two groups of flowering plants, monocotyledons and dicotyledons. The number of flower parts indicates to which group a plant belongs. Generally, monocots have the flower parts in 3s or multiples of 3. Dicots have their parts in 4s or 5s or multiples thereof. Count the number of petals or sepals in the flower you have been examining. Are you studying a monocot or dicot?

II. MICROSPORANGIA AND THE MALE GAMETOPHYTE

(Starr, p. 255)

MATERIALS

Per student:

- prepared slide of young lily anther, cross section
- prepared slide of mature lily anther (pollen grains), x.s.
- *Impatiens* flowers, with mature pollen
- glass microscope slide
- coverslip
- compound microscope

Per student pair:

- 0.5% sucrose, in dropping bottle

PROCEDURE   With the low power objective of your compound microscope, examine a prepared slide of a cross section of an immature anther (Fig. 26-5d). Find

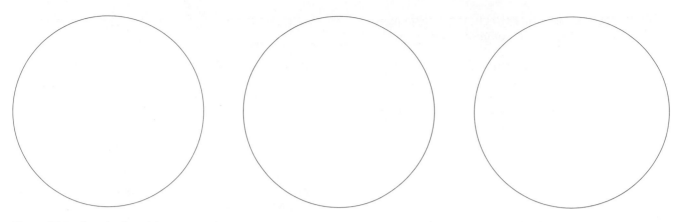

Figure 26-2 Germination of *Impatiens* pollen grain
(_____×)

sections of the four **microsporangia** (also called *pollen sacs*), which appear as four clusters of densely stained cells within the anther. Study the contents of a single microsporangium. Depending upon the stage of development, you will find either diploid *microspore mother cells* (Fig. 26-5e) or haploid *microspores* (Figs. 26-5f, 5g).

Obtain a prepared slide of a cross section of a mature anther. Observe it first with the low power objective, noting that the walls have split open to allow the pollen grains to be released, as illustrated in Fig. 26-5i. Pollen grains are immature *male gametophytes,* and very small ones at that. Switch to the high-dry objective to study an individual pollen grain more closely. The pollen grain consists of only two cells. Identify the large **tube cell** and a smaller, crescent-shaped **generative cell** that floats freely in the cytoplasm of the tube cell (Fig. 26-5h).

Note the ridged appearance of the wall layers of a pollen grain. Within the ridges and valleys of the wall, glycoproteins are present that are believed to play a role in recognition between the pollen grain and the stigma.

Transfer of pollen from the microsporangia to the stigma, called **pollination,** is effected by various means—wind, insects, and birds being the most common carriers of pollen. When a pollen grain lands on the stigma of a compatible flower, it germinates, producing a **pollen tube** that grows down the style (Fig. 26-5j). The generative cell flows into the pollen tube, where it divides to form two *sperm* (Fig. 26-5k). Because it bears two gametes, the pollen grain is now considered to be a *mature* male gametophyte.

Obtain an *Impatiens* flower and tap some pollen onto a clean microslide. Add a drop of 0.5% sucrose, cover with a coverslip, and observe with the medium power objective of your compound microscope. Look for the *pollen tube* as it grows from the pollen grain. (You may wish to set the slide aside for a bit and re-examine it 15 minutes later to see what's happened to the pollen tubes.)

In Fig. 26-2, draw a sequence showing the germination of an *Impatiens* pollen grain.

III. MEGASPORANGIA AND THE FEMALE GAMETOPHYTE (Starr, p. 255)

MATERIALS

Per student:

- prepared slide of lily ovary, x.s., megaspore mother cell
- compound microscope

Per lab room:

- demonstration slide of lily ovary, x.s., seven-celled, eight-nucleate gametophyte
- demonstration slide of lily ovary, x.s., double fertilization

PROCEDURE With the medium power objective of your compound microscope, examine a prepared slide of a cross section of an ovary (Fig. 26-5l). Find the several **ovules** that have been sectioned. One ovule probably will be sectioned in a plane so that the very large, diploid **megaspore mother cell** is obvious (Fig. 26-5m). The megaspore mother cell is contained within the **megasporangium,** the outer cell layers of which form two flaps of tissue called **integuments.** After fertilization, the integuments develop into the **seed coat.** The ovule is attached to the ovary wall in a region known as the *placenta.* Find the opening between the flaps of the inner integument, the **micropyle.** (The micropyle is obvious in Fig. 26-5n.) Remember, the micropyle is an opening in the globose ovule. After pollination, the pollen grain germinates on the surface of the stigma and the pollen tube grows down the style (through the micropyle) and penetrates the megasporangium (Fig. 26-5p).

Considerable variation exists in the next sequence of events, but the most common pattern is that described.

The diploid megaspore mother cell undergoes meiosis, producing four haploid megaspores. Three of the four megaspores disintegrate (Fig. 26-5n). The functional megaspore enlarges as its nucleus undergoes three mitotic divisions (Fig. 26-5o), forming the **female gametophyte** (also known as the *embryo sac*). Thus, initially the female gametophyte is one large cell containing eight nuclei.

Three of the nuclei migrate to the micropylar end of the female gametophyte, and three to the opposite end. Cell walls now form around each of these six nuclei. The remaining two nuclei, called **polar nuclei,** reside in the cytoplasm of the large **central cell** (Fig. 26-5p). Hence, the central cell is binucleate. One of the cells at the micropylar end is the **egg cell.** Now the female gametophyte is a seven-celled, eight-nucleate structure and is ready to be fertilized.

Study the demonstration slide of the seven-celled, eight-nucleate female gametophyte represented in Fig. 26-5p. Identify and label the following: placenta, integuments, micropyle, egg cell, central cell, and polar nuclei.

As the pollen tube penetrates the female gametophyte, it discharges the sperm; one of the sperm nuclei fuses with the haploid egg nucleus, forming the *zygote*. Figure 26-5q represents the female gametophyte after fertilization has occurred.

The zygote is a

_____ (haploid, diploid) cell.

The other sperm nucleus enters the central cell and fuses with the two polar nuclei, forming the **primary endosperm nucleus** (Fig. 26-5q).

Thus, the primary endosperm nucleus is

_____ (haploid, diploid, triploid).

The cell containing the primary endosperm nucleus is called the **endosperm mother cell.** Traditionally, the process of fusion of one sperm nucleus with the egg nucleus and the fusion of the other sperm nucleus and the two polar nuclei has been called **double fertilization.** Observe the demonstration slide illustrating double fertilization (Fig. 26-5q), identifying the zygote, primary endosperm nucleus, and central cell of the female gametophyte.

Numerous mitotic and cytoplasmic divisions of the endosperm mother cell form the **endosperm,** a tissue used for nutrition of the embryo sporophyte as it develops within the seed.

IV. EMBRYOGENY

The zygote undergoes mitosis and cytokinesis to produce a two-celled **embryo** (embryo sporophyte). Numerous subsequent divisions produce an increasingly large and complex embryo.

MATERIALS

Per lab room:

- demonstration slides of *Capsella* embryogeny: globular embryo, emerging cotyledons, torpedo-shaped embryo, mature embryo

PROCEDURE Observe the series of four demonstration slides illustrating the development of the embryo in the female gametophyte of *Capsella*. Figure 26-3 illustrates the stages. The first slide shows the so-called "globular stage" (Fig. 26-3a), in which the *embryo*, a spherical mass of cells, is attached to the wall of the female gametophyte (embryo sac) by a chain of cells (the suspensor). Note the endosperm within the female gametophyte.

The second slide illustrates the "heart-shaped stage" (Fig. 26-3b). Now you can differentiate the emerging **cotyledons** (seed leaves), the "lobes of the heart." In many plants the cotyledons absorb nutrients from the endosperm, serving as a food reserve to be used during seed germination.

Further development of the embryo has occurred in the third slide, the "torpedo stage" (Fig. 26-3c). Notice that the entire embryo has elongated. Find the *cotyledons*. Between the cotyledons, locate the **epicotyl,** which is the *apical meristem of the shoot*. Beneath the epicotyl and cotyledons find the **hypocotyl-root axis.** At the tip of the hypocotyl-root axis, locate the *apical meristem of the root* and the **root cap** covering it.

The final slide (Fig. 26-3d) shows a mature embryo, neatly packaged inside the *seed coat*. Identify the seed coat and other regions previously identified in the torpedo stage.

How many cotyledons were there in the slides you examined?

_____

Thus, *Capsella* is a

_____ (monocot or dicot).

V. FRUIT AND SEED   (Starr, p. 255)

Simply stated, a **fruit** is a matured ovary, while a **seed** is a matured ovule. Fertilization not only causes the integuments of the ovule to develop into a seed coat, it also causes the ovary wall to expand into the fruit.

MATERIALS

Per student:

- bean fruits
- soaked bean seeds
- iodine solution (I_2KI), in dropping bottle

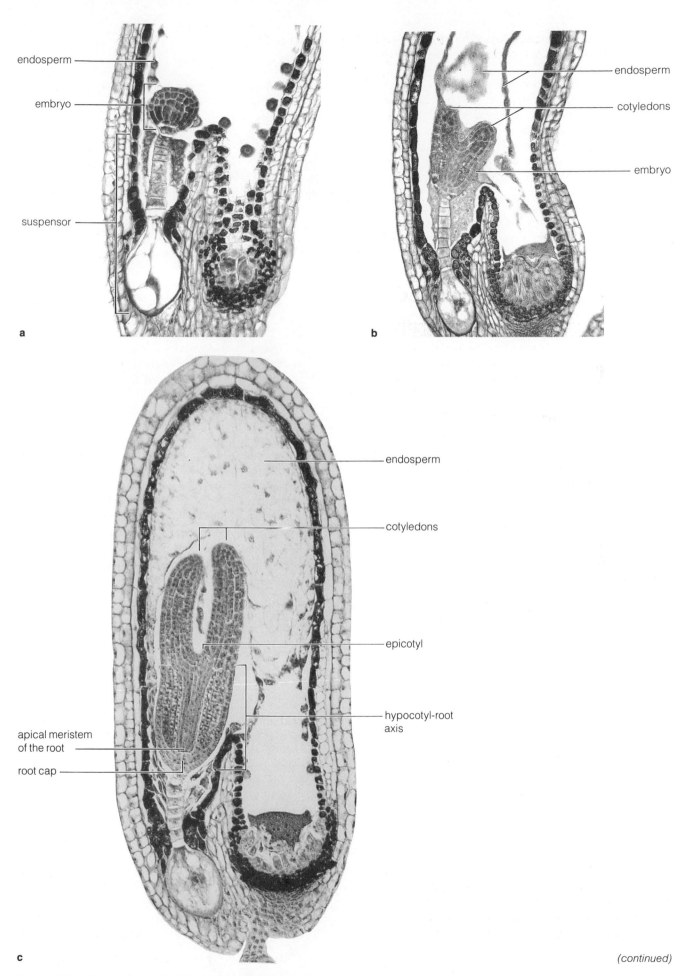

endosperm

embryo

suspensor

a

endosperm

cotyledons

embryo

b

endosperm

cotyledons

epicotyl

hypocotyl-root axis

apical meristem of the root

root cap

c

(continued)

Figure 26-3 Embryogeny in *Capsella* (shepherd's purse). (Photos courtesy Ripon Microslides, Inc.)

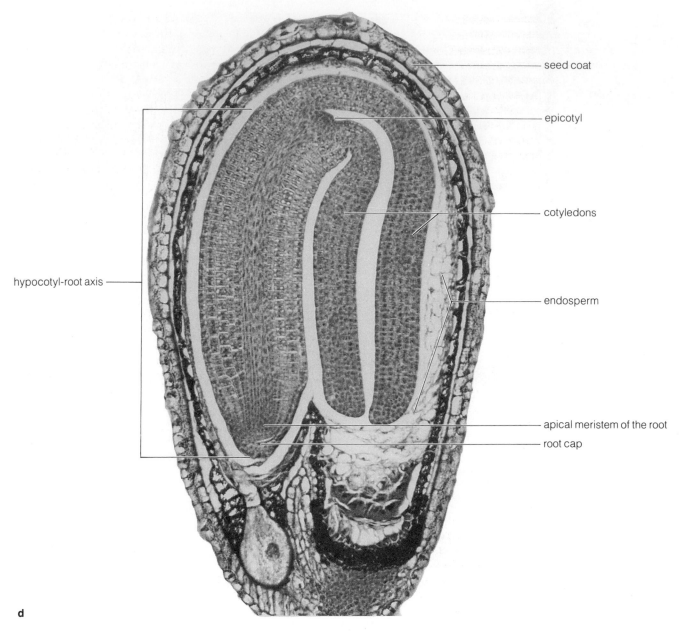

seed coat

epicotyl

cotyledons

hypocotyl-root axis

endosperm

apical meristem of the root

root cap

d

Figure 26-3 *continued*

PROCEDURE Obtain a bean pod and carefully split it open along one seam.

The pod is a matured ovary and thus is a

_____ .

Find the *sepals* at one end of the pod and the shriveled *style* at the opposite end. The "beans" inside are

_____ .

Note the point of attachment of the bean to the pod. What is the point of attachment of the ovule to the ovary wall called?

_____

In Fig. 26-4, draw the split-open bean pod, labeling it with the correct scientific terms.

Figure 26-5r represents a section of a "typical" fruit. Label the fruit wall, seed coat, and the cotyledons and epicotyl of the embryo.

Study more closely one of the beans from within the pod or a bean that has been soaked overnight to soften it. Find the scar left where the seed was attached to the fruit wall. Near the scar, look for the tiny opening left in the seed coat. What is this tiny opening? (*Hint:* The pollen tube grew through it.)

_____

Remove the seed coat to expose the two large *cotyledons*. Split the cotyledons open to find the *epicotyl* and *hypocotyl-root axis*. During maturation of the bean embryo, the cotyledons absorb the endosperm. Thus, bean cotyledons are very fleshy, storing carbohy-

Labels: fruit, seeds, placenta

Figure 26-4   Drawing of an open bean pod

drates that will be used during seed germination. Add a drop of iodine solution (I_2KI) to the cotyledon. What substance is located in the cotyledon?

_____

(*Hint:* Return to Exercise 6 if you've forgotten what is stained by I_2KI.)

VI. SEEDLING

When environmental conditions are favorable for growth (adequate moisture and proper temperatures), the seed **germinates;** that is, growth of the seedling (young sporophyte) begins.

MATERIALS

Per student:

▪ germinating bean seeds

Per table:

▪ a dishpan of water

PROCEDURE   Obtain a germinating bean seed from the culture provided (Fig. 26-5s). Wash the root system in the dishpan provided, *NOT* in the sink.

Identify the **primary root** with the smaller **secondary roots** attached to it. Emerging in the other direction will be the *hypocotyl,* the *cotyledons,* the *epicotyl* (above the cotyledons), and the first *true leaves* above the epicotyl. Label these parts on Fig. 26-5s. Notice that when some seeds (like the pea) germinate, the cotyledons remain *below* ground. Others, like the bean, emerge from the ground by virtue of elongation of the hypocotyl-root axis.

The amount of time between seed germination

and flowering depends largely upon the particular plant. Some plants produce flowers and seeds during their first growing season, completing their life cycle in that growing season. These plants are called **annuals.** Marigolds are an example of an annual. Others, known as **biennials,** grow vegetatively during the first growing season and do not produce flowers and seeds until the second growing season (carrots, for example). Both annuals and biennials die after seeds are produced.

Perennials are plants that live several to many years. The time between seed germination and flowering (seed production) varies, some requiring many years. Moreover, perennials do not usually die after producing seed, but flower and produce seeds many times during their lifetime.

PRE-LAB QUESTIONS

____ **1.** Plants that produce flowers are (a) members of the Anthophyta, (b) angiosperms, (c) seed producers, (d) all of the above.

____ **2.** Collectively, all of the petals of a flower are called the (a) corolla, (b) stamens, (c) receptacles, (d) calyx.

____ **3.** Which of the following refer to the microsporophyll, the male portion of a flower? (a) ovary, stamens, pistil; (b) stigma, style, ovary; (c) anther, stamen, filament; (d) megasporangium, microsporangium, ovule.

____ **4.** A carpel is the (a) same as a megasporophyll, (b) structure producing pollen grains, (c) component making up the anther, (d) synonym for microsporophyll.

____ **5.** The portion of the flower containing pollen grains is the (a) pollen sac, (b) microsporangium, (c) anther, (d) all of the above.

____ **6.** Which of the following is in the correct developmental sequence? (a) microspore mother cell, meiosis, megaspore, female gametophyte; (b) microspore mother cell, meiosis, microspore, pollen grain; (c) megaspore, mitosis, female gametophyte, meiosis, endosperm mother cell; (d) all of the above.

____ **7.** Where would germination of a pollen grain occur in a flowering plant? (a) in the anther, (b) in the micropyle, (c) on the surface of the corolla, (d) on the stigma.

____ **8.** Double fertilization refers to (a) fusion of two sperm nuclei and two egg cells, (b) fusion of one sperm nucleus with one polar nucleus and fusion of another with the egg cell nucleus, (c) maturation of the ovary into a fruit, (d) none of the above.

____ **9.** Ovules mature into _____, while ovaries mature into _____. (a) seeds, fruits; (b) stamens, seeds; (c) seeds, carpels; (d) fruits, seeds.

____ **10.** A bean pod is (a) a seed container, (b) a fruit, (c) a part of the stamen, (d) a and b above.

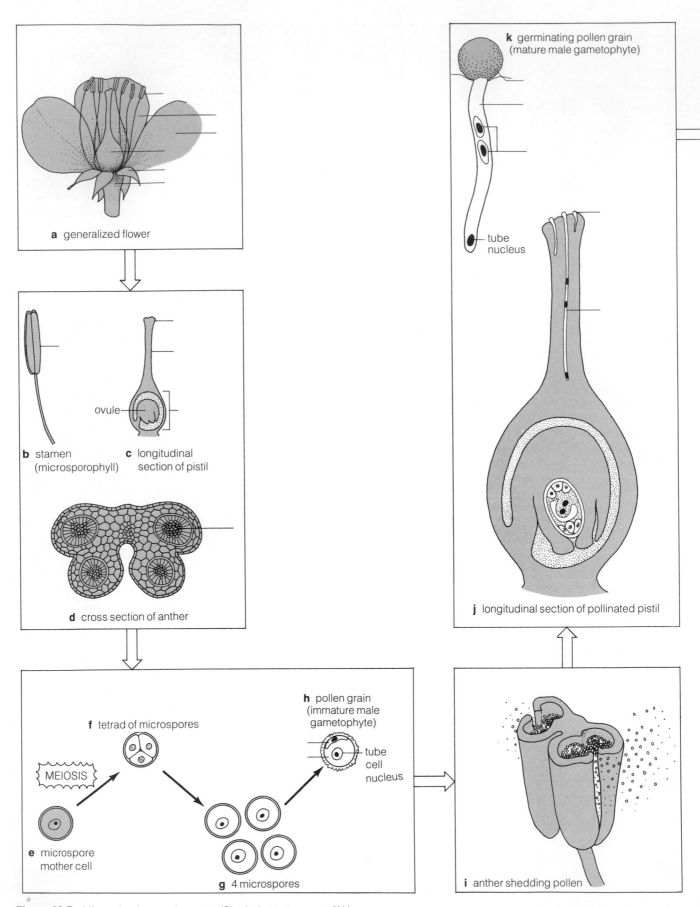

k germinating pollen grain
(mature male gametophyte)

tube
nucleus

a generalized flower

ovule

b stamen
(microsporophyll)

c longitudinal
section of pistil

d cross section of anther

j longitudinal section of pollinated pistil

f tetrad of microspores

h pollen grain
(immature male
gametophyte)

MEIOSIS

tube
cell
nucleus

e microspore
mother cell

g 4 microspores

i anther shedding pollen

Figure 26-5   Life cycle of an angiosperm. (Shaded structures are 2N.)

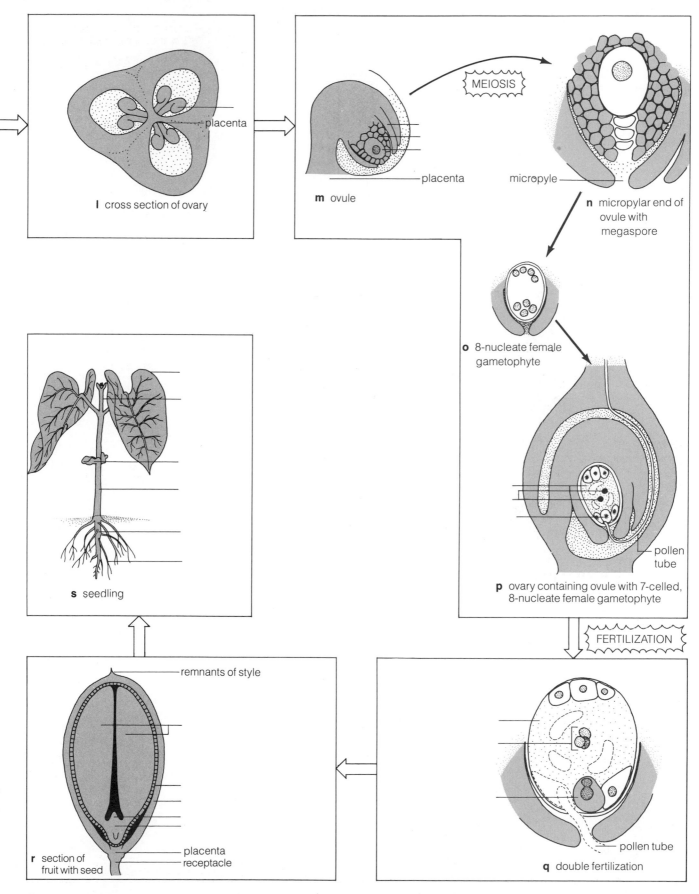

l cross section of ovary

placenta

m ovule

placenta

MEIOSIS

micropyle

n micropylar end of ovule with megaspore

o 8-nucleate female gametophyte

pollen tube

p ovary containing ovule with 7-celled, 8-nucleate female gametophyte

FERTILIZATION

pollen tube

q double fertilization

remnants of style

placenta
receptacle

r section of fruit with seed

s seedling

Labels: receptacle, sepal, petal, filament, anther, microsporangium (pollen sac), stigma, style, ovary, tube cell, generative cell, pollen grain, pollen tube, sperm, ovule, megaspore mother cell, megasporangium, integument,

placenta, female gametophyte (embryo sac), polar nuclei, central cell, egg cell, fruit wall, seed coat, cotyledons, epicotyl, primary root, secondary root, hypocotyl-root axis, true leaf, hypocotyl

EXERCISE 26
SEED PLANTS II: ANGIOSPERMS

POST-LAB QUESTIONS

1. There are two major groups of seed plants, the gymnosperms and angiosperms. What characteristics differentiate these two groups?

2. Some biologists contend that the term "double fertilization" is a misnomer and that the process should be called "fertilization" and "triple fusion." Why do they argue that the fusion of the one sperm nucleus and the two polar nuclei is *not* fertilization?

3. After doing this lab, suppose you and your roommate go to the grocery store. Your roommate says that you need vegetables and asks you to pick up tomatoes. To your roommate's surprise, you say a tomato is not a vegetable, but a fruit. Explain.

4. Compare the source of nutrition for the embryonic sporophyte in an angiosperm with that of a gymnosperm.

5. Distinguish between pollination and fertilization.

6. Why are angiosperms the most successful plants to have evolved?

7. Name a commonly eaten:

   a. leaf

   b. root

   c. stem

   d. flower

   e. fruit

   f. seed

8. Sepals are to calyx as

   _____

   are to corolla.

9. Plants of a particular bamboo species, the preferred food of giant pandas, grow wild in the mountains of southern China. Throughout the country stands of this bamboo, which had been growing vegetatively for nearly 200 years, recently produced flowers and then died, much to the dismay of botanists (and presumably giant pandas). Is this bamboo an annual, biennial, or perennial?

   _____

   Why?

EXERCISE 27

SPONGES, CNIDARIANS, FLATWORMS, ROUNDWORMS, AND ROTIFERS

OBJECTIVES After completing this exercise you will be able to:

1. define spongocoel, osculum, spicules, collar cell, monoecious, budding, spongin, polyp, tentacles, medusa, nematocyst, dioecious, mesoglea, larva, acoelomate, cephalization, bilateral symmetry, phototaxis, regeneration, host, intermediate host, cuticle, scolex, proglottid, bladder worm, pseudocoelomate, cloaca, wheel organ, parthenogenesis;

2. describe the natural history of members of the phyla Porifera, Cnidaria, Platyhelminthes, Nematoda, and Rotifera;

3. identify representatives of the classes Calcarea and Demospongiae of the phylum Porifera, of the classes Hydrozoa, Scyphozoa, and Anthozoa of the phylum Cnidaria, of the classes Turbellaria, Trematoda, and Cestoda of the phylum Platyhelminthes, and of the phyla Nematoda and Rotifera;

4. identify structures (and indicate associated functions) of the representatives of these classes;

5. outline the life cycles of the Chinese liver fluke and the pork tapeworm;

6. outline the life cycle of *Ascaris;*

7. compare and contrast the anatomy and life cycles of free-living and parasitic flatworms and roundworms.

INTRODUCTION Early in our lives, as occurred early in the history of the systematic study of organisms, we learned to identify animals as organisms that behave in certain ways, in particular as organisms that exhibit considerable movement. In his *Scala Naturae,* Aristotle distinguished plants from animals on the basis of the extent of movement and response to stimuli. Until relatively recently, every living organism was assigned to either the plant or animal kingdom, partly on the basis of whether animallike motility was exhibited. Even today, when organisms are divided among five kingdoms in Whittaker's classification scheme (see Exercise 2), creatures outside the animal kingdom (for example, *Euglena;*

Fig. 22-6) are sometimes described as animallike because of the behavior or movement they exhibit.

Evidence accumulated from centuries of study supports the notion that animals evolved from a group of protistans distinct from those that gave rise to the plants and fungi. Current classification schemes rely less heavily on movement as a diagnostic feature of animals because we have learned that some organisms that exhibit considerable motility are unrelated to animals. Animals are now recognized as *multicellular, heterotrophic organisms composed of distinct tissues, organs,* and often *systems. Cell walls are absent,* and *photosynthesis does not occur.* During individual development of an animal, *blastula and larval stages are usually seen.*

I. PHYLUM PORIFERA (THE SPONGES)   (Starr, pp. 261–263)

Aristotle said of sponges, "In the sea there are things which it is hard to label as either animal or vegetable." This was a reference to their relative lack of motility and of sensitivity to stimuli. In fact, sponges are the least complex animals other than mesozoans—rare, minute, wormlike parasitic animals with little tissue differentiation.

Apparently, sponges evolved directly from protistan stock and diverged early from the main line of animal evolution. Most are marine. They are sessile as adults, but many disperse as free-swimming larvae before becoming attached to the substrate. Although sponges exhibit cellular specialization, they are atypical animals in their lack of definite tissues.

Three body forms, or canal systems, exist in sponges (Fig. 27-1).

One form is vase-shaped, in which water enters the animal through pores, flows directly to a large, central, internal cavity, the **spongocoel,** and through a superior exit, the **osculum.** Another form is similar, except that the body wall everts to form *incurrent canals* between, and *radial canals* within, the pockets.

317

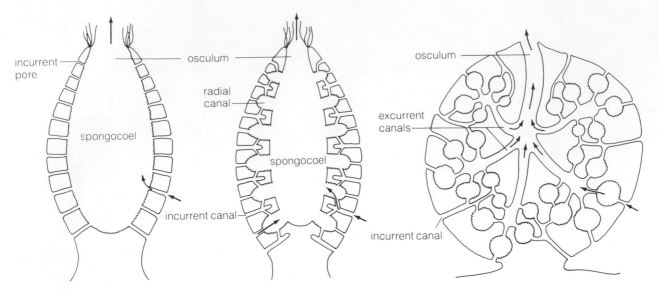

Figure 27-1 Three body forms of sponges. (After Villee, Walker, and Barnes, 1973.)

A third system consists of pores opening into numerous chambers that empty into *excurrent canals* leading to the osculum.

MATERIALS

Per student:

- dissection microscope
- compound microscope
- glass microscope slide
- coverslip

Per student pair:

- preserved (dried) specimen of *Scypha*
- preserved (dried) specimen of *Leucosolenia*
- prepared slide of a longitudinal section of *Scypha*
- prepared slide of a whole mount of *Leucosolenia*

Per student group (4):

- 100-mL beaker of heat-proof glass
- bottle of 5% sodium hydroxide (NaOH) or potassium hydroxide (KOH)
- hot plate or laboratory burner, wire gauze square with a ceramic fiber center, tripod, and flint lighter
- squeeze bottle of distilled water

Per lab room:

- hand lens
- demonstration collection of commercial sponges
- several prepared slides of sponge gemmules
- squeeze bottle of 50% vinegar and water

PROCEDURE

A. Class Calcarea

Members of this class contain **spicules** (skeletal elements) composed of calcium carbonate ($CaCO_3$). All three body plans are represented. You will study the genera *Scypha* and *Leucosolenia*.

1. With the dissection scope, examine the general morphology of an intact sponge, *Scypha*. Note the osculum, an excurrent opening through which water passes from the body wall and central cavity, the spongocoel. Note the pores in the body wall. Observe the long spicules surrounding the osculum and the shorter spicules protruding from the surface of the sponge.

CAUTION *Hot 5% hydroxide solutions are corrosive. Wear safety goggles. Wash exposed skin for 15 minutes in running tap water, then flood the area with a 50% vinegar and water solution.*

---

2. Place a portion of a dried specimen of *Scypha* in a 100-mL heat-proof beaker. Add 5% sodium hydroxide or potassium hydroxide solution to the beaker and bring this to a boil. Allow the beaker to cool.
Your instructor will provide directions for disposing of solutions. Pour off the liquid. Wash the residue by adding water, allowing the remains of the sponge to settle, and then pour off the water. Make

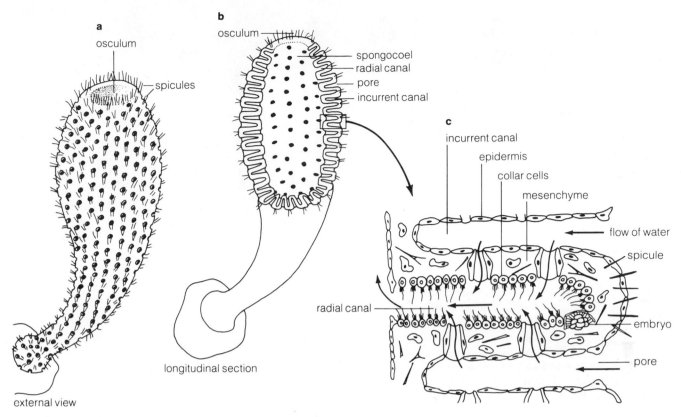

Figure 27-2 External view (**a**) and longitudinal section (**b**) of *Scypha* (magnified in **c** to show detail). (After Lytle and Wodsedalek, 1984.)

a wet mount of the residue and describe what you observe.

_____

_____

_____

_____

3. Examine longitudinal sections of *Scypha* (Fig. 27-2) with a compound microscope.

4. Trace the path of water flowing into and through the sponge. Water enters the incurrent canal through a pore. From the incurrent canal it flows through many smaller pores into the radial canal. From here it flows past the collar cells, through the radial canal into the large central spongocoel. Once inside the spongocoel, the water flows out of the animal through the large, superior opening, the osculum.

5. The **collar cells** will appear as numerous small, dark bodies lining the radial canal. Each bears a flagellum that projects into the radial canal, but it is unlikely that you will be able to see the flagella, even with the high power objective. Examine the structure

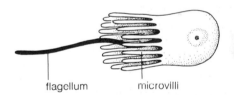

Figure 27-3 Collar cell. (After Starr and Taggart, 1987.)

of a collar cell in Fig. 27-3. What action results from the beating of the flagella of the collar cells?

_____

_____

_____

_____

Embedded in the gelatinous material around the collar cells are amoeboid cells and spicules. Both the collar cells and amoeboid cells engulf and digest food fragments carried into the sponge by the flow of water.

6. Sexual reproduction occurs in sponges. Many sponges are **monoecious** (an individual has both male and female reproductive organs) or *hermaphroditic*, producing both eggs and sperm from certain amoebocytes. Although it is unlikely that you will observe gametes, you may be able to detect embryos surrounded by a shell and suspended in the radial canals of *Scypha*. These appear as relatively large, dark structures (Fig. 27-2) that soon will break loose and exit the sponge through the osculum.

7. Asexual reproduction occurs by **budding.** Look for buds on a specimen of the colonial sponge, *Leucosolenia*. You also may find buds on the prepared slides of a whole mount of *Leucosolenia*.

B. Class Demospongiae

Members of this class of sponges are supported by **spongin,** a flexible substance similar chemically to human hair, by siliceous (silica containing) spicules, or both. Silica is silicon dioxide, a major component of sand and of many rocks like flint and quartz. Sponges in this class are able to hold and pass large volumes of water in and through their bodies, making them commercially valuable as bath sponges. This class contains the majority of sponge species, and most are marine. Commercial sponges have been largely replaced by synthetic sponges in today's marketplace.

1. Examine demonstration specimens of commercial sponges. Freshwater sponges in this class reproduce asexually by forming *gemmules,* resistant reproductive organs. Under favorable conditions, gemmules develop into new individuals.

2. Find a gemmule on the prepared slide of sponge gemmules.

II. PHYLUM CNIDARIA (THE COELENTERATES)   (Starr, pp. 263, 264)

This phylum contains some of the most beautiful organisms in the seas, many of which are brightly colored and plantlike or flowerlike, with tentacles that move in response to the currents. The colonial hydroids are branching organisms, plantlike in appearance and in their lack of motility. Sea anemones, often brilliantly colored, are members of bottom communities. The corals secrete calcareous (chalky) exoskeletons of intricate detail that not only protect the coral organisms but also provide places to live for other marine organisms. Jellyfish are graceful, delicate animals capable of slow movement through the water.

Cnidarians are derived from the protistan ancestors that gave rise to the main line of evolution from which other animals sprang. Cnidarians diverged early from this evolutionary lineage. These animals are found mostly in shallow marine environments, with the notable exceptions being the freshwater

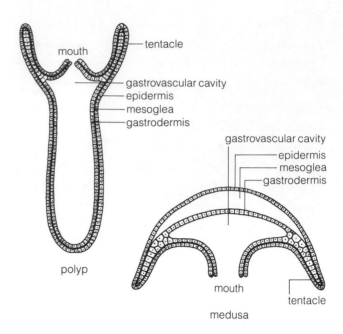

Figure 27-4 Two body forms of Cnidarians. (After Hickman, Hickman, and Hickman, 1978.)

hydras. There are definite tissues. Cnidarians are mostly **radially symmetrical;** that is, any cut from the oral (mouth) to the other end through the center of the animal will yield two like halves of equal size.

Two body forms exist in Cnidarians (Fig. 27-4). The **polyp** is cylindrical in shape, with the oral end and **tentacles** directed upward and the other end attached to the substrate. Hydras, sea anemones, and corals assume this form as adults. The free-swimming **medusa** ("jellyfish") is bell- or umbrella-shaped, with the oral end and tentacles directed downward. Either or both body forms may be present in the Cnidarian life cycle.

The beauty and relative immobility of Cnidarians are deceptive lures to would-be prey; Cnidarians are efficient predators. All are carnivorous and possess tentacles that capture unwary invertebrate and vertebrate prey. Food is captured with the aid of stinging elements called **nematocysts.** The nematocysts are discharged by a combination of mechanical and chemical stimuli. The prey is either pierced by the nematocyst or entangled by its filament and pulled toward the mouth by the tentacles. In some species, nematocysts discharge a toxin that paralyzes the prey.

MATERIALS

Per student:

- dissection microscope
- compound microscope
- microscope slides
- coverslip

Per student pair:

- prepared slide of a whole mount of *Hydra* (showing testes, ovaries, buds, and embryos)
- prepared slide of a transverse section of *Hydra*

Per student group (4):

- small finger bowl
- scalpel or razor blade
- preserved specimen of *Gonionemus*
- preserved specimen of *Aurelia*
- preserved specimen of *Metridium*

Per lab room:

- preserved specimen of *Physalia*
- prepared slide of a scyphistoma
- demonstration collection of corals

Per lab section:

- culture of live *Hydra*
- dropper bottle of glutathione
- culture of live copepods or cladocerans
- dropper bottle of vinegar

PROCEDURE

A. Class Hydrozoa

Members of this class include the hydras (*Hydra*) and the colonial hydroids. This is mostly a marine group, except for the hydras. The polyp, medusa, or both may comprise the life cycle of a hydrozoan. Marine polyps, like *Obelia* (Fig. 27-5), tend to be colonial.

1. *Hydra*

a. Place a live specimen of *Hydra* in a finger bowl. Examine the live *Hydra* with a dissection scope and a prepared slide with the compound microscope, noting the polyp form and structural details of the organism (Fig. 27-6).

b. Observe the tentacles at the oral end. Find the elevation of the body at the base of the tentacles, in the center of which lies a mouth. Observe the swellings on the tentacles; each is a stinging cell that contains a nematocyst. The clear area in the center of the body represents the *gastrovascular cavity*. Look for one to several *testes* or an *ovary* on the body, both of which may occur on the same individual; the testes are conical and are located closer to the mouth than the broad, flattened ovary. You may be able to see a zygote or embryo marginally attached to or detached from the body of the *Hydra*. Is the *Hydra* monoecious or **dioecious** (male and female reproductive organs in separate individuals)?

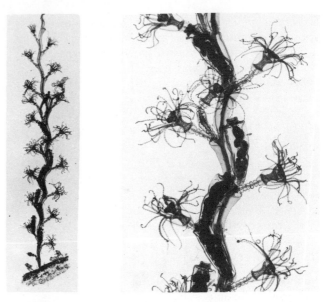

Figure 27-5 The colonial hydrozoan, *Obelia*. (Photos courtesy Triarch, Inc.)

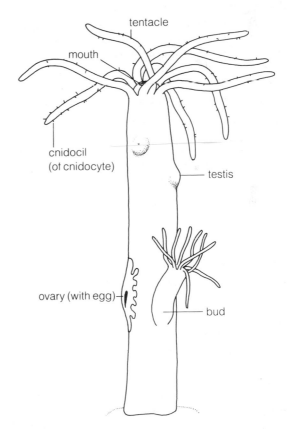

Figure 27-6 External view of *Hydra*. (After Hickman, Hickman, and Hickman, 1978.)

c. *Hydra* also reproduces asexually by budding. Look for buds among your slide-mounted and live specimens.

d. Examine a transverse section of *Hydra* (Fig. 27-7) and locate the *epitheliomuscular cells* and *interstitial cells* comprising the epidermis. An occasional stinging cell

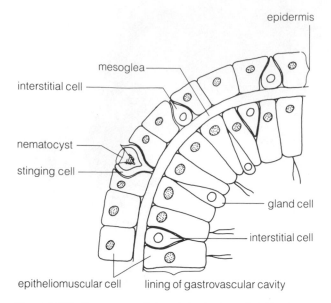

epidermis

mesoglea

interstitial cell

nematocyst

stinging cell

gland cell

interstitial cell

epitheliomuscular cell — lining of gastrovascular cavity

Figure 27-7 Transverse section of the body wall of *Hydra*

may be seen in this layer. The lining of the gastrovascular cavity is also comprised mostly of epitheliomuscular cells and interstitial cells. Between the two is a thin gelatinous membrane called the **mesoglea**. In *Hydra*, the mesoglea is without cells. As their name suggests, epitheliomuscular cells are contractile cells. How might these cells function to help maintain the body shape of the polyp?

_____

_____

_____

What additional function could you suggest for the epitheliomuscular cells of the lining of the gastrovascular cavity?

_____

_____

_____

Gametes arise from gonads that form accumulations of interstitial cells.

e. Place live *Hydra* in a small finger bowl, add a drop of glutathione, and add a number of live copepodan or cladoceran crustaceans. The glutathione slows down the movements of organisms. With the dissection microscope, look for predatory behavior by the *Hydra*. Note the action of the tentacles and the nematocysts. If you do not observe the discharge of nematocysts, place a drop of vinegar in the finger bowl and observe carefully. Describe what you observe.

_____

_____

_____

f. Are any of your *Hydra* green? If so, use a scalpel or razor blade to remove a tentacle. Make a wet mount of the tentacle, being certain to crush it with the coverslip. What do you observe? Can you explain the green color of your *Hydra*?

_____

_____

_____

2. Hydrozoan medusa (jellyfish) and *Physalia*

_____

_____

CAUTION *Preserved specimens are kept in a formalin-based or other preservative solution. Wash any part of your body exposed to this solution with copious amounts of water. If preservative solution is splashed into your eyes, wash them with the safety eyewash bottle for 15 minutes.*

_____

a. Examine the preserved specimen of the hydrozoan medusa or jellyfish, *Gonionemus* (Fig. 27-8). Note the tentacles (which contain nematocysts and adhesive pads) around the margin of the umbrellalike *bell*. Extending downward from the center of the umbrella is a saclike structure containing the gastrovascular cavity, with the mouth at its tip surrounded by four *oral lobes*. Extending inward from the lower margin of the bell is a thin flap, the muscular *velum*. Four *radial canals* extend from the gastrovascular cavity to the *ring canal* at the margin of the bell. Between the bases of the tentacles are the *statocysts*, balancing organs. Gonads, either ovaries or testes, are attached to the radial canals. The medusa moves by "jet propulsion." Contractions of the body bring about a pulsating motion that alternately fills and empties the cavity of the bell.

b. Examine a preserved specimen of *Physalia*, the Portuguese man-of-war. This organism is really a colonial hydrozoan that is composed of several types of individuals. It consists of a gas-filled bladder from which a large number of polyps are suspended. Some of the polyps capture food, others process the food, and still others are reproductive in function. The sting of the Portuguese man-of-war is very painful, and a neurotoxin associated with the nematocysts of the tentacles can be fatal to humans.

B. Class Scyphozoa

This marine class contains the "true" jellyfish. The medusa is the predominant stage of the life cycle and is larger than hydrozoan medusae. The nematocysts in the tentacles and oral arms of the jellyfish can produce painful stings. Like other members of this

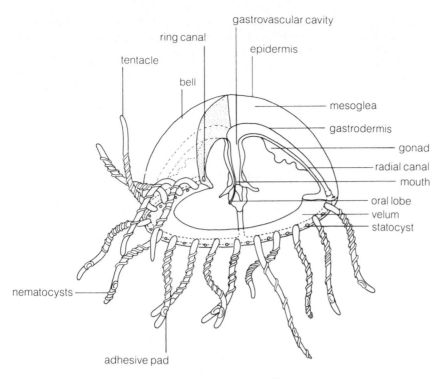

Figure 27-8 The hydrozoan medusa, *Gonionemus*. (After Boolootian and Stiles, *College Zoology*. Copyright © 1981 Macmillan Publishing Company. Reprinted by permission.)

phylum, scyphozoans are carnivorous and feed on a variety of animals, including fish. Some forms are bioluminescent, producing flashes of light as a result of chemical reactions. Bioluminescence may serve to lure prey toward the jellyfish or warn potential predators.

1. *Aurelia*

a. Examine a preserved specimen of the scyphozoan jellyfish, *Aurelia* (Fig. 27-9). The structure of this medusa is very similar to that of the hydrozoan, *Gonionemus*. *Sense organs*—consisting of cells sensitive to light, touch, chemicals, and balance—occur periodically between short tentacles around the muscular margin of the bell. Note the absence of a velum. Four *oral arms*, used in capturing prey, arise from the square mouth. Four horseshoe-shaped, brown gonads surround the mouth. A complex system of branching radial canals radiates from the center of the animal to the ring canal in the margin of the bell.

b. The polyp of *Aurelia* is a larval stage of the life cycle called the *scyphistoma*. A **larva** (plura *larvae*) is a sexually immature, free-living form of an animal that grows and transforms into an adult. Examine the demonstration slide of this stage. This is an active, feeding stage, equipped with tentacles and a mouth.

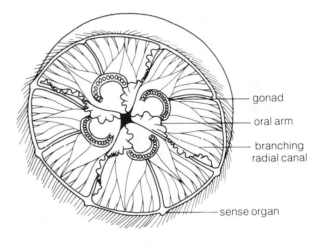

Figure 27-9 The scyphozoan jellyfish, *Aurelia*

C. Class Anthozoa

This class contains the sea anemones and corals. This is a marine group in which all members are polyps. The largest organisms in this class are the anemones. They attach to hard substrates or burrow in the sand or mud. They are heavy-bodied polyps that feed on small fish and invertebrates. Some species form symbiotic relationships with certain small fishes that live

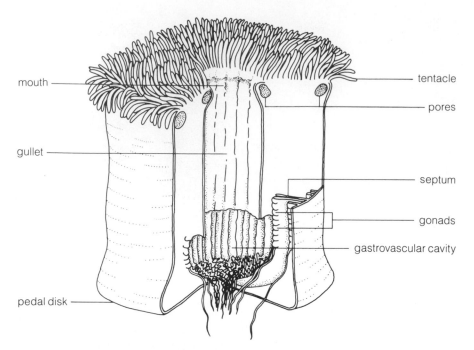

mouth

gullet

pedal disk

tentacle

pores

septum

gonads

gastrovascular cavity

Figure 27-10 Cutaway diagram of the sea anemone, *Metridium*. (After P. Abramoff and R. G. Thomson, *Laboratory Outlines in Biology-III*. Copyright © 1962, 1963 Peter Abramoff and Robert G. Thomson. Copyright © 1964, 1966, 1972, 1982 W. H. Freeman and Company.)

among the tentacles of the anemones. The fish receive protection from the tentacles, which for some reason do not discharge their nematocysts into these fish. In return, when the fish dart from and back to the protective cover of the tentacles to capture prey, the anemones benefit by ingesting food particles not eaten by the fish.

The smaller polyps of corals are similar to those of the anemones. The epidermis of the coral polyp secretes an exoskeleton of calcium carbonate. After a polyp dies, another polyp uses the remaining skeleton of the dead polyp as a foundation and secretes its own skeleton. In this fashion, coral polyps have built up various types of reefs, atolls, and islands.

1. Examine the demonstration exoskeletons of a variety of corals.

2. *Metridium*

a. Examine a preserved specimen of *Metridium*, the common sea anemone (Fig. 27-10). Note its stout, stumplike body, with the mouth and surrounding tentacles at the oral end and the *pedal disk* at the other end. For what is this pedal disk an adaptation?

_____

b. With your scalpel or razor blade, slice through the radius of the body wall from the oral end to the base as though you were cutting a piece of cake. The mouth is continuous with the *gullet*, which leads to the gastrovascular cavity, where digestion occurs. The gullet communicates with the body wall by a series of walls,

or *septa*, that serve to increase the surface area for absorption of nutrients. You should be able to see circular pores in the oral ends of the septa that permit water to pass between adjacent chambers formed by the septa.

Anemones are dioecious. The gonads are a series of spherical structures along the inner margins of the septa. Eggs and sperm are cast out of the body through the mouth, and fertilization is external. The fertilized egg develops into a motile, ciliated larva that settles on the substrate to develop into an adult polyp.

III. PHYLUM PLATYHELMINTHES (THE FLATWORMS) (Starr, pp. 265–268)

All members of this phylum have soft, flattened, wormlike bodies. The flatworms are **acoelomates**—they lack a body cavity. They have three distinct tissue layers: ectoderm, mesoderm, and endoderm. Organizationally, these animals represent an important evolutionary transition because they are the simplest forms to have organ systems, a central nervous system, **cephalization** (a definite "head" with sense organs), and **bilateral symmetry.** This is the type of symmetry exhibited by most advanced animals. In animals with bilateral symmetry, the body can be cut through the center in only one way, parallel to the main axis, to yield two equal halves. This type of symmetry permits greater motility.

MATERIALS

Per student:

- dissection microscope
- scalpel

Per student pair:

- prepared slide of a whole mount of *Dugesia*
- prepared slide of a whole mount of *Dugesia* with the digestive system stained
- prepared slide of a cross section of *Dugesia*
- prepared slide of a whole mount of *Fasciola hepatica*
- prepared slide of a whole mount of *Taenia pisiformis*
- prepared slide of a bladder worm of *Taenia*

Per student group (4):

- penlight or microscope light source
- glass petri dish
- preserved specimen of *Fasciola hepatica*

Per lab room:

- chunk of beef liver
- container of fresh stream water
- container of live planaria
- ice cubes
- package of razor blades
- tray of deciduous leaves
- collection of preserved tapeworms

PROCEDURE

A. Class Turbellaria

Most of the species in this class are free-living flatworms, although a few species may be parasitic or symbiotic with other organisms. They are found in marine and fresh waters on the underside of submerged rocks, leaves, and sticks. They are especially abundant in cool, freshwater streams and along the ocean shoreline. Some live in wet or moist terrestrial habitats.

The freshwater planarians are the representatives of this class typically studied in biology courses. They are relatively small flatworms, usually 2 centimeters or less in length. *Dugesia* is the genus most commonly studied.

1. Examine a whole mount of *Dugesia* and refer to Fig. 27-11. Locate the *head*, the *eyes*, and the *auricles*, lateral projections of the head that are organs sensitive to touch and to molecules dissolved in the water.

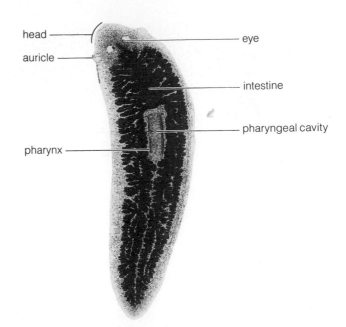

Figure 27-11 Whole mount of *Dugesia*, with the digestive system highlighted (12×). (Photo courtesy Ripon Microslides, Inc.)

2. Obtain a slide with the digestive system specially stained so that it is clearly visible. You should be able to see the protractile *pharynx*, with the mouth at its terminus. The pharynx protrudes from the ventral (belly) side of the animal to suck up the organic morsels (insect larvae, small crustaceans, and other small living and dead animals) that comprise its food. The *gastrovascular (pharyngeal) cavity* within the pharynx is continuous with the cavity of the *intestine*. The intestine has a front and two rear divisions; all are elaborately branched and together occupy much of the animal's body. For what is the extensive branching of the intestine an adaptation?

_____

_____

3. Place a small piece of beef liver in a glass petri dish containing clean, fresh stream water. Place several live planaria in the dish and observe their feeding behavior. Describe what you see.

_____

_____

_____

_____

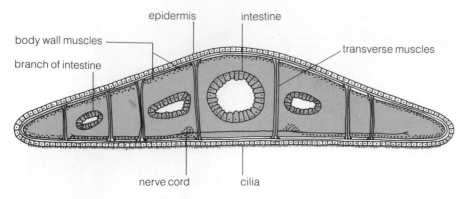

Figure 27-12 Cross section of *Dugesia*

4. Examine a cross section of *Dugesia* (Fig. 27-12). Find the following structures: *intestine, nerve cords, transverse muscles, epidermis,* and the several layers of muscles in the body wall. Along the bottom of the animal are *cilia*. What is the function of these cilia?

_____

_____

_____

The body wall contains three layers of muscle: an outer circular layer, a middle diagonal layer, and an inner longitudinal layer. You may not be able to distinguish all three layers.

5. With the dissection microscope, observe a living specimen in a glass petri dish and note its general size and shape. Touch the animal with a probe. What is its reaction? How does its shape change? Can you explain the change in shape on the basis of its muscular apparatus?

_____

_____

_____

6. **Phototaxis** is movement of an individual in response to light. Use a light source (penlight or light source for a microscope) to determine phototaxis in planarians. In otherwise dark surroundings, shine the light at the head of the planarian. What is its response?

_____

_____

_____

7. From the side of the animal, point the light toward its head. What is the animal's response?

_____

_____

_____

Is the planarian positively or negatively phototactic?

_____

The eyes of a planarian are photoreceptors. They are sensitive to light but do not form images.

Planarians are monoecious but apparently do not self-fertilize. Individuals possess testes, ovaries, a penis, and a vagina, and copulation occurs. Planarians exhibit a high degree of **regeneration,** in which lost body parts are gradually replaced. You can observe regeneration by cutting a planarian and allowing it to regenerate the lost tissue over a period of a couple of weeks.

8. If you wish to observe regeneration, place a planarian on a piece of ice and allow it to cool for at least 5 minutes. Take a clean razor blade or scalpel and cut the animal where your instructor indicates (for example, a portion of the "tail" may be cut off, the head may be split in two, or an auricle may be removed). Place the animal in a container of fresh stream water in a cool, dark place. Place a few leaves in the container to provide an appropriate substrate for the planarian. Check the animal every few days for two weeks to look for evidence of regeneration.

B. Class Trematoda

The flukes are small, leaf-shaped, parasitic forms with complex life cycles. They are generally internal parasites of vertebrates, including humans. They are covered by a thick **cuticle** but have no real epidermis. For what is the cuticle an adaptation?

_____

_____

_____

1. Examine a preserved specimen of a sheep liver fluke, *Fasciola hepatica,* noting its general morphology. Find the *oral sucker* at the front end and the *ventral sucker* just behind it.

2. Examine a whole mount of an adult *Clonorchis* (Fig. 27-13). *Clonorchis sinensis* is the Chinese liver

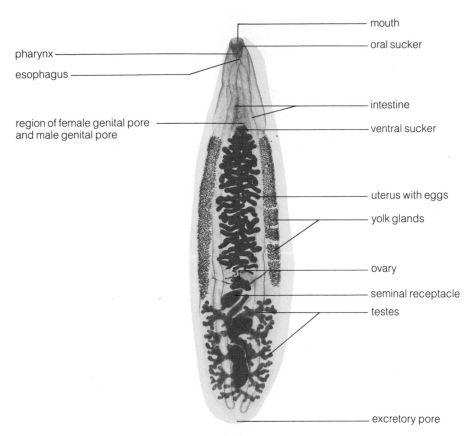

pharynx

esophagus

region of female genital pore
and male genital pore

mouth

oral sucker

intestine

ventral sucker

uterus with eggs

yolk glands

ovary

seminal receptacle

testes

excretory pore

Figure 27-13 Photomicrograph of a whole mount of *Clonorchis sinensis*, the Chinese liver fluke (10×). (Photo courtesy of Triarch, Inc.)

fluke, the adult of which is a common parasite of Asian humans. Another way of putting this is that humans are the **host** or final host for the Chinese liver fluke. *Clonorchis* has two suckers for attachment to its host: an *oral sucker* surrounding the mouth and a *ventral sucker*. The digestive tract begins with the *mouth*. This is followed by a muscular *pharynx*, a short *esophagus*, and the blind two-part *intestine*. At the tip of the back end, locate the *excretory pore* through which nitrogenous wastes are excreted.

Clonorchis is monoecious. Find the following female reproductive structures: *ovary, yolk glands, seminal receptacle*, and *uterus*. Find the following male reproductive structures: *testes* and *seminal vesicle*. During copulation, sperm is conveyed from the testes to the *genital pore* by the paired vasa efferentia, which fuse to form the vas deferens. These ducts cannot be seen in Fig. 27-13 but may be visible in your specimen. Copulating individuals exchange sperm at the genital pore, and the sperm is stored temporarily in the seminal receptacle. The yolk glands provide yolk for nourishment of the developing eggs. *Eggs* can be seen as black dots in the uterus, where they mature.

The life cycle of *Clonorchis sinensis* (Fig. 27-14) involves two *intermediate hosts*, a snail and the golden carp. Fertilized eggs are expelled from the adult fluke into the human host's bile duct and are eventually

voided with the feces of the host. If the eggs are shed in the appropriate moist or aquatic environment, they may be ingested by snails (the first **intermediate host**). A larva emerges from the egg inside the snail. The larva passes through several developmental stages and gives rise to a number of tadpolelike larvae. These free-swimming larvae leave the snail, encounter a golden carp (the second intermediate host), and burrow into this fish to encyst in muscle tissue. If the fish is not properly cooked and is eaten by humans, cats, dogs, or some other mammal, the larvae emerge from the cysts and make their way up the host's bile duct into the liver, where they mature into adult flukes. The cycle is then repeated. What is the infectious stage (for humans) of this parasite?

_____

What two precautions would be effective in reducing the frequency of contact of humans and other mammals with this parasite?

_____

_____

_____

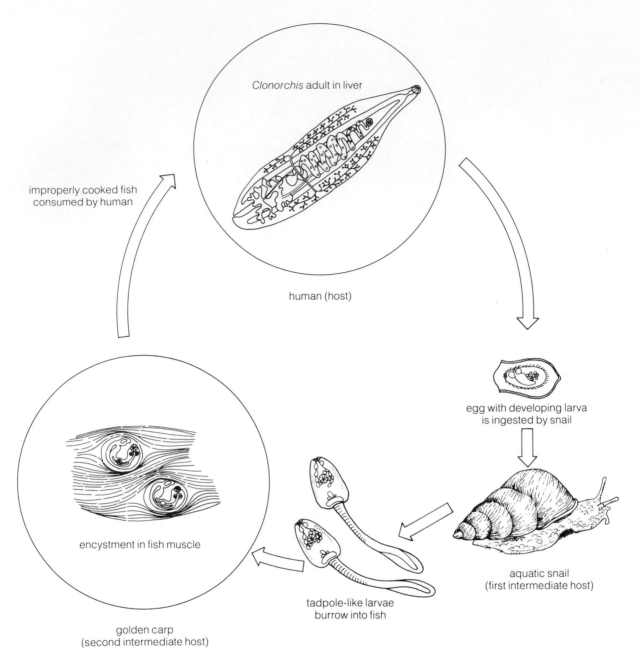

Figure 27-14 Life cycle of the human liver fluke, *Clonorchis sinensis.*

C. Class Cestoda

The tapeworms are flattened internal parasites that inhabit the intestine of vertebrates. Their general appearance resembles that of a segmented egg noodle. Tapeworms have no digestive system; instead, they absorb across their body walls predigested nutrients provided by the host. Their bodies are essentially reproductive machines, with extensive, well-developed reproductive organs occupying much of each mature body segment. The body is covered by a **cuticle.**

1. Examine preserved specimens of tapeworms taken from a variety of vertebrate hosts.

2. Examine the preserved specimens and whole

mounts of *Taenia pisiformis*, a tapeworm of dogs and cats (Fig. 27-15). The flattened body is divided into three general regions: the **scolex,** *neck*, and body. The scolex is an anterior holdfast organ with hooks and suckers. For what are the hooks and suckers used by the tapeworm?

Behind the scolex is a short neck. The body is comprised of successive units, the **proglottids.** Because proglottids are added to the animal at the neck,

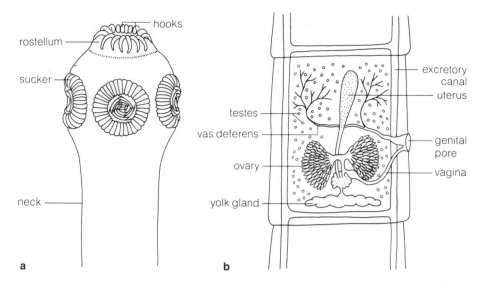

a

b

Figure 27-15 Scolex (**a**) and proglottid (**b**) of the tapeworm, *Taenia pisiformis*. (After Lytle and Wodsedalek, 1984.)

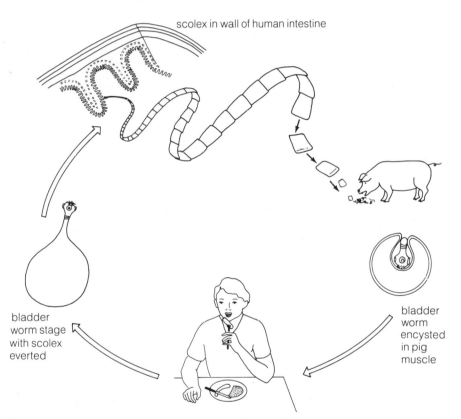

Figure 27-16 Life cycle of the pork tapeworm, *Taenia solium*

the hindmost proglottids are the oldest. Mature proglottids are almost completely filled with reproductive organs and eggs ready for release.

Tapeworms are monoecious. They are one of the few organisms that may self-fertilize, perhaps because of their isolation from each other. Examine the mature proglottids with a dissection microscope. The lateral *genital pore* opens into the *vagina*. The *uterus* should be visible as a fine line running top to bottom

in the middle of the proglottid. The *ovaries* are represented by a dark, diffuse mass of tissue.

The *vas deferens* leads from the genital pore to the *testes*, a paler, more diffuse mass of tissue than the ovaries. Flanking the reproductive organs are the *excretory canals*.

3. The life cycle of the pork tapeworm, *Taenia solium*, involves two host organisms, the pig and a human (Fig. 27-16). Mature proglottids containing embryos are voided with the feces of a human. If a proglottid is ingested by a pig, the embryos escape from the proglottid and bore through the intestinal mucosa into the circulatory or lymphatic system. From there they make their way to muscle where they encyst as larvae known as *bladder worms*. If improperly cooked pork is ingested by a human, a bladder worm is released from its encasement (bladder) and attaches with its scolex to the intestinal mucosa of the human host. Here it matures and adds proglottids to complete the life cycle.

Examine the bladder worm stage of *Taenia* with the dissection microscope.

IV. PHYLUM NEMATODA (Starr, p. 266)

The roundworms are widespread and abundant, occurring in large numbers in marine and freshwater bottom sediments and in water films around soil particles. Most are microscopic, and some are parasitic. Besides being especially troublesome for our pet dogs and cats, members of this phylum parasitize a wide variety of animals and plants. Nonetheless, most roundworms are free-living and beneficial; great numbers are characteristic of rich soils, where they are important in nutrient and mineral cycling.

Roundworms are **pseudocoelomates.** They have a body cavity, or *pseudocoelom*, that is not considered a true body cavity because it is not lined by a derivative of the mesoderm. Unlike flatworms, nematodes have a complete digestive tract. The body is elongated and cylindrical and covered by a cuticle.

Parasitic roundworms include *Ascaris* (see below), the hookworms, *Trichinella* worms of mammals (which cause trichinosis), pinworms of humans, and the filarial worms of humans that cause elephantiasis by obstructing lymphatic vessels.

MATERIALS

Per student:

- dissection microscope
- dissecting needle

Per student pair:

- finger bowl

Per lab room:

- collection of preserved specimens and whole mounts of roundworms
- tray of moist soil
- demonstration dissection of a female *Ascaris*
- demonstration dissection of a male *Ascaris*

PROCEDURE

A. Free-Living Roundworms

1. Examine a variety of demonstration specimens and whole mounts of free-living and parasitic roundworms.

2. Place some of the moist soil your instructor has provided in a finger bowl with some water. Using a dissection microscope, look for translucent roundworms in the soil sample. Describe the behavior of any roundworms you find, and make a note of any visible anatomical features so that you can compare these to the features of the internal parasite, *Ascaris*, described below.

_____

_____

_____

_____

B. *Ascaris*

Ascaris is a common parasite of humans, pigs, horses, and other mammals. Adults range from 15 to 40 cm in length, with males being smaller and having a sharp bend in the posterior part of the body. In this exercise you will study the anatomy of both sexes. Your lab instructor has provided demonstration dissections of both female and male *Ascaris*.

_____

CAUTION *Wash your hands and instruments thoroughly after examining the female* Ascaris; *some eggs may still be alive even though the specimen has been preserved.*

_____

1. Examine a female *Ascaris*. At the front end, the triangular *mouth* is surrounded by three *lips* (Fig. 27-17). The opening on the ventral surface near the other end of the body is the *anus. A genital pore* is located about one-third the distance from the mouth.

The *pseudocoel* is the space containing the internal organs. The lips surround a *mouth* that opens into a muscular *pharynx* (*esophagus*). The esophagus leads into a flat *intestine*, running nearly the length of the body. The intestine ends in a *rectum* and opens into the *anus*.

Find the paired *excretory tubes* on each side of the body. These tubes open to the outside through a

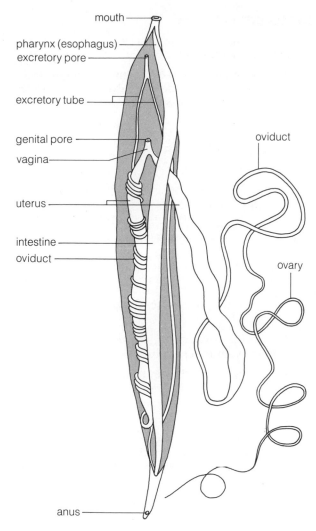

mouth

pharynx (esophagus)

excretory pore

excretory tube

genital pore

vagina

uterus

intestine

oviduct

oviduct

ovary

anus

Figure 27-17 The parasitic roundworm, *Ascaris* (female). (After Boolootian and Stiles, 1981.)

3. After copulation, the eggs are fertilized in the female oviduct. The eggs, surrounded by a thick shell, are expelled through the genital pore to the outside; as many as 200,000 may be shed into the host's intestinal tract per day.

The eggs are eliminated in the host's feces. They are very resistant and may live for years under adverse conditions. Infection with *Ascaris* results from ingesting eggs in food or water contaminated by feces or soil. The eggs pass through the stomach and hatch in the small intestine. The small larvae migrate through the bloodstream, exit the circulatory system into the respiratory system, crawl up the respiratory tubes and out the trachea into the throat. There they are swallowed to re-infect the intestinal tract and mature. *Sometimes larvae exit the nasal passages rather than being swallowed!* The entire journey of the larva from intestine back to intestine takes about 10 days.

V. PHYLUM ROTIFERA (Starr, p. 266)

The rotifers are microscopic, the smallest animals. They are mostly bottom-dwelling, freshwater forms and are important prey of larger animals in aquatic food chains. Their body wall is very thin, usually transparent, and is covered by a cuticle composed of protein. Like the nematodes, the rotifers are pseudocoelomates.

MATERIALS

Per student:

- several microscope slides and coverslips
- compound microscope

Per lab room:

- live culture of rotifers

PROCEDURE

A. Live Rotifers

You will examine rotifers from a living culture. Many of the details of the digestive, excretory, and reproductive systems of these animals will not be distinguishable, and prepared slides are usually no better (and frequently worse) in revealing these details. However, the general body outline and structures described below should be apparent.

1. Because rotifers are microscopic, place a drop from the rotifer culture on a slide, cover with a coverslip, and observe with the compound microscope. The elongate, cylindrical body is divisible into the *trunk* and the *foot* (Fig. 27-18).

The foot bears two to several *toes* for adhering to objects. The superior end of the trunk bears the **wheel**

single *excretory pore* located near the mouth on the ventral surface. The *ovaries* are the threadlike ends of a Y-shaped reproductive tract. The ovary leads to a larger *oviduct* and a still larger *uterus*. The two uteri converge to form a short, muscular *vagina* that opens to the outside through the *genital pore*.

2. Examine a male *Ascaris*. The lips and mouth are as in the female. However, the male has a *cloacal opening* instead of an anus. A **cloaca** is an exit for both reproductive and digestive products. A pair of *copulatory spicules* protrudes from this opening.

Excretion is as in the female. The reproductive system superficially resembles that of the female. It consists of a *single* tubular structure that includes a threadlike *testis* joined to a larger *vas deferens* (*sperm duct*), an even larger *seminal vesicle,* and terminally a short, muscular *ejaculatory duct* that empties into the cloaca.

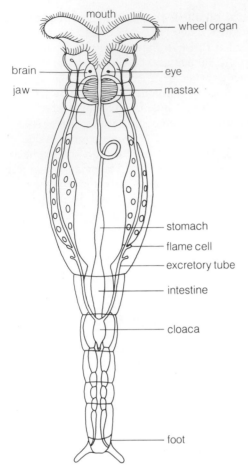

Figure 27-18 Dorsal view of a rotifer, showing the internal anatomy. (After Boolootian and Stiles, 1981.)

Labels in figure: mouth, wheel organ, brain, eye, jaw, mastax, stomach, flame cell, excretory tube, intestine, cloaca, foot

organ, which is encircled by cilia. These cilia beat in such a fashion as to give the appearance of a wheel (or two) turning, hence the common reference to these animals as the "wheel animals." The cilia are responsible for creating a current that sweeps organic food particles into the mouth at the center of the wheel organ. The cilia also enable the organism to move.

2. Inside the trunk, below the wheel organ, you may be able to see the movement of a grinding organ, the *mastax*. This structure is equipped with jaws made hard by a substance called chitin. For what is the mastax an adaptation?

_____

_____

_____

_____

B. Life Cycle of Rotifers

Rotifers are dioecious. However, many species have no males, and the eggs develop parthenogenetically. **Parthenogenesis** is the development of unfertilized eggs, generally in response to seasonal chemical or temperature changes in the aquatic environment.

PRE-LAB QUESTIONS

_____ **1.** Which of the following is characteristic of animals? (a) organs and organ systems, (b) cell walls, (c) autotrophy, (d) photosynthesis.

_____ **2.** Sponges are considered to be among the least complex animals because they (a) reproduce sexually, (b) are sessile (stationary) as adults, (c) lack definite tissues, (d) lack a "head."

_____ **3.** Cnidarians evolved from an ancestral group of (a) sponges, (b) protists, (c) fungi, (d) none of the above.

_____ **4.** The free-swimming jellyfish is known as a (a) polyp, (b) velum, (c) medusa, (d) hydra.

_____ **5.** Coral reefs, atolls, and islands are the result of the buildup of coral (a) molts, (b) calcareous exoskeletons, (c) wastes, (d) chitin.

_____ **6.** Cephalization is (a) the division of the trunk into similar segments, (b) the presence of a head with sense organs, (c) sexual reproduction involving self-fertilization, (d) the ability to regenerate lost body parts.

_____ **7.** Tapeworms have no (a) reproductive organs, (b) nervous system, (c) excretory system, (d) digestive system.

_____ **8.** Nematodes and rotifers have a body cavity that (a) comprises the lumen of the digestive tract, (b) is termed a eucoelom, (c) is filled with blood, (d) is not lined by mesoderm.

_____ **9.** Filarial roundworms cause elephantiasis by (a) encysting in muscle tissue, (b) promoting the growth of fatty tumors, (c) obstructing lymphatic vessels, (d) laying numerous eggs in the joints of their hosts.

_____ **10.** The rotifers obtain their food by the action of (a) tentacles, (b) cilia on the wheel organ, (c) a muscular pharynx, (d) pseudopodia.

EXERCISE 27
SPONGES, CNIDARIANS, FLATWORMS, ROUNDWORMS,
AND ROTIFERS

POST-LAB QUESTIONS

1. List five characteristics of members of the animal kingdom.

2. Define cellular specialization and indicate how sponges exhibit this phenomenon.

3. Name and illustrate the two body forms of Cnidarians.

   a.   **b.**

4. Describe the means by which the Cnidarians seize and ingest organisms that are faster than they are.

5. Define bilateral symmetry.

6. Describe several adaptations of free-living flatworms to their external environment.

7. Describe several adaptations of parasitic flatworms to their external environment.

8. The relatively simple animals, *Hydra* and *Dugesia*, can regenerate lost body parts, but humans generally cannot. Discuss this in terms of tissue differentiation and the comparative levels of structural complexity of these organisms.

9. **a.** Provide an explanation for the evolution of hermaphroditism in flatworms.

   b. *Explain* the relationship between the tremendous numbers of eggs produced by flukes, tapeworms, and parasitic roundworms and the complexity of their life cycles.

10. The nematodes and rotifers are pseudocoelomates, even though they have a body cavity. Why is this cavity not considered to be a "true body cavity"?

MOLLUSKS, SEGMENTED WORMS, AND JOINT-LEGGED ANIMALS

OBJECTIVES After completing this exercise you will be able to:

1. define protostome, deuterostome, coelomate, coelom, foot, mantle, somite, metamerism, peritoneum, setae, clitellum, cocoon, exaskeleton, hemolymph, chitin, carapace, sexual dimorphism, gills, tracheae;

2. differentiate between protostomes and deuterostomes;

3. describe the natural history of members of the phyla Mollusca, Annelida, and Arthropoda;

4. identify representatives of the classes Pelecypoda, Gastropoda, and Cephalopoda of the phylum Mollusca;

5. outline the life cycle of a freshwater clam or mussel;

6. identify representatives of the classes Oligochaeta, Polychaeta, and Hirudinea of the phylum Annelida;

7. outline the life cycle of an earthworm;

8. identify representatives of the classes Crustacea, Insecta, Chilopoda, Diplopoda, and Arachnida of the phylum Arthropoda;

9. identify structures (and indicate their associated functions) of the representatives of these phyla and classes.

INTRODUCTION Some time after the origin of the first bilaterally symmetrical animals, two divergent paths were taken by the animals that subsequently evolved. The mollusks, annelids, and arthropods followed one of these paths, and the echinoderms and chordates the other. The mollusks, annelids, and arthropods comprise a collection of animals known as **protostomes,** animals whose blastopore (the first opening) in the gastrula stage of development ultimately becomes a mouth. The anus develops later. The echinoderms and chordates comprise the **deuterostomes,** whose blastopore develops into an anus. A second opening becomes the mouth.

Animals in both groups are **coelomates,** possessing a true **coelom** that is lined by tissue derived from the

mesoderm. This exercise describes the protostomes. The next exercise covers the deuterostomes.

I. PHYLUM MOLLUSCA   (Starr, pp. 269–271)

This is the second largest animal phylum and includes snails, slugs, nudibranchs, clams, mussels, oysters, octopuses, and squids. The body is typically soft (*mollusc* in Latin means "soft"). Apart from the squids and octopuses, and the land snails and slugs, which have adopted a terrestrial existence, mollusks inhabit shallow marine and fresh waters, where they crawl along or burrow into the soft substrate. Many species are valued as food for humans.

Two characteristics distinguish the mollusks from other animals. They possess a ventral, muscular **foot** for movement and a dorsal integument, the **mantle,** that secretes the shell and functions in gaseous exchange. Even forms without a shell usually have a mantle.

MATERIALS

Per student:

- scalpel
- blunt probe or dissecting needle

Per student pair:

- preserved freshwater clam or mussel
- dissection microscope
- glass petri dish
- prepared slide of *glochidia*

Per lab room:

- labeled demonstration dissection of a freshwater clam or mussel
- collection of pelecypod shells and preserved specimens
- freshwater aquarium with snails (for example, *Physa*)

- fluorescent or gooseneck lamp
- black construction paper
- aquarium lid
- collection of gastropod shells and preserved specimens
- several preserved squids
- several preserved octopuses and chambered nautiluses
- large plastic bag for disposal of dissected specimens

PROCEDURE

A. Class Pelecypoda

These are the *bivalves*, so named because they have shells comprised of two halves or valves. Included in this class are the clams, mussels, oysters, and scallops. They have a hatchet-shaped (laterally flattened) body and foot for burrowing in soft substrates. *Growth ridges* on the shell's outer surface represent periods of restricted winter growth and thus the age of the organism.

CAUTION *Preserved specimens are kept in a formalin based or other preservative solution. Wash any part of your body exposed to this solution with copious amounts of water. If preservative solution is splashed into your eyes, wash them with the safety eyewash bottle for 15 minutes.*

---

1. Dissection of a freshwater clam or mussel.
a. Examine a freshwater clam or mussel. Each valve has a hump on the dorsal surface (back), the *umbo*, that points toward the front of the organism. The umbo is the oldest part of the valve. Encircling the umbo are concentric rings of annual growth, some of which form the growth ridges mentioned previously. To determine the age of your mussel, count the number of ridges formed by the growth rings.

---

b. Remove the left valve by cutting the *anterior* and *posterior adductor muscles* that hold the valves together (Fig. 28-1). To do this, slip a scalpel between the

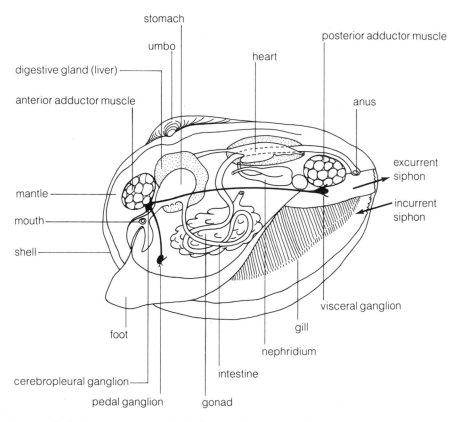

Figure 28-1 Internal anatomy of a freshwater clam or mussel. (After Boolootian and Stiles, 1981.)

valves and cut back and forth in the areas where the two valves come together at the dorsal surface. You should be able to feel the scalpel cutting through the adductor muscles.

The relatively simple body plan includes a ventral **foot** and a dorsal *visceral mass.* The membrane that lines the inner surface of each valve is the **mantle,** which secretes the shell. The shell is composed of three layers, each formed of calcium carbonate. Pearls are formed as a result of irritations (usually grains of sand or small pebbles) in the mantle and are sometimes found embedded in the *pearly layer* of the shell. Note the *incurrent* and *excurrent siphons* at the rear end. These represent extensions of the mantle. What are their functions?

_____

_____

_____

c. Remove the mantle on the left side and observe the *gills,* one on each side of the *visceral mass.* The gills function in gaseous exchange. Water enters the incurrent siphon and flows over the gills, ultimately leaving through the excurrent siphon. Gills also trap food particles contained in incoming water; these are transported to the mouth by the cilia on the gills.

d. The circulatory system is *open,* so the blood passes from arteries leading from the heart through body spaces called *sinuses.* Find the *heart* in the pericardial cavity between the visceral mass and the hinge between the valves. Dispose of your dissected specimen in the large plastic bag provided for this purpose.

e. Examine the labeled dissection prepared by your instructor. Note the *esophagus, stomach, digestive gland* (liver), *anus, nephridium* (kidney), and *gonad.*

f. Fertilization is internal in bivalves. Sperm are taken in through the female's incurrent siphon and fertilize eggs in the gills. Larvae (*glochidia*) develop from the fertilized eggs and after several months, leave the gill area via the excurrent siphons to parasitize fish by clamping onto the gills or fins with their jawlike valves. The fish grows tissue over the parasite, which remains attached for two to three months. The glochidium then breaks out of the fish to develop into an adult on the lake or river bottom. Look at *glochidia* in a prepared slide.

2. Assorted pelecypods. Examine an assortment of pelecypod shells and preserved specimens. List the features they have in common.

_____

_____

_____

List the differences.

_____

_____

_____

_____

How can you tell which live on sandy bottoms?

_____

_____

B. Class Gastropoda

The snails, slugs, and nudibranchs (sea slugs) comprise this class of mollusks (Fig. 28-2).

Slugs and nudibranchs have no shell; the former are terrestrial, the latter marine. Nudibranchs are very ornate gastropods, often brightly colored and frequently adorned with numerous, fleshy appendages.

1. Place a live freshwater snail (for example, *Physa*) in a glass petri dish. Once the snail has attached itself to the glass, invert the petri dish and use the dissection microscope to observe it moving along a slimy path. A *slime gland* in the front of the foot secretes a mucus. Note the muscular contractions of the foot that allow the animal to glide over this mucus. You can see the *mouth* in this position.

2. In snails with spirally coiled shells, the shell either spirals to the snail's right or to the snail's left. Which way does your snail's shell spiral?

_____

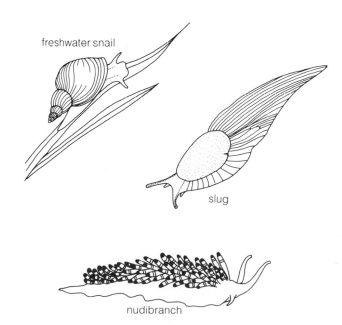

Figure 28-2   Assorted members of the class Gastropoda

3. Observe the feeding behavior of snails in a freshwater aquarium that has sides coated with algae. Are most of the snails oriented with their heads directed upward or downward?

_____

4. At this time, your instructor will place a light over one half of the aquarium and cover the other half with dark paper or a lid. At the end of the lab period check to see if there has been any movement of the snails toward one side or the other. What is their response to light?

_____

_____

5. Gastropods, especially snails, are important intermediate hosts for trematode parasites of vertebrates (see Exercise 27). Examine the variety of gastropod shells and preserved specimens on display.

C. Class Cephalopoda

This class includes the squids, octopuses, and nautiluses (Fig. 28-3). Cephalopods have several significant modifications of the basic molluskan body plan.

They are adapted for swimming and a carnivorous habit.

1. Examine the external morphology of a squid. The foot of the squid has become divided into four pairs of arms and two tentacles, each with suckers, and a _siphon_ (_funnel_). The tentacles, with their terminal suckers, shoot out to grasp prey. The siphon functions in jet propulsion by forcing jets of water outward through this opening to propel the animal through the water. Two large eyes superficially resemble those of vertebrates. The shell of the squid lies dorsally beneath the mantle. Located posteriorly is a pair of _lateral fins_. What might be the function of these fins?

_____

_____

2. Note octopuses have no shell.

3. The chambered nautilus has a compartmentalized, coiled shell. The body of the animal occupies a chamber, outgrows it, and secretes a new chamber. The old chambers are filled with air to allow the animal to remain buoyant. Examine specimens of octopuses and nautiluses.

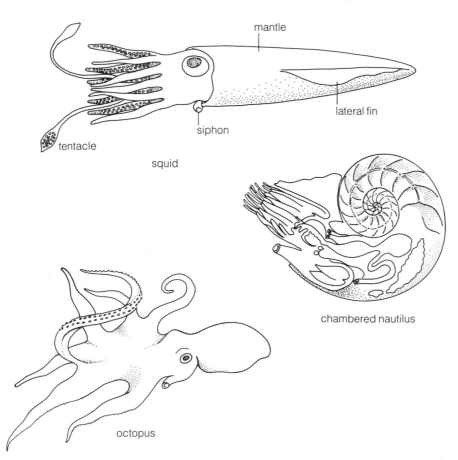

Figure 28-3 Assorted members of the class Cephalopoda

II. PHYLUM ANNELIDA (Starr, p. 272)

You are probably more familiar with the phylum Annelida—through its representative, the earthworm—than with the other animal phyla previously described. Probably the most striking annelid characteristic is the division of the cylindrical trunk into a series of similar segments, the **somites**. Annelids were the first animals to evolve the condition of **metamerism** (segmentation). Segmentation is internal as well as external, with segmentally arranged components of various systems and the body cavity. Unlike the proglottids of the cestodes, a somite cannot function independently of other somites. As protostomes, the annelids have a true coelom lined by the shiny **peritoneum**. *The circulatory system is closed*, entirely contained in vessels.

MATERIALS

Per student:

- blunt probe or dissecting needle

Per student pair:

- preserved earthworm
- dissection microscope

Per lab room:

- labeled demonstration dissection of an earthworm
- collection of preserved polychaetes
- collection of preserved leeches
- live freshwater leeches in an aquarium

PROCEDURE

A. Class Oligochaeta

1. You may know several characteristics of the bristleworms because of past associations with the earthworm (*Lumbricus terrestris*), especially if you go fishing. Recalling past experiences with this organism, list as many features of the earthworm as you can.

_____

_____

_____

_____

_____

2. Let's see how you did! Obtain a preserved earthworm and have a dissection microscope handy. Is the body bilaterally or radially symmetrical?

_____

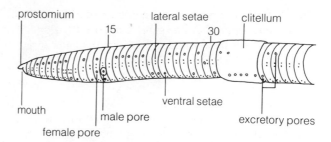

Figure 28-4 External anatomy of the earthworm (numbers refer to the segment numbers)

3. Examine the external anatomy (Fig. 28-4). Note the somites. Count the segments and record the number.

_____

Is the number of somites the same in all specimens?

_____

The *prostomium* is a fleshy lobe that hangs over the *mouth*, the first segment. Take the specimen between your fingers and feel the **setae**, tiny, bristlelike structures. What is their function?

_____

_____

_____

4. Although you will not separate the following layers of the body wall of the earthworm, it is important that you are aware of their relationship to each other. A noncellular secretion, the *cuticle*, covers the body. Beneath the cuticle is the epidermis, and internal to this are an outer circular layer and an inner longitudinal layer of muscle.

5. Use the dissection microscope to help you find the following. A pair of small excretory pores is found on the ventral surface of each segment, except the first few and the last. On the sides of segment 14, the openings (*female pores*) of the oviducts can be seen with the microscope. The openings (*male pores*) of the sperm ducts, with their swollen lips, can be found on segment 15. The **clitellum** is the enlarged ring beginning at segment 31 or 32 and ending with 37. This glandular structure secretes a slimy band around two copulating individuals. The *anus* is a vertical slit in the terminal segment.

6. Examine the labeled dissection prepared by your instructor. Note the *buccal cavity, pharynx, esophagus, crop, gizzard, intestine, seminal vesicles, testes, vas deferens* (sperm duct), *seminal receptacles, oviducts, ovaries, dorsal blood vessel, hearts, ventral blood vessel, brain, ventral nerve cord,* and *nephridia* (see Fig. 28-5).

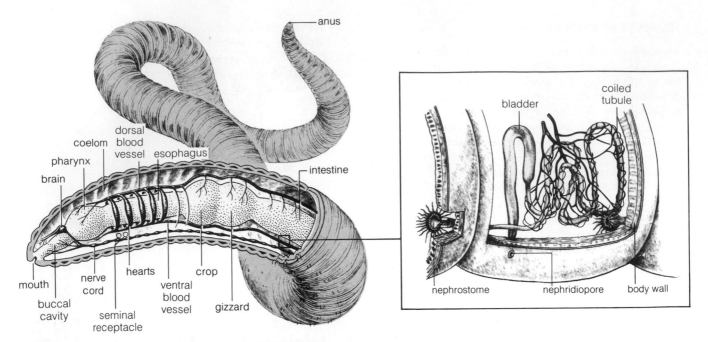

Figure 28-5 Internal anatomy of the earthworm. Other than the seminal receptacles, the reproductive tracts are not included in the diagram. (After Starr and Taggart, 1987.)

7. The earthworm is monoecious. Copulation in earthworms usually takes two to three hours. Individuals face belly to belly in opposite directions and come together at their clitellums. These organs secrete a slimy substance that forms a band around the worms. Sperm are transferred between the participants, and the sperm are stored temporarily in the seminal receptacles. When the individuals separate, part of the slimy band formed by the clittellums remains with each worm to form a **cocoon.** The cocoon receives eggs and stored sperm from the seminal receptacles. The eggs are fertilized, and the cocoon moves forward as the worm backs out, slips off the front end of the worm, and is deposited in the soil. In this species, a single young earthworm completes development using the other fertilized eggs as food, eventually breaks free of the cocoon, and becomes an adult in several weeks. As long as the stored sperm lasts, the earthworm continues to form cocoons.

B. Class Polychaeta

The polychaetes are abundant marine annelids with dorsoventrally flattened bodies (Fig. 28-6).

They are most common at shallow depths in the intertidal zone at the seashore. They are prey for a variety of marine invertebrates.

1. Examine preserved polychaetes. Although they are similar to oligochaetes in many ways, notice that

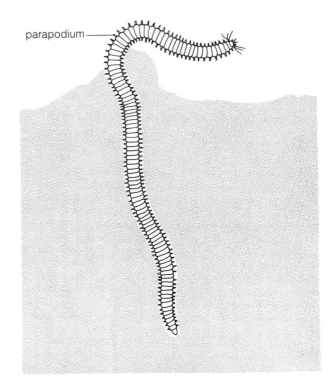

Figure 28-6 A polychaete in its burrow

the polychaetes have fleshy, segmental outgrowths called *parapodia*. These structures are equipped with numerous setae, the characteristic for which they get their class name, Polychaeta, meaning "many setae." Notice also the definite head.

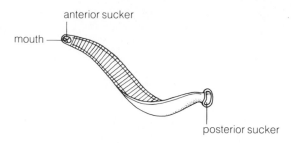

Figure 28-7 A leech (class Hirudinea)

C. Class Hirudinea

The leeches (Fig. 28-7) are dorsoventrally flattened annelids that are usually found in fresh water.

1. Examine preserved specimens of leeches. They lack parapodia and setae. They are predaceous or parasitic; parasitic species suck blood with the aid of a muscular pharynx. The anterior somites are modified into a small *sucker* surrounding the mouth. The mouth is supplied with jaws made hard with the substance chitin used to wound the host and supply the leech with blood. Parasitic leeches secrete an anticoagulant into the wound to prevent the blood from clotting. A *posterior sucker* helps the leech attach to the host or prey.

2. If living freshwater leeches are available, observe the action of their oral sucker by allowing them to briefly attach to and hang from your hand or arm. Observe their movement as they swim through the water in an aquarium. Explain this movement in terms of the contraction and relaxation of circular and longitudinal layers of muscle that comprise part of their body wall.

III. PHYLUM ARTHROPODA (Starr, pp. 273–278)

In numbers of both species and individuals, this is the largest animal phylum. There are more than 1 million species of arthropods, more in fact than in all the other animal phyla combined. Arthropods are believed to be evolved from annelid ancestors, the annelid cuticle having become thicker and harder to form an **exoskeleton** (external skeleton) hardened with **chitin** (chitin is a structural polysaccharide). Like the annelids, the arthropods are segmented, with a pair of jointed appendages per body segment.

The body is typically divided into three regions: the *head, thorax,* and *abdomen.* The mouthparts are modified appendages (legs). The coelom is filled with the blood, or **hemolymph,** because of the *open cir-* *culatory system.* Arthropods possess a highly developed central nervous system and complex sense organs and behavior, features that are partly responsible for their evolutionary success and current abundance.

MATERIALS

Per student:

- dissecting scissors
- blunt probe or dissecting needle
- scalpel

Per student pair:

- dissection microscope
- preserved crayfish
- grasshopper (*Romalea*)

Per student group:

- living crayfish
- five-gallon aquarium filled with pond or aquarium water
- black construction paper
- gooseneck lamp with dim light bulb
- several live millipedes (optional)

Per lab room:

- collection of preserved and mounted crustaceans
- collection of preserved and mounted insects
- collection of preserved centipedes
- collection of preserved millipedes
- collection of preserved and mounted arachnids
- large plastic bag for disposal of dissected specimens

PROCEDURE

A. Class Crustacea

Some of our most prized food items are animals that belong to the class Crustacea. This group contains shrimps, lobsters, crabs, and crayfish, and of considerably less culinary interest, water fleas, sand fleas, isopods (sow or pill bugs), ostracods, and barnacles. Most species are marine but some live in fresh water. Most isopods occupy moist areas on land. For the most part, crustaceans are carnivorous, scavenging, or parasitic.

The head of crustaceans has two pairs of antennae and three pairs of mouth parts, one pair of mandibles, and two pairs of maxillae. The exoskeleton is hardened dorsally and sometimes laterally to form a **carapace.** The appendages of the body are specialized for a wide variety of functions, and the eyes may be simple or compound.

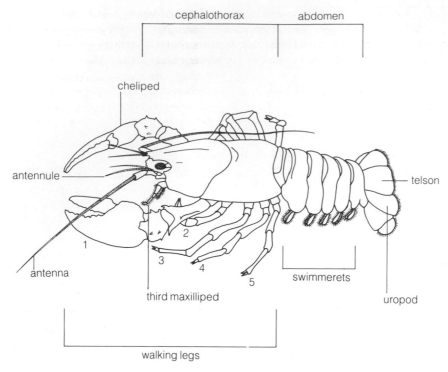

Figure 28-8 Dorsolateral view of the external anatomy of a crayfish. (After Boolootian and Stiles, 1981.)

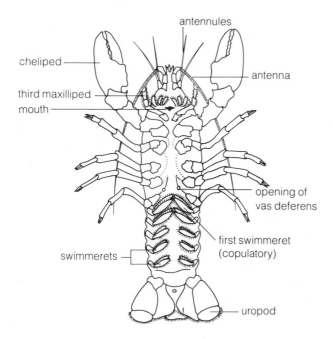

Figure 28-9 Ventral view of a male crayfish

Larger crustaceans often can adapt to the color of their background because of special pigment-containing cells, the *chromatophores*, in the outer body wall.

1. *Dissection of a crayfish.* The crayfish is a relatively large representative of the phylum Arthropoda whose internal anatomy is readily apparent in preserved specimens.

a. Examine the demonstration specimens of crustaceans. You will study in detail the external features of the crayfish (*Cambarus* or *Procambarus*) as a representative of the class Crustacea. Obtain a specimen and rinse it in water before you begin your examination.

b. Notice that the body is divided into two major regions: an anterior *cephalothorax,* comprised of a fused head and thorax, and a posterior *abdomen* (Figs. 28-8, 28-9). The hard outer cover of the *cephalothorax* is the protective **carapace.**

c. Note the large *compound eyes* on movable stalks, two large *antennae,* and two pairs of smaller *antennules.* The *mouth* is ventral, surrounded by several pairs of appendages modified for handling food. There are five pairs of large *walking legs;* the first, called the *chelipeds,* bear large pincers. There are five pairs of abdominal appendages called *swimmerets.*

The *uropods* are a pair of large, flattened lateral appendages near the end of the abdomen. The *telson* is an extension of the last abdominal segment, bearing the *anus* ventrally.

d. Crayfish exhibit **sexual dimorphism** (morphological differences between male and female). Females have a broad abdomen, five pairs of swimmerets of roughly the same size on the abdominal segments, and a *seminal receptacle* ventrally between the bases of the fourth and fifth pairs of walking legs. Males have a narrower abdomen, the front two pairs of swimmerets enlarged for copulation and transferring sperm to the female, and openings of the *vasa deferentia* (sperm ducts) at the base of each fifth walking leg. Observe both sexes.

e. Examine the appendages on one side of the body, listed from anterior to posterior below with their functions.

---

i. *antennule*—contains a statocyst, an organ of balance, in the basal segment.

ii. *antenna*—for chemoreception.

iii. *mandible*—modified as a functional jaw.

iv. *maxillae* (first and second)—modified for handling food, the second for creating a current for gaseous exchange in the gills.

v. *maxillipeds* (first, second, and third)—modified for handling food.

vi. *walking legs* (one through five)—for defense (the cheliped), movement.

vii. *swimmerets* (one through five)—the first two are modified in males for copulation and transferring sperm to the female; young are brooded among these structures on the female.

viii. *uropod*—modified to form (with the *telson*) the tail fan, used for swimming backward.

---

f. Refer to Fig. 28-10 to study the internal anatomy of the crayfish. Locate the **gills** within the branchial chambers by carefully cutting away the lateral flaps of the carapace with your scissors. The gills are feathery structures containing blood channels that function in gaseous exchange. The second maxilla creates a current of water that flows past the gills, bringing oxygen in contact with the blood supply and carrying away carbon dioxide.

g. With scissors, superficially cut forward from the rear of the carapace to just behind the eyes. With your scalpel, carefully separate the hard carapace from the thin, soft, underlying tissue. Next, remove the gills to reveal the internal organs.

h. Note the two longitudinal bands of *extensor muscles* that run dorsally through the thorax and abdomen. What is their function?

---

i. In the abdomen, find the large *flexor muscles* lying below the extensor muscles. These muscles bend the abdomen to provide a quick backward thrust.

j. Locate the small, membranous heart just posterior to the stomach. Remember, the circulatory system is open; from the arteries, blood flows into open spaces, or *sinuses*, before returning to the heart through openings in the wall of this organ.

k. The *mouth*, hidden by several oral appendages, leads to the tubular *esophagus*, which leads to the *cardiac stomach*. At the border between the cardiac stomach and the *pyloric stomach* is the *gastric mill*, a grinding apparatus composed of three chitinous teeth. What role in digestion do you think this structure plays?

---

l. Locate the large *digestive gland*. This organ secretes digestive enzymes into the cardiac stomach and takes up nutrients from the pyloric stomach. Absorption of nutrients from the tract continues in the intestine, which runs from the abdomen to the *anus*.

m. Locate the *green glands*, the excretory organs situated ventrally in the head region near the base of the antennae. A duct leads to the outside from each green gland.

n. The gonads lie beneath the heart. They will be obvious if your specimen was obtained during the reproductive season. If not, they will be small and difficult to locate. In a female, find the *ovaries* just beneath the heart. In the male, two white *testes* will occupy a similar location.

o. Remove the organs in the cephalothorax to expose the *ventral nerve cord*. Observe the segmental ganglia and their paired lateral nerves. Trace the nerve cord forward to locate the *brain*. Note the nerves leading from the brain toward the eyes, antennules, antennae, and mouth parts. Dispose of your dissected specimen in the large plastic bag provided for this purpose.

2. *Crayfish behavior.*

a. Place a crayfish in a small aquarium containing pond or aquarium water. Cover the aquarium to prevent light from entering, except for one end. Shine a dim light toward the uncovered end of the aquarium and describe the crayfish's response.

---

b. Remove the crayfish from the aquarium. Threaten the animal with a large object (for example, a notebook) and describe its behavior.

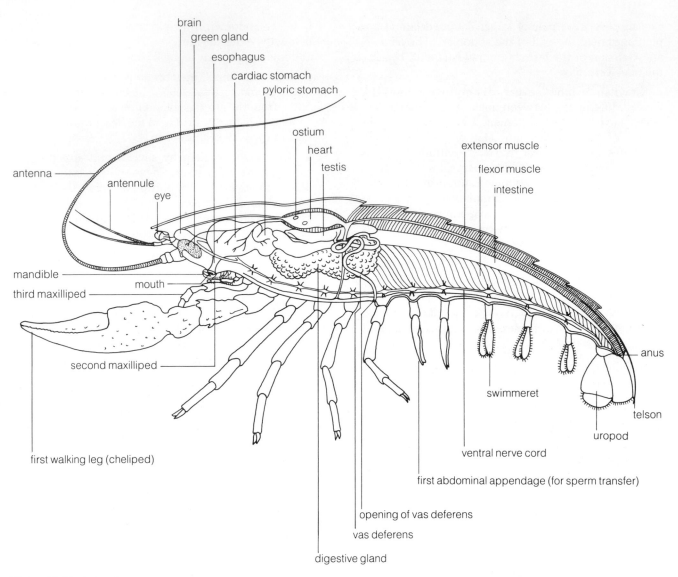

brain
green gland
esophagus
cardiac stomach
pyloric stomach
ostium
heart
testis
extensor muscle
flexor muscle
intestine
antenna
antennule
eye
mandible
mouth
third maxilliped
second maxilliped
first walking leg (cheliped)
anus
swimmeret
telson
uropod
ventral nerve cord
first abdominal appendage (for sperm transfer)
opening of vas deferens
vas deferens
digestive gland

Figure 28-10 Internal anatomy of a male crayfish. (After Boolootian and Stiles, 1981.)

Approach the animal from a variety of angles, including from the side and behind it. Based on the animal's response, would you say that its eyes are effective in detecting movement from all directions? Explain.

_____

_____

c. Repeatedly place the crayfish on a moderate incline, and observe its response and the orientation or movement it exhibits. Is its response positive or negative to gravity?

_____

_____

B. Class Insecta

Of the more than 1 million species of arthropods, more than 850,000 of them are insects. These animals occupy virtually every kind of terrestrial habitat, and they have invaded fresh waters as well. A number of the typical arthropod adaptations have contributed to their success, including the chitinous exoskeleton that physically protects the internal organs, keeps foreign substances out of the body, and prevents loss of water. Moreover, they have an additional adaptation that many biologists believe has been especially

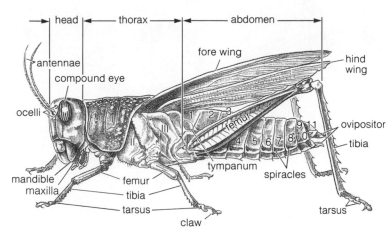

Figure 28-11 External anatomy and body form of a grasshopper, a typical insect. (After Jensen, Heinrich, Wake, and Wake, 1979.)

responsible for their proliferation—flight. A currently popular hypothesis suggests that wings became adapted for flight secondarily, with their initial function as heat-absorbing devices. Whatever the initial function of wings, insects were the first organisms to fly and thus were able to exploit a variety of opportunities not available to other animals.

1. Examine members of the class Insecta on demonstration in the lab.

2. Examine the grasshopper, *Romalea*, described here as a representative insect. The general structure of insects is relatively uniform. The body consists of a *head*, *thorax*, and *abdomen* (Fig. 28-11).

An insect's body is covered by an exoskeleton. The thorax is composed of three segments, each of which bears a pair of legs. The middle and back segments can each bear a pair of wings. The abdomen, unlike that in crustaceans, bears no appendages. Evidence of the respiratory system is located on the sides of the thorax and abdomen in the form of *spiracles*, tiny openings into the **tracheae,** or breathing tubes, that course throughout the body and connect directly with the tissues.

3. The head of the grasshopper contains a pair of compound eyes between which are three simple eyes, the *ocelli*, light-sensitive organs that do not form images (Fig. 28-12).

4. A single pair of antennae distinguishes the insects from the crustaceans (that have two) and the arachnids (that have none). Feeding appendages consist of a pair of *mandibles* and two pairs of *maxillae*, the second pair fused together to form the lower lip, the *labium*. The upper lip, the *labrum*, covers the mandibles. Near the base of the labium is a process called the *hypopharynx*. Mouthparts are extremely variable in in-

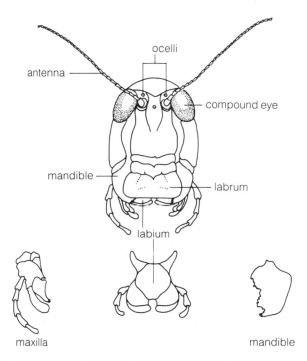

Figure 28-12 Structure of the head region of a grasshopper

sects, with modifications for chewing, sucking, or piercing. With the dissection microscope, examine the mouthparts of a variety of specimens your instructor has provided, and speculate on the feeding behaviors of these insects.

C. Class Chilopoda

The centipedes are predaceous arthropods adapted for running. The body is flattened dorsoventrally, with one pair of legs per segment except for the head and the rear two segments (Fig. 28-13).

a   b   c

Figure 28-13 Examples of members of the classes Chilopoda (the centipedes) (**a**), Diplopoda (the millipedes) (**b**), and Arachnida (the spiders, scorpions, etc.) (**c**). (Photos courtesy Carolina Biological Supply Company.)

The first segment of the body bears a pair of legs modified as *poison claws* for seizing and killing prey. In some tropical species as long as 20 cm, the poison can be dangerous to humans.

Centipedes are swift arthropods and live under stones or the bark of logs. They may be abundant on the forest floor in your area. Their prey consists of other arthropods, worms, and mollusks. The common house centipede eats roaches, bedbugs, and other insects.

Examine several different species of centipedes.

D. Class Diplopoda

Although superficially similar, the millipedes and centipedes differ in many ways. The millipedes are typically cylindrical in body form. All segments, except those of the short thorax, bear two pairs of legs and two sets of spiracles, and there are no poison claws.

Millipedes are very slow arthropods that live in dark, moist places. You may find them in the same habitat as centipedes. They scavenge on decaying organic matter. When handled or disturbed, they often curl up into a ball and secrete a noxious fluid from their scent glands as a means of defense. Examine the demonstration specimens of millipedes in your lab.

1. Examine several different species of millipedes.

2. If your instructor has provided live specimens, hold one to your nose to see if you can detect a smelly secretion.

E. Class Arachnida

This class includes spiders, scorpions, ticks, and mites. The body of spiders and scorpions consists of a *cephalothorax* and *abdomen.* In ticks and mites, these parts are fused. The cephalothorax bears six pairs of appendages, the rear four of which are walking legs. How many pairs of walking legs do insects have?

_____

Arachnids have no true mandibles or antennae, and the eyes are simple. In spiders, the first pair of appendages bear fangs. Each fang has a duct that is connected to a poison gland. The second pair of appendages are used to chew and squeeze their prey. They are also sensory and are used by males to store and transfer to females their sperm.

Most, but not all, spiders spin webs to catch their prey. They secrete digestive enzymes into the prey's (usually an insect's) body, and the liquified remains are sucked up.

Examine the assortment of arachnids on display in the lab. How do ticks and mites obtain nourishment?

_____

_____

_____

_____

PRE-LAB QUESTIONS

_____ **1.** The protostomes are animals whose (a) stomach is in front of the crop, (b) mouth is covered by a fleshy lip, (c) blastopore becomes a mouth, (d) digestive tract is lined by mesoderm.

_____ **2.** One of the two major distinguishing characteristics of mollusks is (a) the presence of three body regions, (b) the mantle, (c) segmentation of the body, (d) jointed appendages.

_____ **3.** Earthworms exhibit metamerism, defined as the (a) division of the body into a series of similar segments, (b) presence of a true coelom, (c) difference in size of the male and female, (d) presence of a "head" equipped with sensory organs.

_____ **4.** The copulatory organ of the earthworm is the (a) penis, (b) clitellum, (c) gonopodium, (d) vestibule.

_____ **5.** The classes (Oligochaeta, Polychaeta, and Hirudinea) of the phylum Annelida are distinguished partly on the basis of the extent or number of (a) segments of the body, (b) eyes, (c) antennae, (d) setae.

_____ **6.** The animal phylum with the most species is (a) Mollusca, (b) Annelida, (c) Arthropoda, (d) Onychophora.

_____ **7.** Sexual dimorphism, as seen in the crayfish, is (a) the presence of male and female individuals, (b) the production of eggs and sperm by the same individual, (c) another term for copulation, (d) the presence of observable differences between males and females.

_____ **8.** The grinding apparatus of the digestive system of the crayfish is the (a) oral teeth, (b) gizzard, (c) pharyngeal jaw, (d) gastric mill.

_____ **9.** The insects were the first organisms to (a) bury their dead, (b) exhibit segmentation, (c) fly, (d) produce lungs.

_____ **10.** Like the insects, the arachnids (spiders, etc.) have (a) three pairs of walking legs, (b) one pair of antennae, (c) true mandibles, (d) an abdomen.

EXERCISE 28
MOLLUSKS, SEGMENTED WORMS, AND
JOINT-LEGGED ANIMALS

POST-LAB QUESTIONS

1. Indicate the differences between protostomes and deuterostomes, and list the phyla of animals in each group.

2. Explain what causes the growth ridges found on the shells of bivalves.

3. Define metamerism, and indicate how this phenomenon is exhibited by the annelids, arthropods, and *your* body.

4. Define sexual dimorphism, describe it in the crayfish, and list two other animals that exhibit this phenomenon.

5. How has flight been at least partly responsible for the success of the insects?

6. List several differences between insects and arachnids.

ECHINODERMS, HEMICHORDATES, AND CHORDATES

OBJECTIVES After completing this exercise you will be able to:

1. define endoskeleton, water vascular system, madreporite, tube foot, notochord, gill slits, gill arches, dorsal hollow nerve cord, lateral line, placoid scale, operculum, atrium, ventricle, artery, vein, cranium, ectothermic, endothermic, viviparous;

2. describe the natural history of members of the phyla Echinodermata, Hemichordata, and Chordata;

3. identify representatives of the classes Asteroidea, Ophiuroidea, Echinoidea, Holothuroidea, and Crinoidea;

4. compare and contrast representatives of the phylum Hemichordata and the chordate subphyla, Cephalochordata and Urochordata;

5. compare and contrast representatives of the vertebrate classes Agnatha, Chondrichthyes, Osteichthyes, Amphibia, Reptilia, Aves, and Mammalia;

6. identify structures (and indicate associated functions) of the representatives of these phyla, subphyla, and classes;

7. associate structural features of vertebrate skeletons with various locomotory behaviors.

INTRODUCTION The echinoderms and chordates are deuterostomes, animals whose blastopore develops into an anus. Although echinoderms and chordates look different, the chordates appear to have evolved from echinoderm ancestors some 600 million years ago. The most popular hypothesis explaining the origin of the chordates states that these animals evolved from the bilaterally symmetrical larvae of ancestral echinoderms. This idea is also supported by similarities between echinoderms and chordates in blood serum proteins and in the way mesoderm forms during development.

I. PHYLUM ECHINODERMATA

(Starr, pp. 278, 279)

Echinoderm means "spiny skin." Members of this phylum are the spiny-skinned animals—the sea stars, brittle stars, sea urchins, sea cucumbers, sea lilies, and feather stars. These animals are all marine, living on the bottom of both shallow and deep seas. Their feeding methods range from trapping organic particles and plankton (sea lilies and feather stars) to scavenging (sea urchins) and predatory behavior (sea stars).

The echinoderms exhibit five-part radial symmetry and a calcareous **endoskeleton** (internal skeleton containing calcium salts) composed of many small plates. Part of the coelom is taken up by the **water vascular system,** important to movement, attachment, respiration, food handling, and sensory perception. You will examine the sea star as a representative echinoderm.

MATERIALS

Per student:

- dissecting scissors
- blunt probe or dissecting needle

Per student pair:

- dissection microscope
- preserved sea star
- wax dissection pan
- dissection pins

Per lab room:

- collection of preserved or mounted brittle stars
- collection of sea urchins and sand dollars (preserved specimens and shells)
- collection of preserved sea cucumbers

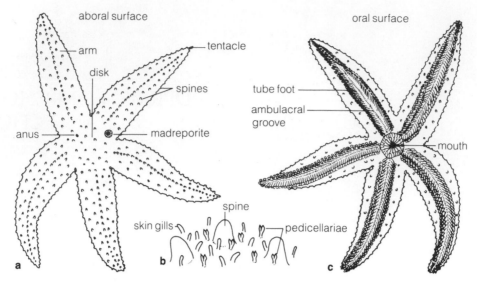

Figure 29-1 Aboral (**a**) and oral (**c**) views of a sea star. (**b**) shows aboral features. (After Abramoff and Thomson, 1982.)

- collection of preserved or mounted feather stars and sea lilies

- large plastic bag for disposal of dissected specimens

PROCEDURE

A. Class Asteroidea

The sea stars are familiar occupants of the sea along shores and coral reefs. They are slow-moving animals that sometimes gather in large numbers on rocky substrates. Many are brightly colored. Large sea stars may be more than 1 meter in size from the tip of one arm to the tip of another.

CAUTION *Preserved specimens are kept in a formalin-based or other preservative solution. Wash any part of your body exposed to this solution with copious amounts of water. If preservative solution is splashed into your eyes, wash them with the safety eyewash bottle for 15 minutes.*

---

1. Obtain a preserved specimen of a sea star and keep your specimen moist with water in a wax dissection pan. Note the central disk on the upper side (the side without the mouth), the five *arms,* and the **madreporite,** a light-colored calcareous sieve near the edge of the disk between two arms (Fig. 29-1a). The madreporite is the opening of the water vascular system.

2. With a hand lens or dissection microscope, note the many spines scattered over the surface of the body. Near the base of the spines are many small pincerlike structures, the *pedicellariae.* These struc-

tures grasp objects that land on the surface of the body. For what might these be an adaptation?

Also among the spines are many soft, hollow skin gills that communicate directly with the coelom and function in gaseous exchange.

3. The *mouth* is located on the *oral* side. An *ambulacral groove* extends from the mouth down the middle of the oral side of each arm. Numerous **tube feet** extend from the water vascular system and occupy this groove (Fig. 29-1c). Each tube foot consists of a bulb-like structure attached to a sucker (Fig. 29-2). The amount of water in the bulb of a tube foot determines whether the tube foot applies suction to the substrate or releases from the substrate. By alternating this suction and release, the animal moves across the substrate. The suction created by the tube feet is also used to adhere to the shells of bivalves as the sea star uses muscular action to pry the shells open to get at their soft insides. The tube feet also function in gaseous exchange.

4. With your dissecting scissors cut across the top of an arm about 1 cm from the tip. Next, cut out a rectangle of spiny skin by carefully cutting along each side of the arm to the central disk and then across the top of the arm at the margin (edge) of the disk. Observe the hard, calcareous plates of the skeleton as you cut. From the upper surface, observe within the arm the large *coelom,* which contains the internal organs (Fig. 29-2).

5. Cut around the madreporite to remove the upper portion of the body wall of the central disk. The

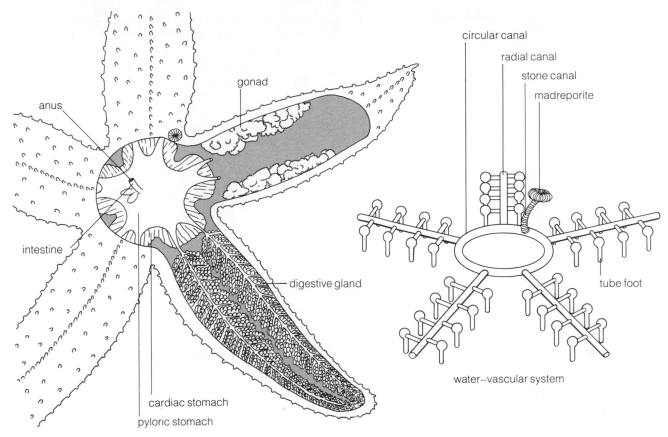

Figure 29-2 Internal anatomy of a sea star. (After Lytle and Wodsedalek, 1984, and Starr and Taggart, 1987.)

mouth connects with an extremely short esophagus that leads to the pouchlike *cardiac stomach*. The cardiac stomach opens into the upper *pyloric stomach*. A slender, short *intestine* leads from the upper side of the stomach to the *anus*. The two green, fingerlike bodies in each arm are the *digestive glands* that produce digestive enzymes and deliver them to the pyloric stomach.

6. Dark *gonads* are located near the base of each arm. The sexes are separate but are difficult to distinguish, except by microscopic examination.

7. The water–vascular system is unique to echinoderms. The madreporite leads to a short *stone canal*, which in turn leads to the *circular canal* encircling the mouth. Five *radial canals* lead from the circular canal into the ambulacral grooves. Each radial canal connects by short side branches with many pairs of *tube feet*.

8. The nervous system (not shown in Fig. 29-2) is simple. A circular *nerve ring* surrounds the mouth, and a *radial nerve* extends from this into each arm, ending at a light-sensitive eyespot. There are no excretory organs. Excretion is accomplished by wandering amoebocytes in the coelomic fluid. Dispose of your dissected specimen in the large plastic bag provided for this purpose.

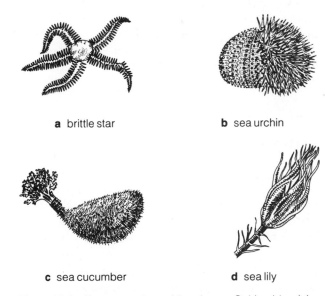

a brittle star   **b** sea urchin

c sea cucumber   **d** sea lily

Figure 29-3 Representatives of the classes Ophiuroidea (**a**), Echinoidea (**b**), Holothuroidea (**c**), and Crinoidea (**d**)

B. Classes Ophiuroidea, Echinoidea, Holothuroidea, Crinoidea

Examine representatives of the other echinoderm classes—Ophiuroidea, Echinoidea, Holothuroidea, and Crinoidea (Fig. 29-3).

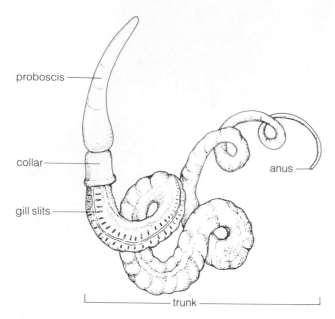

proboscis

collar

gill slits

anus

trunk

Figure 29-4 General body structure of an acorn worm

II. PHYLUM HEMICHORDATA

The acorn worms are related to the chordates, but just what the relationship is or how close it is is not known. The key chordate characters are little developed in the hemichordates. These marine animals are long, slender, wormlike organisms found in tidal flats. They are stationary mud burrowers and filter feeders.

MATERIALS

Per student:

- blunt probe

Per student pair:

- dissection microscope
- preserved acorn worm

PROCEDURE

1. Using the dissection microscope, study a preserved acorn worm, *Balanoglossus*. In front, the acorn worms have a tough, flexible, muscular snout, or *proboscis*, that is used as a burrowing organ (Fig. 29-4).

2. Behind the snout is the thick *collar*, the "cap" of the acorn, that contains a *dorsal, hollow nerve cord* and also a ventral nerve cord. Behind the collar is the trunk bearing two rows of **gill slits.** The *anus* opens at the end of the body.

III. PHYLUM CHORDATA   (Starr, pp. 280–288)

The word *chordate* refers to one of the three major characteristics of members of this phylum, the **notochord.** This structure is a dorsal, flexible rod that pro-

vides support for most of the length of the body during at least some part of the life cycle. In addition to the notochord, chordates possess **gill slits** during at least some portion of their life cycle. These structures persist through adulthood in the protochordates and fishes, but the tissues between the gills (**gill arches**) undergo rearrangement to form jaw-support and other structures in the individual development of amphibians, reptiles, birds, and mammals. A third feature of the chordates is the **dorsal, hollow nerve cord.**

MATERIALS

Per student:

- scalpel
- several microscope slides and coverslips
- compound microscope
- forceps
- dissecting scissors
- blunt probe or dissecting needle
- dissection pins

Per student pair:

- dissection microscope
- prepared slide of whole mount of amphioxus
- preserved leopard frog
- wax dissection pan

Per student group (4):

- preserved sea squirt
- plastic mount of sea squirt
- preserved *amphioxus*
- plastic mount of *amphioxus*
- preserved sea lamprey
- preserved hagfish
- preserved dogfish
- preserved yellow perch

Per lab room:

- collection of preserved cartilaginous fishes
- collection of preserved bony fishes
- collection of preserved amphibians
- collection of preserved reptiles
- skeleton of snake
- turtle shell
- an assortment of feathers
- collection of stuffed birds (study skins)
- several field guides to the birds
- collection of stuffed mammals (study skins)
- prepared skeleton of a frog

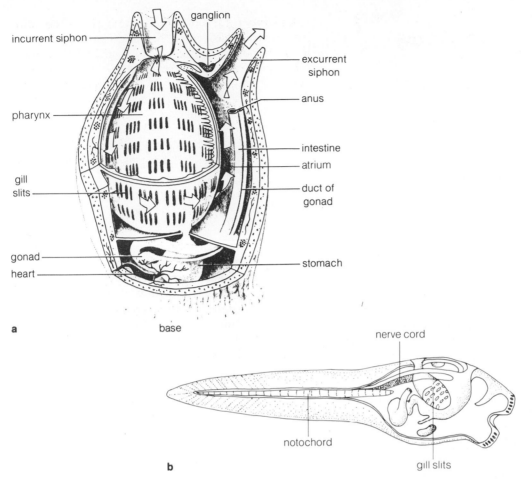

incurrent siphon
ganglion
excurrent siphon
anus
intestine
atrium
duct of gonad
pharynx
gill slits
gonad
heart
stomach

a

base

nerve cord
notochord
gill slits

b

Figure 29-5 Cutaway view of an adult sea squirt (**a**), and the tadpolelike larva (**b**). (After Jensen, Heinrich, Wake, and Wake, 1979.)

- collection of mammalian placentas and embryos (in utero)
- large plastic bag for disposal of dissected specimens

PROCEDURE

A. Subphylum Urochordata

The protochordates include members of the subphyla Urochordata and Cephalochordata. The urochordates are the tunicates, or sea squirts. Both common names are descriptive of obvious features of these organisms. A leatherlike "tunic" covers the adult tunicate, and water is expelled ("squirted") from the animal out an *excurrent siphon*.

1. Examine preserved specimens and a plastic mount of the sea squirt with a dissection microscope (Fig. 29-5a). These animals are small, inactive, and very common marine organisms that inhabit coastal areas of all oceans. The adult is a filter feeder, capturing organic particles by ciliary action into an *incurrent siphon*. Individuals either float freely in the water, sin-

gly or in groups, or are stationary, attached to the bottom as branching individuals or colonies.

2. Water enters a sea squirt through the incurrent siphon. It travels into the *pharynx*, where it seeps through gill slits to reach a chamber, the *atrium*. The water eventually exits through the excurrent siphon. Back in the pharynx, food particles are trapped in sticky mucus and then passed to the digestive tract. Undigested materials are discharged into the atrium to be expelled with water out the excurrent siphon.

3. Urochordate larvae (Fig. 29-5b) resemble tadpoles in general body form and have a notochord confined to the tail; the notochord degenerates when the tadpole becomes an adult sea squirt. The larvae also have a hollow, dorsal nerve cord with a rudimentary brain and sense organs. These structures undergo reorganization into a nerve ganglion and nerve net in the adult.

B. Subphylum Cephalochordata

The lancelets are distributed worldwide and are especially abundant in coastal areas with warm, shallow waters. *Amphioxus* is the commonly studied represen-

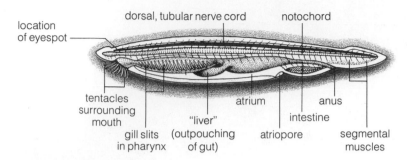

location of eyespot — dorsal, tubular nerve cord — notochord

tentacles surrounding mouth — gill slits in pharynx — "liver" (outpouching of gut) — atrium — atriopore — intestine — anus — segmental muscles

Figure 29-6 *Amphioxus*, showing internal features

tative of this subphylum. The word *amphioxus* means "sharp at both ends," and this is descriptive of the elongate, fishlike body form of cephalochordates.

1. Examine a preserved *amphioxus* (Fig. 29-6). The lancelets are translucent, with a low, continuous *dorsal* and *tail fin*. They reach 50 to 75 mm in length as adults. Despite their streamlined appearance, these organisms are not very active. They spend most of their time buried in the sand with their heads projecting into the water while they filter organic debris from the water.

2. Feeding is similar to that described above for the urochordates. Obtain a plastic mount of *amphioxus*. Referring to Fig. 29-6, trace the flow of water from the mouth to the *atriopore*. Then describe the capture of food and the path it takes through the digestive system, as was done previously for the sea squirt.

3. Find the flexible notochord extending nearly the full length of the individual; this structure persists into the adult stage. There is a small "brain," with a dorsal, hollow nerve cord that bears light-sensitive cells (the *eyespot*) at its front end. About 150 gill slits occur in the pharynx.

C. Subphylum Vertebrata

This is the largest chordate subphylum. The vertebrates have the three primary characteristics of chordates mentioned earlier plus a **vertebral column.** The nerve cord has differentiated into a *brain* and a *spinal cord*. Both of these structures are protected by bones, the brain by the *cranium* and the spinal cord by the *vertebral arches*.

Vertebrates are bilaterally symmetrical, with a fundamental metamerism reflected by segmentation of muscles, vertebrae, and ribs. The body is typically divided into a *head, neck, trunk,* and *tail*. If present, the lateral appendages are paired *thoracic* (pectoral fins, forelimbs, wings, and arms) and *pelvic* (pelvic fins, hindlimbs, and legs) *limbs* that support the body or serve in movement.

1. Class Agnatha. The agnathans were the first vertebrates to evolve. They include the present-day lampreys and hagfishes.

a. Examine a preserved sea lamprey, *Petromyzon marinus* (Fig. 29-7). The word *agnatha* means "without jaws." As you can see from examining the lamprey, there are no jaws but rather a round, sucker-like mouth within which are circular rows of horny, rasping teeth and a deep, rasping tongue. There also are no paired appendages (fins), although there are two dorsal fins and a tail fin. The gill slits and notochord persist in the adult, although some vertebral elements are present.

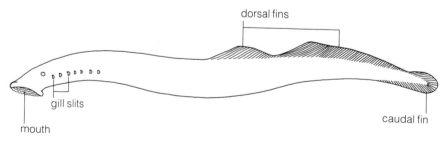

dorsal fins

gill slits

mouth

caudal fin

Figure 29-7 External appearance of the sea lamprey, *Petromyzon marinus*. (After Kent, 1983.)

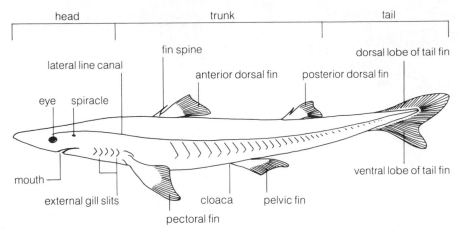

head   trunk   tail

fin spine
lateral line canal
anterior dorsal fin   posterior dorsal fin   dorsal lobe of tail fin
eye   spiracle

mouth
external gill slits   cloaca   pelvic fin   ventral lobe of tail fin
pectoral fin

Figure 29-8 External appearance of the dogfish (shark), *Squalus acanthias*. (After S. Wischnitzer, *Atlas and Dissection Guide for Comparative Anatomy*, Third Edition. Copyright © 1967, 1972, 1979 W. H. Freeman and Company. Used by permission.)

b. Examine a preserved hagfish, *Myxine glutinosa*. The hagfishes are marine, and the lampreys are represented by both marine and freshwater species. The marine lampreys spawn in fresh waters. Both hagfishes and lampreys feed on the tissue fluids of fishes, rasping wounds in the sides of their prey. A landlocked population of the sea lamprey is infamous for having nearly ruined the commercial fish industry in the Great Lakes until that industry was restored recently by controlling this pest. The hagfish is well known for its habit of burrowing inside the body of a fish prey and "eating its way out."

2. Class Chondrichthyes. The first vertebrates to evolve jaws and paired appendages, according to the fossil record, were the heavily armored *placoderms*. These fishes arose about 420 million years ago. The evolution of jaws and paired appendages was a critical event in the history of vertebrate evolution, allowing predation on larger and more active prey. Current hypotheses suggest that the placoderms ultimately gave rise to the cartilaginous and bony fishes.

The cartilaginous fishes are jawed fishes with cartilaginous skeletons. This mostly marine group includes the carnivorous skates, rays, and sharks. Paired appendages are present in these organisms.

a. Examine the assortment of cartilaginous fishes on display.

b. Examine a preserved dogfish (shark), *Squalus acanthias* (Fig. 29-8). How many dorsal fins are there?

---

c. Notice the front paired appendages, the *pectoral fins*, and the back pair, the *pelvic fins*. There are five to seven pairs of naked gill slits in cartilaginous fishes, six in the dogfish. The most anterior gill slit is called a

spiracle and is located just behind the eye. The *tail fin* has a dorsal lobe larger than the ventral one. In the male, the pelvic fins bear *claspers*, thin processes for transferring sperm to the oviducts of the female. Examine the pelvic fins of a female and a male. Also examine the nearby *cloaca*. The **cloaca** is the terminal chamber that receives the products of the digestive, excretory, and reproductive systems.

d. The *nostrils* open into blind olfactory sacs; they do not connect with the pharynx, as your nostrils do. They function solely in olfaction (smell) in the cartilaginous fishes. The *eyes* are effective visual organs at short range and in dim light. There are no eyelids.

e. Find the dashed line that runs along each side of the body. This is called the **lateral line** and functions in the detection of vibrations in the water. It consists of a series of minute canals perpendicular to the surface that contain sensory hair cells. When the hairs are disturbed, a nerve impulse is initiated that ultimately results in the detection and interpretation of the disturbance.

f. The body is covered by **placoid scales,** toothlike outgrowths of the skin. Run your hand from head to tail along the length of the animal. How does it feel?

Now run your hand in the opposite direction along the animal. How does it feel this time?

Cut out a small piece of skin with your scalpel and examine it with the dissection microscope. What do the scales look like? Now you can understand why the skin feels as it does!

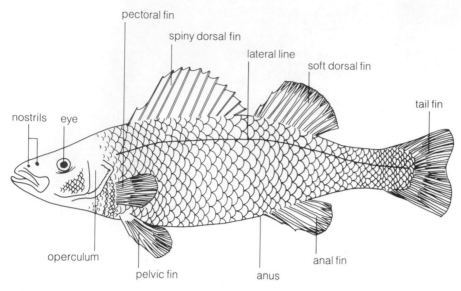

nostrils

eye

pectoral fin

spiny dorsal fin

lateral line

soft dorsal fin

tail fin

operculum

pelvic fin

anus

anal fin

Figure 29-9 External anatomy of a bony fish, *Perca flavescens*

3. Class Osteichthyes. These are the bony fishes, the fishes with which you are most familiar. These fishes inhabit virtually all the waters of the world. They are economically important, commercially and as game species. The skeleton is at least partly ossified (bony), and the flat scales that cover at least some of the surface of most bony fishes are bony as well.

a. Examine the assortment of bony fishes on display.

b. Examine a yellow perch, *Perca flavescens*, as a representative advanced bony fish (Fig. 29-9). Notice that the *nostrils* are double, leading into and out of *olfactory sacs*. The eyes are similar to those in sharks, with no eyelids. Gill arches and slits are housed in a common chamber covered by a bony, movable flap, the **operculum.** A spiny *dorsal fin* is in front of a soft *dorsal fin.* Paired *pectoral* and *pelvic fins* are present, and behind the *anus* is an unpaired *anal fin.* The *tail fin* consists of dorsal and ventral lobes of approximately equal size. The fins, as in the cartilaginous fishes, are used for braking, steering, and maintaining an upright position in the water.

c. Find the **lateral line.** This functions similarly to that of the shark.

d. Bony scales cover the body. Remove a scale, make a wet mount, and examine it with the compound microscope. Locate the *annual rings* that indicate the age of the fish. How are the annual rings of the fish scale analogous to the annual ridges of the mussel or clam shell?

4. Class Amphibia. The amphibians were the first vertebrates to assume a terrestrial existence, having evolved from a group of "lobe-finned fishes." The paired appendages are modified as legs that support the individual during movement on land. Respiration is by lungs, gills, and the highly vascularized skin and lining of the mouth. Reproduction requires water, or at least moist conditions on land. The larvae generally live in water. The skeleton is more bony than that of the bony fishes, but a considerable proportion of it remains cartilaginous. The skin is usually moist, with mucous glands; scales are usually absent. This group of vertebrates includes the familiar frogs, toads, and salamanders.

a. Examine preserved specimens of a variety of amphibians.

b. The leopard frog, *Rana pipiens*, illustrates well the general features of the vertebrates and the specific characteristics of the amphibians. Study a preserved specimen after rinsing it in fresh water. Examine the external anatomy of the frog (Fig. 29-10).

c. Find the two nostrils at the tip of the head. These are used for inspiration and expiration of air. Just behind the eye is a disklike structure, the *tympanum*, the outer wall of the middle ear. There is no external ear. The tympanum is larger in the male than in the female. Examine the frogs of other students in your lab. Is your frog a male or female?

d. At the back end of the body locate the *cloacal opening*.

e. The forelimbs are divided into three main parts: the *upper arm, forearm,* and *hand.* The hand is divided into a *wrist, palm,* and *fingers* (digits). The three divisions of the hindlimbs are the *thigh, shank* (lower leg),

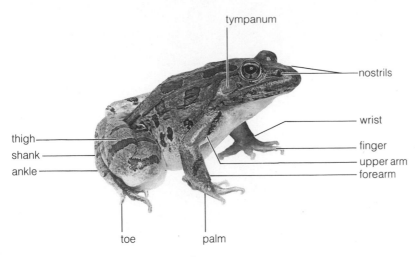

Figure 29-10   External anatomy of the frog. (Photo courtesy Carolina Biological Supply Company.)

and *foot*. The foot is further divided into three parts: the *ankle*, *sole*, and *toes* (digits).

f. Fasten the frog, ventral side up, with pins to the wax of a dissection pan. Lift the skin with forceps. Then make a superficial cut with your scissors from the end of the trunk forward, and just left or right of center, to the tip of the lower jaw. Pin back the skin on both sides. Refer to Fig. 29-11 to study the internal anatomy.

g. Note the white line along the midventral line and the large abdominal muscles you have exposed. Lift these muscles with your forceps, and cut through the body wall with your scissors from the end of the trunk to the tip of the lower jaw, cutting through the *sternum* (breastbone) but not damaging the internal organs. Pin back the body wall as you did the skin to expose the internal organs.

h. Locate the *spleen* and the following organs of the digestive system: *mouth, pharynx* (throat), *esophagus, liver, gall bladder, stomach, pancreas, small intestine, large intestine* (colon), and *cloaca*. Go back to the esophagus and find the *bronchi* (the singular is *bronchus*) that lead toward the *lungs*. Are the bronchi dorsal or ventral to the esophagus?

_____

i. Amphibians have a three-chambered *heart*, with two thin-walled **atria** and a thick-walled **ventricle**. In the adult frog, most of the venous (deoxygenated) blood moves from the heart to the *lungs* and back before it is pumped to the various regions of the body. There is a *pulmocutaneous circulation*, involving the lungs and the skin, and a *systemic circulation*, involving the rest of the body. Blood is carried from the ventricle to the systemic circulation by a large artery, the *aorta*.

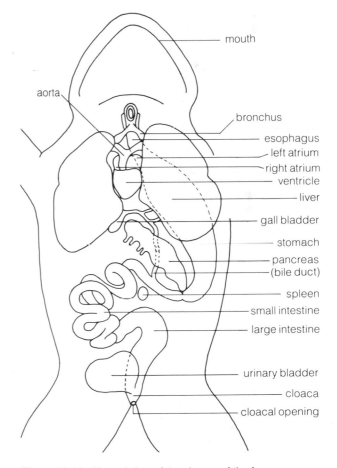

Figure 29-11   Ventral view of the viscera of the frog

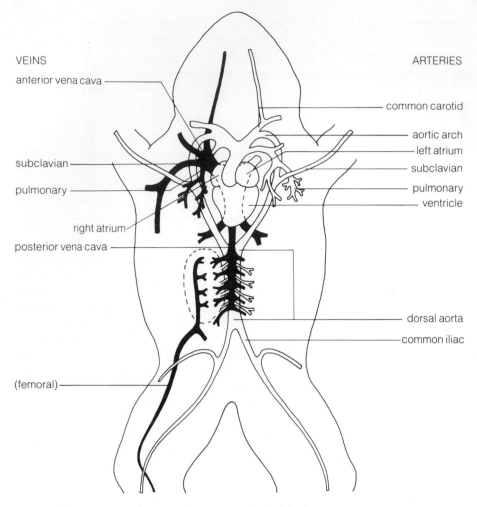

anterior vena cava

common carotid

aortic arch

left atrium

subclavian

subclavian

pulmonary

pulmonary

ventricle

right atrium

posterior vena cava

dorsal aorta

common iliac

(femoral)

Figure 29-12 Ventral view of the major blood vessels of the frog

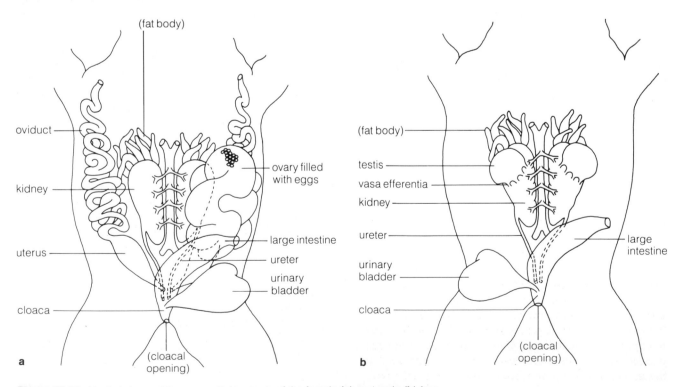

(fat body)

oviduct

ovary filled
with eggs

kidney

uterus

large intestine

ureter

urinary
bladder

cloaca

a

(cloacal
opening)

(fat body)

testis

vasa efferentia

kidney

ureter

urinary
bladder

large
intestine

cloaca

b

(cloacal
opening)

Figure 29-13 Ventral views of the urogenital systems of the female (**a**) and male (**b**) frog

j. Arteries carry blood (usually oxygenated) away from the heart, and **veins** carry blood (usually deoxygenated) toward the heart from the tissues. Refer to Fig. 29-12 and find the following major arteries: *pulmonary, aortic arch, subclavian, dorsal aorta,* and *common iliac.*

Where does blood in the dorsal aorta go?

_____

_____

Does the pulmonary artery contain oxygenated or deoxygenated blood?

_____

k. Locate the following major veins: *anterior* and *posterior vena cavae, subclavian, femoral,* and *pulmonary.* Find the lungs and describe the path of the blood from the tissues of the body, through the major veins to the heart, and through the major arteries ultimately back to the tissues.

_____

_____

_____

_____

l. The excretory and reproductive organs together comprise the *urogenital system.* Locate the urine-producing excretory organs, the pair of *kidneys* located dorsally in the body cavity. A duct, the *ureter,* leads from the kidney to the *urinary bladder,* an organ that stores urine for resorption of water from the urine into the circulatory system. The bladder empties into the cloaca.

m. In the female, find the *ovaries;* these organs expel eggs into the *oviducts* (Fig. 29-13). The oviducts lead into the *uterus.* The reproductive tract ends in the cloaca.

n. In the male, locate the *testes;* these organs produce sperm that are carried to the kidneys through tiny tubules, the *vasa efferentia* (Fig. 29-13). The ureters serve a dual function in male frogs, transporting both urine and sperm to the cloaca.

o. The nervous system of vertebrates is comprised of two divisions: (1) the *central nervous system,* the brain and spinal cord, and (2) the *peripheral nervous system,* nerves extending from the central nervous system. Turn your frog over and remove the skin from the dorsal surface of the head between the eyes and along the vertebral column. With your scalpel, shave thin sections of bone from the skull, noting the shape and size of the **cranium** (braincase), until you expose the *brain.* Pick away small pieces of bone with your forceps to expose the entire brain. Use the same procedure to expose the vertebrae and *spinal cord.* Note the *cranial nerves* coming from the brain and the *spinal nerves*

from the spinal cord. Dispose of your dissected specimen in the large plastic bag provided for this purpose.

p. The skeleton of vertebrates consists of the following: (1) the *axial skeleton* and its *skull, sternum,* and *vertebral column,* and (2) the *appendicular skeleton* and its *girdles* (support structures) and their appendages. Find the bones illustrated in Fig. 29-14 on the prepared skeleton of a frog.

5. Class Reptilia. This diverse group of vertebrates includes the turtles, lizards, snakes, crocodiles, and alligators. These animals, like the fishes and reptiles, are largely **ectothermic,** meaning that without the capability of maintaining their body temperatures physiologically their body temperature is more likely to vary with the outside temperature. There is no larval stage. The skeleton is well ossified. The skin is dry and covered by epidermal scales. There are virtually no skin glands. This is the first group of vertebrates that adapted to strictly living on land, although many do live in fresh or sea water. In most reptiles, the heart consists of two atria and a partially divided ventricle, so there is still some mixing of oxygenated and deoxygenated blood. The nervous system is more highly developed than that of amphibians.

The *amniotic egg,* first developed by the reptiles and characteristic of the birds and mammals as well, is an adaptation to an entire life cycle on land. The embryo is suspended in an internal aquatic environment, and the shell prevents its drying out.

Examine an assortment of preserved reptiles and note their diversity as a group. Look at a skeleton of a snake and the inside of the *carapace* (the upper portion) of a turtle shell. What portions of the individual are incorporated into the shell?

_____

_____

6. Class Aves. The birds are **endothermic** vertebrates, capable of maintaining their body temperatures physiologically. Their body is covered with *feathers,* and scales, reminiscent of their reptilian heritage, are present on the feet. The front limbs are modified as *wings* for flight in most birds. An additional internal adaptation for flight is the well-developed *sternum* (breastbone), with a *keel* for the attachment of powerful muscles for flight. The major bones of birds are hollow and contain air sacs connected to the lungs. Birds have a four-chambered heart, with two atria and two ventricles. This permits the complete separation of oxygenated and deoxygenated blood. Why is this circulatory arrangement an advantage for birds compared to an arrangement similar to that found in amphibians and reptiles? (*Hint:* Recall that birds are endothermic and can fly.)

_____

_____

_____

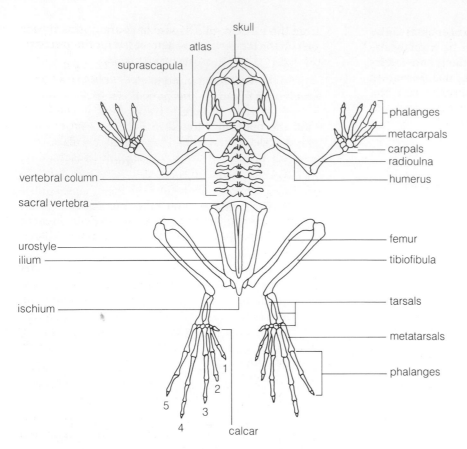

Figure 29-14 Dorsal view of the skeleton of a frog

The nervous system is similar to that found in reptiles. However, the brain is larger, permitting more sophisticated behavior and muscular coordination; the optic lobes are especially well developed in association with a keen sense of sight.

a. Examine a feather with the dissection microscope, and note the *rachis* (shaft), *vane,* and *barbs* and *barbules* that comprise the vane (Fig. 29-15). What else does the bird use its feathers for besides flight?

_____

_____

_____

b. The upper and lower jaws are modified as variously shaped *beaks* or *bills,* with the shapes reflecting the feeding habits of the species. No teeth are present in adults. Examine the assortment of birds, and speculate on their food preferences by studying the configurations of their bills. Check your conclusions with a field guide or other suitable source that describes the food habits.

7. Class Mammalia. Like the birds, these vertebrates are endothermic. The term *Mammalia* stems from the

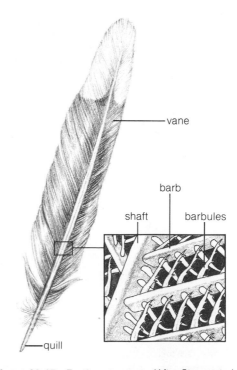

Figure 29-15 Feather structure. (After Storer *et al.,* 1979.)

mammary glands that all female mammals possess. The heart is four-chambered, as in birds. The brain is relatively larger than that of other vertebrates, and its surface area is increased by grooves and folds. The outer ear frequently has a cartilaginous portion called the *pinna*. Mammals are equipped with modifications and outgrowths of the skin, in addition to hair. There are four types of skin glands: *mammary, sebaceous, sweat,* and *scent glands. Hair* is present during some portion of the life cycle.

a. Study the assortment of mammals on display, and list below as many functions for hair as you can.

_____

_____

_____

b. Most mammals are **viviparous,** the female bearing live young after supporting them in the uterus with a *placenta,* a nutritive connection between the mother and the embryo. Examine the placentas and embryos in uteri of mammals on display in the lab. What relationship between the early development of mammals and the comparative sophistication of the nervous system and behavior of adult mammals can you suggest?

_____

_____

PRE-LAB QUESTIONS

_____ **1.** The deuterostomes are animals whose (a) blastopore becomes a mouth, (b) blastopore becomes an anus, (c) digestive system opens into a cloaca, (d) circulatory systems are open.

_____ **2.** The skin gills of sea stars are organs (a) for gaseous exchange, (b) of defense, (c) for movement, (d) that produce the skeletal elements.

_____ **3.** The tube feet of a sea star (a) bear tiny toes, (b) are located only at the tips of the arms, (c) function in gaseous exchange, (d) protect the organism from predatory fish.

_____ **4.** The madreporite, stone canal, circular canal, and radial canals of a sea star are components of its (a) nervous system, (b) water–vascular system, (c) digestive system, (d) excretory system.

_____ **5.** The tunicates and lancelets have an inner chamber from which water is expelled. This chamber is the (a) intestine, (b) bladder, (c) atrium, (d) nephridium.

_____ **6.** Which of the following is *not* a major characteristic of the chordates? (a) dorsal hollow nerve cord, (b) notochord, (c) gill slits, (d) vertebral column.

_____ **7.** The lampreys and hagfishes are unusual vertebrates in that they have no (a) eyes, (b) jaws, (c) gill slits, (d) mouth.

_____ **8.** The structure of the cartilaginous and bony fishes that detects vibrations in the water is the (a) anal fin, (b) operculum, (c) nostrils, (d) lateral line.

_____ **9.** Amphibians have (a) a two-chambered heart, (b) a three-chambered heart, (c) a four-chambered heart, (d) none of the above.

_____ **10.** The amniotic egg of reptiles, birds, and mammals is an adaptation to (a) carnivorous predators, (b) a life on land, (c) compensate for the short period of development of the young, (d) protect the young from the nitrogenous wastes of the mother during the formation of the embryo.

EXERCISE 29
ECHINODERMS, HEMICHORDATES, AND CHORDATES

POST-LAB QUESTIONS

1. Explain how the tube feet of the water–vascular system of a sea star function in movement and predatory behavior.

2. List similarities and differences between the hemichordates and chordates.

3. List the three major structural features of chordates, and describe how these are exhibited in the chordate sub-phyla: Urochordata, Cephalochordata, and Vertebrata.

4. List the ways in which agnathans differ from typical vertebrates.

5. List the three divisions (and their components) of the nervous system of vertebrates.

6. Compare the structure (compartmentalization) of the hearts of typical amphibians, reptiles, birds, and mammals, and suggest associations between the several structural patterns and the metabolic demands on members of these vertebrate classes.

ECOLOGY: LIVING ORGANISMS IN THEIR ENVIRONMENT

OBJECTIVES After completing this exercise you will be able to:

1. define ecology, *oikos*, population, habitat, community, ecosystem, producer, trophic level, primary consumer, secondary consumer, decomposer, herbivore, carnivore, omnivore;

2. construct a food web;

3. determine the trophic level to which an organism belongs;

4. identify the three basic types of survivorship curves, and describe the trends exhibited by each.

INTRODUCTION "Ecology." The word probably brings to mind the activities of groups maintaining a clean environment, or lobbying the government to prevent loss of natural areas, or protesting the policy and/or actions of some other groups. To the biologist, this is *not* what ecology as a science is all about. Rather, these activities are environmental concerns that may have grown out of ecological studies.

The word *ecology* is derived from the Greek *oikos*, which means "house." Broadly defined, **ecology** is the study of the interactions between living organisms and their environment. Individuals, whether they are bacteria or humans, do not exist in a vacuum. Rather, they interact with one another in complex ways. As such, ecology is an extremely diverse and complex study. There are so many aspects of ecology that it is difficult to pick one, or even a few, to represent the entire science.

Let's look at the levels of organization considered in an ecological study.

A **population** of organisms is a group of individuals *of the same species* occupying a given area at a given time. For example, you and your colleagues in the laboratory at the moment are a population; you are all *Homo sapiens*. Your population consists of humans in biology class right now. The place you are occupying, the lab room, is your **habitat**. Of course, you are sharing your habitat with other organisms; unseen bacteria and possibly insects are present in the room as

well. Maybe there are plants decorating the lab room; if so, there are other populations in your habitat.

A **community** consists of all those populations of *all* species occupying a given area at a given time. To use our simple example, our laboratory community consists of humans, bacteria, insects, and plants.

Our community exists in a physical and chemical environment. The combination of the community and its environment comprise an **ecosystem.** Our laboratory ecosystem consists of humans, bacteria, insects, plants, an atmosphere, lab benches, walls, floors, light, heat, etc. As you see, an ecosystem has both living and nonliving components.

In this exercise, we will introduce some of the principles of ecology. We will study three concepts to illustrate how ecological studies are performed: food webs, survivorship curves, and population density and distribution.

I. FOOD WEBS

(Starr, pp. 525–527)

Let's consider the community of organisms in any environment. Each population plays a different role within the structure of the community. Each community consists of producers, consumers, and decomposers.

Producers are autotrophic organisms, such as most plants and a few microorganisms. They utilize the energy of the sun or chemical energy to synthesize organic (carbon-containing) compounds from inorganic compounds.

Consumers are heterotrophic organisms. In ecological studies, consumers are usually classified into feeding levels (**trophic levels**) by what they eat.

- **Primary consumers** are **herbivores,** eating plant material.

- **Secondary consumers** are organisms that eat primary consumers or other secondary consumers. These organisms are **carnivores.** Sometimes **parasites**—organisms that utilize living tissue for their nourishment—are considered secondary consumers as well.

Table 30-1   A Simplified Food Web

Trophic Level	Organism(s)
Decomposers	
Secondary consumers (parasites)	
Secondary consumers (carnivores)	
Omnivores	
Primary consumers (herbivores)	
Producers	

- **Decomposers** include fungi, most bacteria, and some protistans that break down organic material into smaller molecules, which are then recycled in the ecosystem.

Sometimes feeding strategies cross trophic levels. **Omnivores** are organisms that eat either producers or consumers.

How would you classify yourself with respect to your feeding strategy?

_____

PROCEDURE   Below you will find a list of organisms in a forest community and their source(s) of energy. In the blank beside each, list the *most specific* trophic level from the descriptions presented above.

- Human (black raspberries, hickory nuts, deer, rabbits) _____
- Black raspberry (sun) _____
- Deer (plants) _____
- Bear (black raspberries, deer) _____
- Coyote (deer, rabbits, black raspberries) _____
- Nematode (living hickory tree roots) _____
- Bacterial species 1 (black raspberries) _____
- Bacterial species 2 (deer) _____
- Bacterial species 3 (dead oak trees, black raspberries, dead humans, dead black bears, dead deer)

_____

- Mosquito (blood of living humans, deer, and bears)

_____

- Hickory tree (sun) _____
- Cyanobacterium (sun) _____
- Fungal species 1 (living black raspberries) _____
- Fungal species 2 (dead oak trees, dead black raspberries) _____
- Rabbit (black raspberries) _____

Now, in Table 30-1, construct a **food web** by placing the names of the organisms in their appropriate trophic levels. Then connect arrows to the organisms to complete the food web.

II. THE FLOW OF ENERGY THROUGH AN ECOSYSTEM

(Starr, pp. 528–530)

Ecologists have learned through careful measurement that the amount of energy captured decreases at each succeeding trophic level. A green plant is much more efficient at capturing the energy of the sun than is a machine at capturing the energy stored in fossil fuels to create motion. Nonetheless, only a small fraction of the light energy falling on the surface of a leaf is converted to the chemical energy of ATP. A primary consumer may only capture about 10% of the energy stored in a green plant. A secondary consumer may capture only 10% of the energy stored in the body of the animal that *it* eats.

_____

Suppose another secondary consumer eats the first (herbivore-eating) secondary consumer. Assuming a similar flow of energy, how much of the sun's original energy does this secondary consumer gain?

_____

Suppose an omnivore can obtain all the nutritional requirements necessary for life by either eating plant material or animal material. From an energetics standpoint, by which route will the greater amount of sun's energy be captured?

_____

By what different routes could this organism obtain equal amounts of the sun's energy?

_____
_____
_____

III. SURVIVORSHIP[1]   (Starr, pp. 494–503)

Within a population, some individuals die very young while others live into old age. To a large extent, the *pattern* of survivorship is species-dependent. Generally, three patterns of survivorship have been identified. These three have been summarized by *survivorship curves*, graphs that indicate the pattern of mortality (death) in a population.

Humans in highly developed countries with good health-care services are characterized by a Type I curve, where there is high survivorship until some age, then high mortality. The insurance industry has generated information to determine "risk groups." The premiums they charge are based upon the risk group to which an individual belongs.

While survivorship curves for humans are relatively easy to generate, information about other species is more difficult to determine. It can be quite a trick to simply determine the age of an individual plant or animal, not to mention watching an entire population over a period of years. However, the principle of determining survivorship can be demonstrated in the laboratory using nonliving objects.

In this exercise we will study the populations of dice and soap bubbles, using them as models of real populations to construct survivorship curves. We will subject these populations to different kinds of stress to determine the effects upon survivorship curves.

_____

1. Adapted from an exercise courtesy J. Shepherd, Mercer University.

MATERIALS

Per student:

- 15-cm ruler

Per student group (4):

- bucket containing 50 dice
- soap bubble solution and wand
- stopwatch or digital watch

Per lab room:

- overhead transparency of Fig. 30-2 and Fig. 30-3
- 1 set of red, blue, black, and green pens (Sharpies or similar marker)
- overhead projector
- projection screen

PROCEDURE

A. Dice Survivorship

Work in a group of four for this experiment. One person should be assigned to dump the dice, another to record data, while the other two count.

Population 1.

1. Empty the bucket of 50 dice onto the floor.

2. Assume that all individuals that come up as "ones" die of heart disease. All others survive.

3. Pick up all the "ones," set them aside (in the cemetery), and count the number of individuals who have survived. Record the number of *survivors* in this generation (Generation 1) in Table 30-2.

4. Return the survivors to the bucket.

5. Dump the survivors onto the floor again, and remove the deaths ("ones") that occurred during this second generation.

6. Count and record the number of survivors.

7. Continue this process until all the dice have died from heart disease.

Population 2.

1. Start again with a full bucket of 50 dice. Assume that "ones" die of heart disease and "twos" die of cancer. Proceed as described for Population 1, recording numbers in Table 30-2, until all dice are dead.

2. Now determine the percentage of survivors for each generation with the following formula:

$$\text{percentage survivors} = \frac{\text{number surviving}}{50} \times 100\%$$

Table 30-2 Dice Population Data

Generation	Population 1 (heart disease only)		Population 2 (cancer & heart disease)	
	Number Surviving	Percentage Surviving	Number Surviving	Percentage Surviving
0	50	100%	50	100%
1				
2				
3				
4				
5				
6				
7				
8				
9				
10				
11				
12				
13				
14				
15				
16				
17				
18				
19				
20				
21				
22				
23				
24				
25				

B. Soap Bubble Survivorship

Work in groups of four for this experiment. One person will blow bubbles, a second group member will serve as the timer, a third observes survivorship, and the fourth records data in Table 30-3.

Three different populations of soap bubbles will be formed based upon actions of group members,

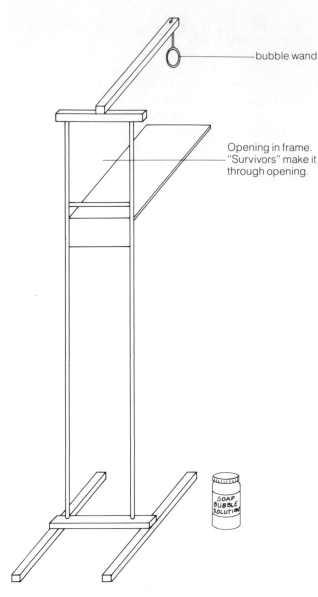

Figure 30-1 Frame for Population 3

with each group being assigned one population to work with during this exercise:

Population 1: Once the bubble leaves the wand, group members wave, blow, or fan in an effort to keep the bubble in the air and prevent it from breaking ("dying").

Population 2: Group members do nothing to interfere with the bubbles or keep them up in the air.

Population 3: This group uses a wand mounted on a wooden frame (Fig. 30-1). The group member blowing bubbles tries to blow the bubbles through the opening in the frame. Bubbles that break rather than passing through the frame are timed and included in the data. Bubbles that fall without passing through or breaking on the frame are ignored (not counted in the data). *Do **not** attempt to manipulate the frame in any way so as to increase the chances that the bubbles will pass through it.*

Age at Death (seconds)	Put a Check for Each Bubble Dying	Total Number Dying at This Age	Number Surviving to This Age	Percentage Surviving to This Age
0	_____	_____	50	100%
1	_____	_____	_____	_____
2	_____	_____	_____	_____
3	_____	_____	_____	_____
4	_____	_____	_____	_____
5	_____	_____	_____	_____
6	_____	_____	_____	_____
7	_____	_____	_____	_____
8	_____	_____	_____	_____
9	_____	_____	_____	_____
10	_____	_____	_____	_____
11	_____	_____	_____	_____
12	_____	_____	_____	_____
13	_____	_____	_____	_____
14	_____	_____	_____	_____
15	_____	_____	_____	_____
16	_____	_____	_____	_____
17	_____	_____	_____	_____
18	_____	_____	_____	_____
19	_____	_____	_____	_____
20	_____	_____	_____	_____
21	_____	_____	_____	_____
22	_____	_____	_____	_____
23	_____	_____	_____	_____
24	_____	_____	_____	_____
25	_____	_____	_____	_____
26	_____	_____	_____	_____
27	_____	_____	_____	_____
28	_____	_____	_____	_____
29	_____	_____	_____	_____
30+	_____	_____	_____	_____

Table 30-3 Soap Bubble Population Data

1. Practice blowing bubbles for a few minutes until they can be generated with the single end of the wand.

2. Once the bubble is free of the wand, the timer should start the watch. When the bubble bursts, the timer notes the time and puts a check mark next to the appropriate age at "death" in Table 30-3.

3. Obtain data on 50 bubbles.

4. Summarize your data as follows:

a. Count the number of checks (the number of bubbles "dying") at each age. Record the number in the column marked "Total Number Dying at This Age."

b. By subtracting the number dying at each age from 50, determine and record the surviving at each age. For example, if 5 bubbles broke (died) at age 1 second, then 50 − 5 = 45 survived at least 1 second.

c. Calculate the percentage surviving at each age. Since at "birth" (moment the bubble left the wand) 50 bubbles were "alive," 100% were alive at age 0. Use the following formula:

$$\frac{\text{percentage surviving}}{\text{to this age}} = \frac{\text{number surviving}}{50} \times 100\%$$

d. Plot the percentage surviving on the graphs in Figs. 30-2 and 30-3. (See Section C.)

C. Plotting Survivorship Curves

1. Use Fig. 30-2 to make an *arithmetic plot* for survivorship of the dice and soap bubble populations. Note that the horizontal lines are all the same distance apart on this graph.

Use open circles to plot percentage of dice surviving at each age (throw of the dice). Use closed circles to plot the percentage of soap bubbles surviving at each age, transposing data from the "Percentage Surviving" columns of Tables 30-2 and 30-3. Connect the points from each population by drawing straight lines with the aid of a ruler.

2. We will use Fig. 30-3 to make a *logarithmic plot* of survivorship. Note that on this *semi-log* paper, the horizontal lines become closer together toward the top of the page.

Perhaps you've never used semi-log paper. Examine Fig. 30-3 more closely. Note that the scale on the vertical axis runs from 1 at the bottom to 9 near the middle, then from 1 to 9 near the top; there is a 1 at the topmost line. The lines are spaced according to the *logarithms* of the numbers.

The 1 at the top represents 100%, the 9 below it 90%, and so on. This means that the 1 in the middle of the page is 10%, while the 9 below it represents 9%; the 8, 8%; and so forth until the 1 at the bottom, which represents 1%.

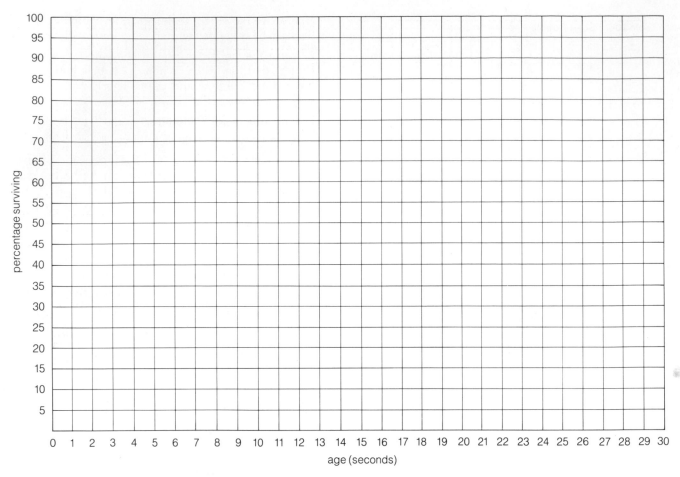

Figure 30-2 Arithmetic plot of survivorship

Plot the survivorship of the dice and soap bubbles that your group generated, transferring the numbers from the "Percentage Surviving" columns from Tables 30-2 and 30-3. Use open circles to plot the percentage of dice surviving and closed circles for the percentage of soap bubbles surviving. Use a ruler to connect the points, forming a curve for each.

After you have made your own plots, using the pens provided, copy the plot from your population onto the overhead transparency your instructor has for this purpose. Different colored pens are available so that each population will have a separate color.

D. Interpretation of the Survivorship Curves

Examine first the plots from the dice populations. In the first population (heart disease only), a constant ⅙, or 17%, of the population dies at each age. In the second population, ⅓, or 33%, dies at each age. As you see, on the *arithmetic plot* these data form a smooth curve, while on the *logarithmic plot* they form a straight line.

Both types of plots provide useful information. A straight line on a logarithmic plot indicates the death rate is *constant*. In the arithmetic plot it is easier to see that more individuals die at a young age than at older ages.

In natural populations, three basic trends of survivorship affecting population size have been identified. These are represented in Fig. 30-4 and summarized in Table 30-4.

Now examine the survivorship curves for the soap bubble population.

How do they compare with the Type I, II, III curves in Fig. 30-4?

_____

_____

Do any of the soap bubble populations show constant death rate for at least part of their lifespan? If so, which?

_____

_____

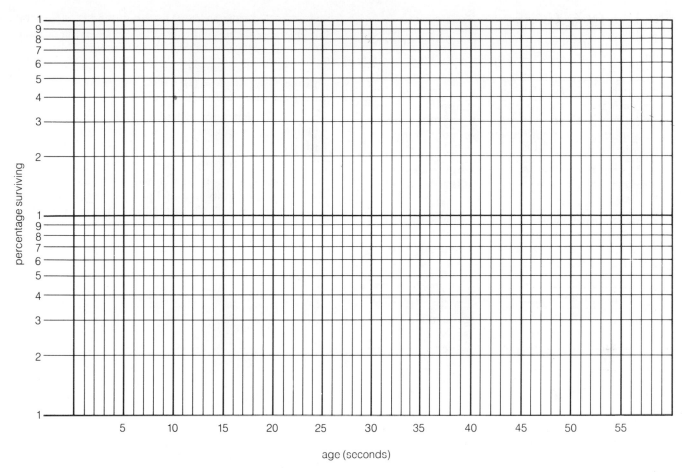

Figure 30-3 Logarithmic plot of survivorship

percentage surviving

age (seconds)

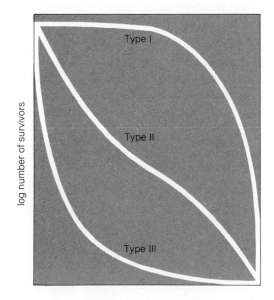

age (percent of lifespan)

Figure 30-4 Three generalized types of survivorship curves. (After Starr and Taggart, 1987.)

Table 30-4 Survivorship Curves	
Type	Trend
I	Low mortality early in life, most deaths occurring in a narrow time span at maturity.
II	Rate of mortality fairly constant at all ages.
III	High mortality early in life.

How did the treatments that Populations 1, 2, and 3 were subjected to affect the shape of the curves?

_____

_____

_____

PRE-LAB QUESTIONS

_____ **1.** The Greek word *oikos* means (a) environment, (b) habitat, (c) house, (d) community.

_____ **2.** A population includes (a) all organisms of the same species in one place at a given moment, (b) the organisms and their physical environment, (c) the physical environment in which an organism exists, (d) all organisms within the environment at a given moment.

_____ **3.** The place that an organism resides is its (a) community, (b) habitat, (c) population, (d) all of the above.

_____ **4.** An ecosystem (a) includes populations but not habitats; (b) is the same as *oikos;* (c) consists of a community and the physical and chemical environment; (d) is none of the above.

_____ **5.** Which of the following is *not* a consumer? (a) bacterium, (b) human, (c) bean plant, (d) fungus.

_____ **6.** The trophic level that an organism belongs to (a) is determined by what it uses as an energy source, (b) determines the population to which it belongs, (c) can be thought of as its feeding level, (d) a and c above.

_____ **7.** A decomposer (a) is exemplified by a fungus, (b) is the same as a parasite, (c) feeds on living organic material, (d) feeds on primary consumers only.

_____ **8.** An herbivore (a) eats plant material, (b) eats primary producers, (c) would be exemplified by a deer, (d) all of the above.

_____ **9.** Which term would best describe a human being? (a) omnivore, (b) herbivore, (c) carnivore, (d) parasite.

_____ **10.** A survivorship curve shows the (a) number of individuals surviving to a particular age, (b) cause of death of an individual, (c) place an organism exists in its environment, (d) trophic level in which an organism is found.

EXERCISE 30
ECOLOGY: LIVING ORGANISMS IN THEIR ENVIRONMENT

POST-LAB QUESTIONS

1. Define the terms *herbivore, carnivore, omnivore.*

2. Define the terms *primary producer, primary consumer, secondary consumer.*

3. Characterize yourself, using the terms listed in questions 1 and 2.

4. Which type of survivorship curve describes a population of organisms that produces a very large number of offspring, most of which die at a very early age, only a few surviving to old age? Give an example of a population of this type.

5. Would you expect a population in which most members survive for a long time to produce few or many offspring? Which would be most advantageous to the population as a whole?

6. Suppose a human population exhibits a Type III survival curve. What would you expect to happen to the curve over time if a dramatic improvement in medical technology takes place?

7. What would you expect to happen to a population where the birth rate is about equal to the death rate?

8. How many humans presently occupy our planet?

9. Is our population increasing, remaining stable, or decreasing?

DISSECTION OF THE FETAL PIG: INTRODUCTION, EXTERNAL ANATOMY, AND THE SKELETAL MUSCLE SYSTEM

OBJECTIVES After completing this exercise you will be able to:

1. define fetus, ungulate, digitigrade locomotion, plantigrade locomotion, antagonistic muscles;

2. locate and describe the external features of a fetal pig;

3. determine the sex of a fetal pig;

4. describe the function of the umbilical cord;

5. define the origin, insertion, and action of a skeletal muscle;

6. identify some of the major skeletal muscles of a mammal.

INTRODUCTION In this and the following two exercises, you will examine in some detail the external and internal anatomy of a fetal (**fetus,** an unborn mammal) pig. As the pig is a mammal, many aspects of its structural and functional organization are identical with those of other mammals, including humans. Thus, a study of the fetal pig is in a very real sense, a study of ourselves.

The specimens you will use in the laboratory were purchased from a biological supply house, which obtained them from a plant where pregnant sows were slaughtered for food. On the average a sow may produce 7–12 offspring per litter. The period of development in the uterus (*gestation period*) is approximately 112–115 days. Generally, lab specimens are approximately 20–30 cm (8–12 inches) long and their age between 100 days to nearly full term.

At the slaughterhouse the fetuses are quickly removed from the sow, cooled, and embalmed with formalin or another preservative, which is injected through one of the umbilical arteries. Following this, the arterial and venous systems are injected under pressure with latex or a rubberlike compound. The arteries are injected with red latex through the umbilical artery, and the venous system is injected with blue latex through one of the jugular veins at the base of the throat.

During the fetal pig exercises keep several points in mind. First, be aware that "to dissect" does not mean "to cut up," but rather primarily "to expose to view." Thus, proceed carefully and never cut or move more than is necessary to expose a given part. Second, for each structure or organ that you identify in the pig, ask yourself if an equivalent one is present in your body. If so, where is it located and is its function similar to that in the fetal pig? Finally, pay particular attention to the spatial relationships of organs, glands, and other structures as you expose them. Realize that their positions in the body are not random. Carefully identify each structure and determine to what organ system it belongs, its relationship to that organ system, and what its general function is. Then determine how it is related to the other organ systems in the body. By proceeding in this manner, you will greatly enhance your understanding of the structure and function of the mammalian body.

To understand the dissection directions, you will need to become familiar with the terms used in virtually all anatomical work (see the appendix on p. 419). Spend a few minutes relating each of these terms to the fetal pig body and to your body as well. Refer to Fig. 31-1 to aid you in this exercise.

I. EXTERNAL ANATOMY OF THE FETAL PIG

Before you begin your examination of the internal structure of the fetal pig, you will examine the external features of its body. This will give you an opportunity to compare the body of the pig with other mammals. Remember, it is the external surface of an organism that has the greatest amount of contact with the environment. Thus, the greatest differences between two organisms may be their external features rather than their internal features.

MATERIALS

Per student pair:

- one fetal pig preserved in formalin or phenol-based chemical and injected with red and blue latex

- dissecting tray

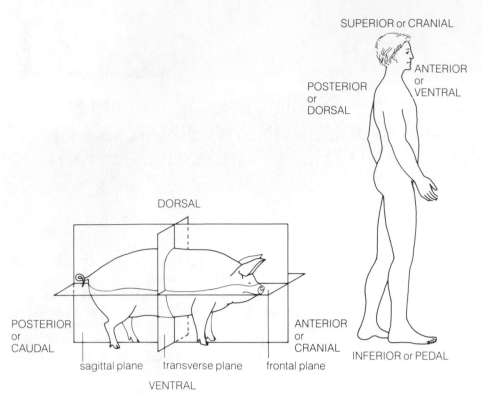

Figure 31-1 Lateral view of pig and human

- 4 large rubber bands *or* 2 pieces of string each 2 feet long
- plastic gloves and/or lanolin hand creme
- goggles if you wear contact lenses
- paper towels

Per lab room:

- liquid waste disposal bottle
- sink
- box of name tags (tags may be provided with fetal pigs)

PROCEDURE

A. Preparing the Fetal Pig for Observation

CAUTION *Preserved fetal pigs are kept in a formalin-based or other preservative solution. Wash any part of your body exposed to this solution with copious amounts of water. If the formalin solution is splashed into your eyes, wash them with the safety eyewash bottle for 15 minutes.*

1. The fetal pigs have been preserved in formalin or a phenol-based chemical, which may be irritating to hands and eyes. Therefore, wear protective plastic gloves or apply lanolin cremes to your hands before proceeding with the dissection. Students wearing contact lenses should wear protective goggles.

2. At the sink, remove your pig from its plastic bag and place on a dissecting tray lined with paper towels. If the plastic bag contains any excess preservative, pour the liquid into the waste bottle provided by your instructor, rinse out the bag with water, and save. As you will be using the same fetal pig for several days, you should place it in the plastic bag at the end of each day's exercise so it does not dry out.

For easy identification, tie a name tag to a hindleg, the bag, or both, *according to the instructor's wishes.* Use a permanent pen or pencil to fill in the tag.

B. General Observations

Identify the four regions of the fetal pig body: the large, compact **head;** the **neck;** the **trunk** with four **appendages** (the *limbs*) and the *tail* (see Fig. 31-2). The trunk may be divided further into the **thoracic region, lumbar region,** and **sacral region,** which also describe the vertebral regions of the spine. We will return to this region during the study of the nervous system.

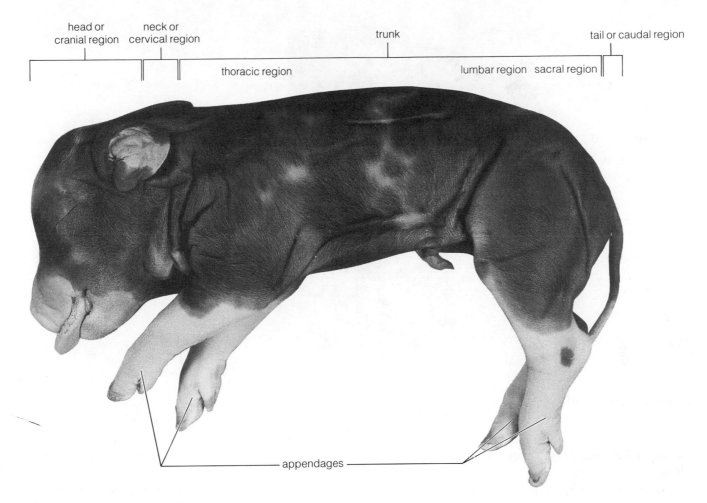

head or
cranial region   neck or
cervical region

trunk

tail or caudal region

thoracic region

lumbar region   sacral region

appendages

Figure 31-2 Lateral view of a fetal pig with the four major body regions indicated. (Courtesy D. Fox.)

C. Head and Neck Region

1. Examine the head in more detail (Fig. 31-3) and identify the **eyes** with **upper** and **lower lids,** the **external ears,** the **mouth,** and the characteristic **nose** or *snout*. Note the position of the **nostrils** or *external nares* on the snout. Feel the texture of the snout. It is composed of bone, cartilage, and other tough connective tissue and as such allows the pig to root and push soil and debris in its search for food.

2. Open the pig's mouth and note the **tongue** with its covering of **papillae,** which contain *taste buds*. Papillae are especially concentrated and prominent along the posterior edges and tip of the tongue. Also notice if any *baby teeth* are present. Like humans, pigs are omnivores; that is, they eat both animal and plant matter. We will return to the structure and placement of teeth in the section on the mouth.

D. Trunk

1. Place the pig on its back (dorsal surface) and examine its *abdomen* (belly). The most prominent feature of the underside (ventral surface) of the fetal pig is the **umbilical cord** seen near the posterior end of the abdomen (see Fig. 31-4). Is this structure present in the adult pig?

(Yes or no) _____

Adult human?

(Yes or no) _____

During its development, the fetus was connected to the placenta on the uterine wall of its mother's reproductive system via the umbilical cord. The cord contains two *arteries* (red), a large *vein* (blue), and a fourth vessel, usually collapsed, the *allantoic duct*. The

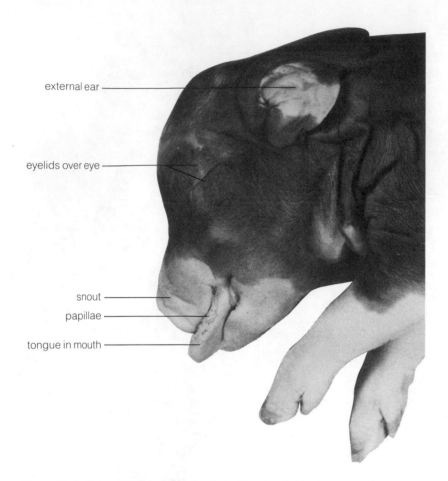

Figure 31-3 External features of the fetal pig. (Courtesy D. Fox.)

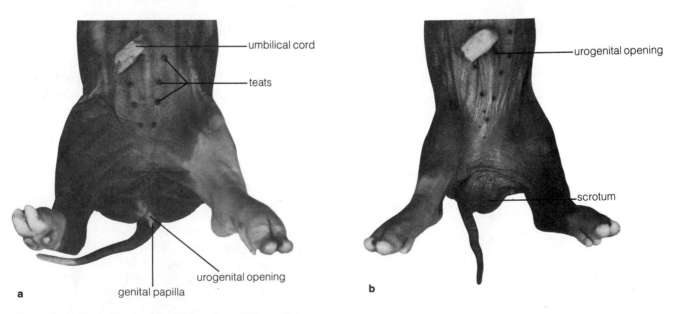

Figure 31-4 Ventral body of the (**a**) female and (**b**) male fetal pig. (Courtesy D. Fox.)

blood in the umbilical vein carries nutrients and oxygen from the mother to the fetus, and blood in the umbilical arteries carries waste materials and carbon dioxide from the fetus to the mother. We will study the umbilical cord in more detail during the exercises on the urogenital (excretory and reproductive) and circulatory systems.

2. Note on the ventral surface of the pig the pairs of **nipples** or *teats*. Both male and female pigs may have from 5–8 pairs of these structures situated in two parallel rows on the **thoracic region** (chest) and **abdominal region** of the body. Finally, locate the **anus,** the posterior opening of the digestive tract. The anus is situated immediately under (ventral to) the tail.

E. Determining the Sex of Your Fetal Pig

1. The male is identified by (1) the presence of a single *urogenital opening* to the excretory and reproductive systems just behind the umbilical cord (see Fig. 31-4b), and (2) the presence of a swelling on the posterior portion of the abdomen between the upper ends of the hindlimbs. The swelling is the **scrotum,** which contains the *testes,* a pair of small, oval structures, that are part of the male reproductive system. These are generally easy to locate in older fetuses. Identify the *penis,* a large, tubular structure immediately under the skin posterior to the urogenital opening.

2. A female fetal pig can be identified by the presence of a single urogenital opening immediately ventral to the anus (see Fig. 31-4a). A small fleshy piece of tissue, the *genital papilla,* projects from the urogenital opening.

3. Note that in both male and female fetal pigs, there is a common urogenital opening shared by the urinary (excretory) system and the genital (reproductive) system. Both adult male pigs and human males have a similar structure. In adult female pigs and humans, however, there are separate openings to the excretory and reproductive systems.

F. Appendages

1. Examine carefully the feet and legs of your pig. The first *toe,* or *digit,* which corresponds to your big toe or thumb, is absent in both forelimbs and hindlimbs of the pig. Furthermore, the second and fifth digits are reduced in size, and the middle two digits, the third and fourth, are flattened or *hoofed.* Pigs and other hoofed animals, referred to as *ungulates,* walk with the weight of their body borne on the tips of the digits. This type of walking is referred to as **digitigrade locomotion.** By contrast, humans use the entire foot for walking and have **plantigrade locomo-**

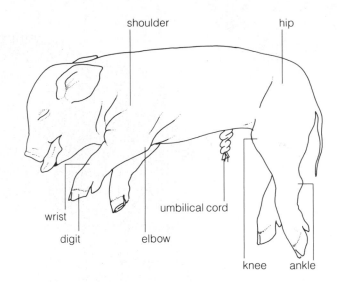

Figure 31-5 Lateral view of external features of the fetal pig

tion. Compare the structure of your hands and feet with the foot of a pig.

2. Complete your study of the appendages of the fetal pig by using Fig. 31-5 to locate and identify the following structures and joints: **wrist, elbow, shoulder, ankle, knee,** and **hip.**

II. SKELETAL MUSCLE SYSTEM   (Starr, p. 253)

The contractions of *skeletal muscle* organs enable the body to move. Through the voluntary contractions of skeletal muscles, you may wink your eye, wave at a friend, or tap your foot in time to the beat of music. In this portion of the exercise, you will examine the structure of skeletal muscles and learn to identify some of them and describe their functions. The movement produced by a skeletal muscle is called its **action.**

Skeletal muscles are attached to the various parts of the skeleton by tough strips of dense fibrous connective tissue called *tendons.* The three parts of a typical muscle are the **origin** (the end attached to the less mobile portion of the skeleton), the **insertion** (the end attached to the portion of the skeleton that moves when the muscle contracts), and the *belly* (the middle portion between the points of attachment).

Skeletal muscles that move an appendage one way usually have an opposing muscle that moves it in the opposite direction. Muscles with such opposite actions are called **antagonistic muscles.** For example, the biceps brachii is responsible for flexing the forelimb (in the pig) or the forearm (in humans), and the triceps brachii (its opposing muscle) straightens or extends them.

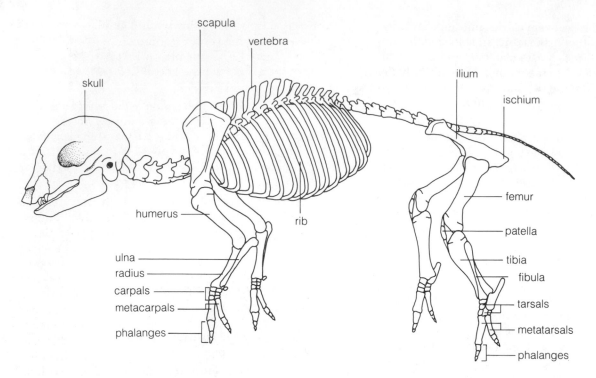

Figure 31-6 Skeleton of the fetal pig. The hyoid bone, located in the upper neck, and the sternum, to which many of the ribs are attached midventrally, are not included in this illustration. (After Gilbert, 1966.)

Realize that the attachments of the muscles can only be understood with reference to the skeleton, and that the shapes of the bones are meaningless when considered apart from the leverage they provide for muscles. Although we will not examine the skeleton in detail, it will be necessary to refer to its various parts during this portion of the exercise. Thus, during your dissection, use the illustration of the fetal pig skeleton (Fig. 31-6) as a reference.

MATERIALS

Per student pair:

- one fetal pig preserved in formalin or phenol-based chemical and injected with red and blue latex
- dissecting tray
- paper towels
- one dissecting kit including the following: scalpel, blunt probe, dissecting needle, scissors, forceps
- 4 large rubber bands *or* 2 pieces of string each 2 feet long
- plastic gloves and/or lanolin hand creme
- goggles if you wear contact lenses

PROCEDURE

A. Directions for Dissection of the Skeletal Muscles

1. Place your specimen on its dorsal side in a dissecting tray lined with paper towels and secure the feet with rubber bands or string as follows:

Tie one end of a string to the left forelimb at the wrist, pass the string underneath the tray, and then tie the other end of the string to the right wrist so that the legs are spread apart under tension. Repeat with the hindlimbs, using the second piece of string. If rubber bands are available, tie two rubber bands together to make them longer. Then loop one end of the band around the right forelimb close to the foot. Bring the rubber band under the dissecting tray and loop it around the other forelimb to anchor the feet securely. Repeat this procedure with another set of rubber bands, and anchor the hindlimbs to the tray.

2. For the following dissection, refer to the drawing in Fig. 31-7. The numbers in the drawing refer to the incisions to be made in the following dissection.

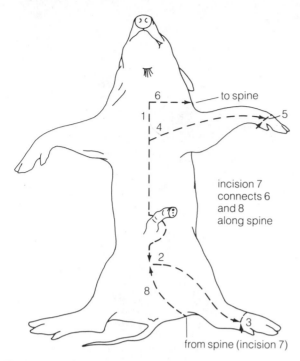

Figure 31-7 Cuts needed to expose muscles

Using a scalpel, make an incision on the ventral side of the pig at the base of the neck (1). *Make sure you cut only through the skin and not the underlying tissues.* With your scissors continue the incision posteriorly to the umbilical cord, then around the cord on the left side (the pig's left side) and to the region between the hindlegs. From this point cut down the medial surface of the left hindleg toward the foot (2) and around the pig's ankle (3). Make a similar cut from the midventral incision in the thorax down the medial surface of the left forelimb (4) and around the pig's wrist (5).

3. Return to the original incision on the ventral surface of the neck, and extend it around the left side of the pig's neck dorsally toward the spine (6). To do this you will need to remove the rubber bands or string and place the pig on its right side in the dissecting pan. When you reach the spine, continue the incision posteriorly along the spine to the base of the tail (7). Finally, connect the posterior ends of incisions 7 and 1 (8). After completing the incisions, place the pig on its back and secure the legs with rubber bands or string as you did earlier.

4. To skin your specimen, grasp the cut edge of the skin at the base of the throat and begin easing the skin loose from the underlying tissues. Use your blunt probe between the skin and underlying connec-tive tissues, working slowly until you have removed the skin from the ventral portion of the trunk and the limbs. Then, turn your pig on its right side and re-move the skin along the lateral (left side) and dorsal surfaces to the spine.

5. You may notice in the region of the neck, shoulder, and trunk, a layer of light brown muscle fibers adher-ing to the skin. These fibers comprise the *cutaneous maximus* muscle that is responsible for the twitching of the skin that gets rid of insects and other irritants. Humans do not have this layer of muscle.

6. After your specimen is skinned, the muscles will not appear as clearly defined as in the illustrations. This is because they are covered by adipose tissue (fat) and two layers of relatively loose connective tissue or *fascia*. The first layer or *superficial fascia* con-nects the skin to the muscles and is relatively easy to remove. The second layer or *deep fascia* connects one muscle to another and maintains them in their proper position to one another. It is considerably tougher than the superficial layer. It will be necessary to break through the deep fascia as you proceed through the exercise. Remember, however, that you are working with a fetus and as such the structures are immature and can be easily torn. Therefore, proceed with care.

7. As you attempt to identify the various muscles, you may find that the boundary of a given muscle is readily apparent, whereas in others it seems to blend with those around it. In order to define the limits of a muscle, use a blunt probe to tease away the overlying adipose and connective tissues until you can see the direction of the muscle fibers. Look for changes in the direction of the muscle fibers, and attempt to slip the blunt probe or flat edge of your scalpel handle be-tween the two separate layers at this point. If the two layers separate readily from one another, you have lo-cated two different muscles.

Now proceed with the exercise and attempt to iden-tify some of the major muscles in the fetal pig *as di-rected by your instructor.* Obviously, not all of the skele-tal muscles have been included, but only those that are relatively easy to identify and will illustrate the general principles of skeletal muscular action.

B. Muscles of the Shoulder and Back

1. The **latissimus dorsi** (Fig. 31-8) is a broad muscle wrapped around the sides of the thoracic region and chest. Carefully pick away the adipose tissue and con-nective layer from the sides of the chest until its fibers are apparent. The origin of this muscle is the lumbar vertebrae, some of the posterior thoracic vertebrae, and the **lumbodorsal fascia** (Fig. 31-8). It is inserted

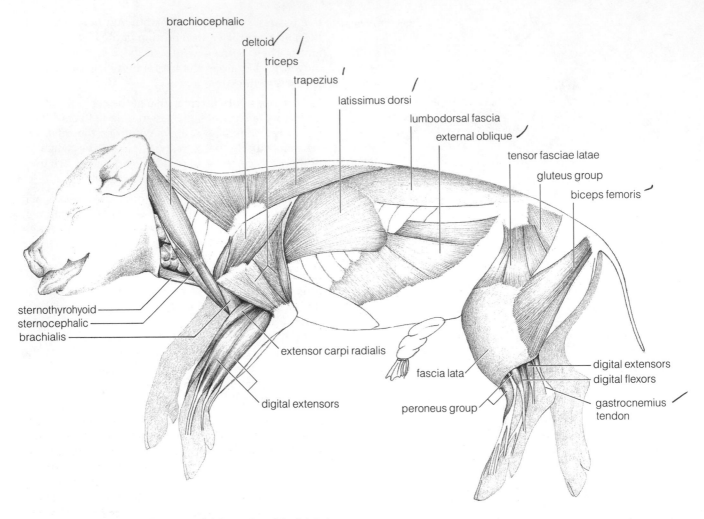

Figure 31-8 Lateral view of the superficial muscles of the fetal pig

by a tendon into the proximal end of the humerus (the major bone of the upper forelimb; see Fig. 31-6). The action of the latissimus dorsi is to move the forelimb dorsally and posteriorly.

2. The **trapezius** (Fig. 31-8) is a broad muscle anterior to the latissimus dorsi. Its origin is from the base of the skull and from the first 10 thoracic vertebrae, and it is inserted on the scapula (the shoulder blade). Its action is to draw or pull the shoulder medially.

3. The **deltoid** (Fig. 31-8) is a relatively broad muscle between the triceps brachii and supraspinatus. It originates from the scapula or shoulder blade and is inserted into the humerus. Its action flexes the forelimb.

4. If you have not done so, carefully remove the cutaneous muscle and gelatinous connective tissues covering the side of the neck. You should now be able to observe a broad, flat strip of muscle, the **brachiocephalic** (Figs. 31-8 and 31-9), extending from the back of the skull to the foreleg. This muscle is inserted in the distal end of the humerus, and its action flexes the forelimb.

C. Muscles of the Forelimb

1. The **triceps brachii** (Figs. 31-8 and 31-9) is a large muscle that virtually covers the entire outer and posterior surface of the upper forelimb. Its origin is from the scapula and proximal third of the humerus, from which it extends posteriorly, and it is inserted on the proximal end of the ulna. Its action is to extend the forelimb.

2. To locate the **biceps brachii** (Fig. 31-9), carefully place your pig on its dorsal side and secure the legs with string or rubber bands. The biceps are a rather small, spindle-shaped muscle extending along the anterior surface of the humerus. This muscle originates on the scapula and inserts on the radius and ulna. Its action is to flex the forelimb and act antagonistically with the triceps. In order to see it clearly, you will need to cut through the overlying muscle (the **brachialis,** see Fig. 31-9) and reflect the cut edges back. The brachialis is a small muscle that also flexes the forelimb. Its origin is the proximal humerus, and it inserts on the ulna.

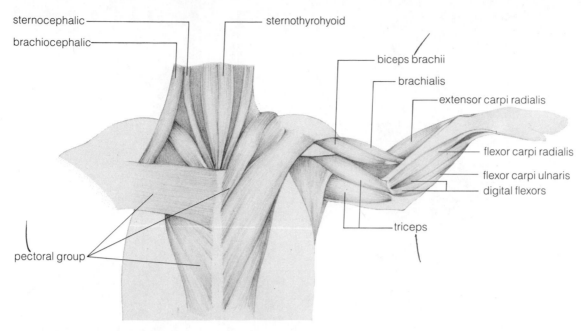

sternocephalic
brachiocephalic
sternothyrohyoid
biceps brachii
brachialis
extensor carpi radialis
flexor carpi radialis
flexor carpi ulnaris
digital flexors
triceps
pectoral group

Figure 31-9 Muscles of the ventral thoracic region and forelimb. Some of the superficial muscles of the pig's left side have been removed. (After Gilbert, 1966.)

3. There are numerous other muscles in the lower forelimb, most of which are concerned with extending or flexing the foot and digits (see Figs. 31-8 and 31-9).

D. Muscles of the Throat and Chest

1. Continue working from the ventral side of your specimen and locate the **sternocephalic** (Figs. 31-8, 31-9) and **sternothyrohyoid** (Figs. 31-8, 31-9). The former is a long muscle lying ventral to the brachiocephalic. Its origin is at the sternum, and it is inserted into the lateral portion of the skull just behind the external ear known as the mastoid process (see Fig. 31-6). Its action flexes the head.

2. The sternothyrohyoid consists of two long flat muscles extending from the sternum (origin) to the hyoid (insertion). The action of these muscles is to retract and depress the base of the tongue and the larynx, as, for example, when swallowing.

3. The **pectoral group** (Fig. 31-9) originates on the ventral side of the sternum and inserts on the humerus. The pectoral group draws, or adducts (moves the appendage medially), and retracts the forelimb toward the chest. If you wish, you may carefully cut the belly of the superficial pectoral and bend it back to examine the underlying muscles more closely.

E. Muscles of the Abdominal Region

The major lateral abdominal muscles consist of the outer **external oblique,** the **internal oblique** (the midlayer), and the **transversus** (the inner layer). To locate these muscles, turn your specimen on its side. Now find the fibers of the external oblique (Figs. 31-8 and 31-10) and observe that they run diagonally, so that their ventral ends are posterior to their dorsal ends. By carefully cutting a "window" approximately 1 cm square in the external oblique, you will expose a portion of the internal oblique. (Remember that the muscles of the fetus are extremely thin. Use care not to cut into the body cavity at this time, as you will release a large amount of fluid.) Notice that the fibers of the internal oblique run at nearly right angles to the direction of those of the external oblique. Now, using the same careful technique as above, remove a small portion of the internal oblique and attempt to reveal the fibers of the transversus, which as the name suggests run transversely. Collectively, the actions of these abdominal muscles, together with the **rectus abdominus** (Fig. 31-10), is to flex the trunk and compress the abdominal viscera to aid expiration or defecation.

F. Muscles of the Hip and Thigh

1. Begin your dissection of the muscles of the hip and thigh by locating the **biceps femoris** (Fig. 31-8). This conspicuous superficial muscle covers most of the caudal half of the lateral surface of the thigh and originates from the ischium of the pelvic girdle. It is inserted in the femur and the upper part of the tibia. Its action extends and abducts (moves the appendage laterally) the hindlimb.

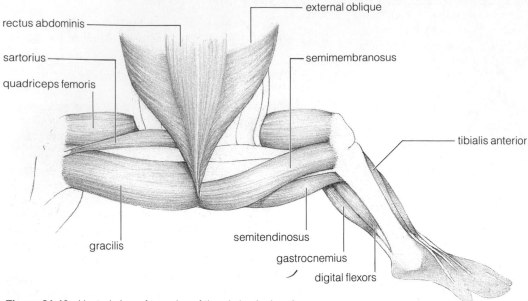

Figure 31-10 Ventral view of muscles of the abdominal region and hindlimb. Some superficial muscles on the pig's left side have been removed. (After Gilbert, 1966.)

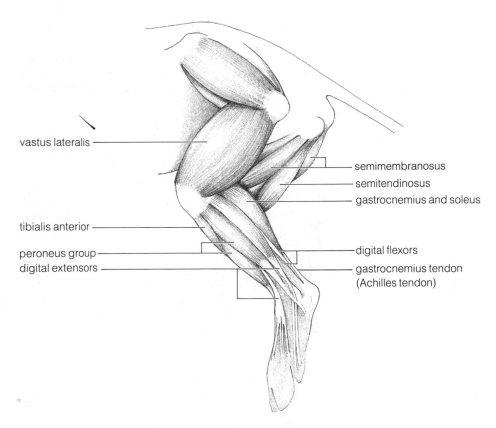

Figure 31-11 Lateral view of the muscles of the hindlimb. (After Gilbert, 1966.)

2. Next, locate the **tensor fasciae latae** (Fig. 31-8), the most anterior of the thigh muscles. It is short and triangular-shaped and continues as a sheet of connective tissue, the *fascia lata*, which attaches to the tibia. The action of the tensor fasciae latae is to tense the fascia lata, flex the hip, and extend the knee.

3. Now carefully free the tensor fasciae latae at its insertion, but leave the origin and its medial portion intact. Cut the biceps femoris at its insertion near the tibia, and peel it back to expose the deeper muscles of the thigh and foot. Attempt to identify the *vastus lateralis*, one of the four **quadriceps femoris** muscles, and the **semitendinosus** (Figs. 31-10 and 31-11), which extend the knee and hip, respectively. Between these two muscles are several arteries and nerves that must be carefully removed to expose the *semimembranosus* (Fig. 31-11).

4. To locate other muscles of the thigh, place your pig on its dorsal side and tie back the legs with string or rubber bands. Referring to Figs. 31-10 and 31-11, locate the **sartorius, gracilus,** and **semimembranosus.** The actions of these muscles are to adduct the hindlimb, flex the hindlimb (sartorius), and extend the hindlimb (semimembranosus). The biceps femoris, semitendinosus, and semimembranosus are collectively called the hamstring muscles.

G. Muscles of the Hindlimb

1. The **gastrocnemius** (Figs. 31-10 and 31-11) and *soleus* (Fig. 31-11) originate, respectively, from the distal end of the femur and the head of the fibula. They are both inserted by the Achilles tendon to the calcaneus (heel bone). These muscles extend the foot, and the gastrocnemius helps flex the knee.

2. Other muscles of the lower hindlimb flex the foot (*tibialis anterior*, Figs. 31-10 and 31-11; *peroneus group*, Figs. 31-8 and 31-11), while others flex and extend the digits (*digital flexors*, Figs. 31-8 and 31-11; *digital extensors*, Figs. 31-8 and 31-11). Spend a few minutes examining the lower hindlimb, and attempt to identify several of these muscles.

When you have completed the laboratory, carefully place your pig in the original plastic bag to prevent it from drying out. Dispose of any paper towels that contain formalin or preservative as directed by your instructor. Clean your dissecting tools and the laboratory table.

PRE-LAB QUESTIONS

_____ **1.** A fetus is (a) a newborn pig, (b) a newborn human, (c) an unborn mammal, (d) all of the above.

_____ **2.** "To dissect" means primarily (a) to cut open, (b) to remove all internal organs, (c) to expose to view, (d) none of the above.

_____ **3.** When the directions for a fetal pig dissection refer to the left, they are referring to (a) your left, (b) the pig's left, (c) it depends on the orientation of the pig, (d) none of the above.

_____ **4.** In a fetal pig, dorsal and ventral refer to (a) the head and tail regions of the body, respectively; (b) the tail and the head regions of the body, respectively; (c) the upper (i.e., back) portion and the lower (i.e., underside) portion of the body, respectively; (d) the lower (i.e., underneath) portion and the upper (i.e., back) portion of the body, respectively.

_____ **5.** The umbilical cord functions to (a) carry waste products in the blood from the fetus to the mother, (b) carry waste products in the blood from the mother to the fetus, (c) carry oxygen in the blood from the mother to the fetus, (d) a and c.

_____ **6.** The female *fetal* pig is similar to the male *fetal* pig in that its body (a) has separate openings for the excretory system and the reproductive system, (b) has a common opening for the excretory system and the reproductive system, (c) has an opening for the excretory system but none for the reproductive system, (d) none of the above.

_____ **7.** The pig and the horse are called ungulates because they (a) walk on their ankles, which are modified as hooves, (b) walk on their feet, which are modified as hooves, (c) walk on the tips of their toes, which are modified as hooves, (d) none of the above.

_____ **8.** Pigs and humans are similar in that they are omnivores. That is, they eat (a) only animal matter, (b) only plant matter, (c) both plant and animal matter, (d) none of the above.

_____ **9.** The insertion of a muscle refers to the part (a) attached to the less mobile portion of the skeleton, (b) attached to the portion of the skeleton that moves when the muscle contracts, (c) attached between the origin and belly, (d) none of the above.

_____ **10.** The biceps brachii is responsible for flexing the forearm while the triceps brachii extends the forearm. Collectively, these opposing muscles are called (a) cooperative, (b) antagonistic, (c) involuntary, (d) sensory.

DISSECTION OF THE FETAL PIG: DIGESTIVE, RESPIRATORY, AND CIRCULATORY SYSTEMS

OBJECTIVES   After completing this exercise you will be able to:

1. define diaphragm, thoracic cavity, abdominopelvic cavity, exocrine gland, endocrine gland;

2. describe and give the functions of the organs of the digestive, respiratory, and circulatory systems;

3. explain the importance of the digestive, respiratory, and circulatory systems to a living mammal;

4. trace the pathway of oxygen and carbon dioxide into and out of the lungs of a mammal.

INTRODUCTION   You will begin your study of the mammalian digestive, respiratory, and circulatory systems by carefully opening the ventral body cavities of your fetal pig to expose the internal organs for further examination. Remember that dissecting does not primarily mean "cutting up" but rather "exposing to view." Thus, proceed carefully, as the internal organs are fragile. And, work closely with your partner, making sure that each structure is fully identified and studied before proceeding to the next step.

DIRECTIONS FOR DISSECTION

MATERIALS

Per student pair:

- one fetal pig preserved in formalin or phenol-based chemical and injected with red and blue latex
- dissecting tray
- paper towels
- dissecting kit
- pins
- bone shears
- 4 large rubber bands *or* 2 pieces of string each 2 feet long
- piece of string 8 inches long

- plastic gloves and/or lanolin hand creme
- goggles if you wear contact lenses

Per lab room:

- liquid waste disposal bottle
- sink

PROCEDURE   For the following dissection, use Fig. 32-1 as a guide for making the various incisions. The numbers in the figure correspond to the numbers in the following directions.

1. Place the pig, ventral side up, in the dissecting tray and restrain it using rubber bands or string as you did in the previous exercise. Begin your incision at the small tuft of hair on the upper portion of the throat (1), and continue the incision posteriorly to approximately 1.5 cm anterior to the umbilical cord. You should cut through the muscle layer but not too deeply or the internal organs may be damaged.

2. You determined the sex of your specimen in the previous exercise. If it is a *female,* make the second incision (2F) completely around the umbilical cord and proceed with a single incision posteriorly for approximately 3 cm between the hindlimbs.

3. If your fetal pig is a *male,* make the second incision (2M) as a half circle anterior to the umbilical cord and then proceed with *two incisions* posteriorly to the region between the hindlimbs. The two incisions are necessary to avoid cutting the *penis,* which lies under the skin just posterior to the umbilical cord. The incisions made in the region of the *scrotum* should be made carefully so as not to damage the testes, lying just under the skin.

4. Deepen incisions 1 and 2 until the body cavity is exposed. Proceed carefully, however, as the body cavity may be filled with a dark fluid. In order to make lateral flaps of the muscle tissue, which can be folded out of the way, make a third (3) and fourth (4) incision as illustrated in Fig. 32-1. Now, carefully open the body cavity. If it is filled with fluid, pour the fluid into

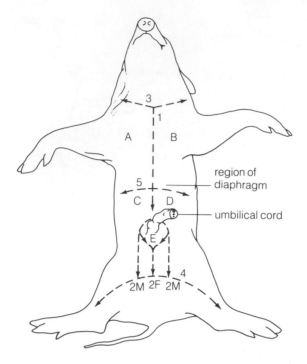

Figure 32-1 Ventral view of fetal pig with the position of incisions indicated. Specific incisions for male and female specimens are marked M and F, respectively.

the waste container provided (not in the sink!), and carefully rinse out the cavity with water.

5. Just below the lower margin of the *rib cage* (use your fingers to determine this), make a fifth (5) incision laterally in both directions from the first incision (1). In this region is the **diaphragm,** a sheet of tissue connected to the body wall and separating the two major body cavities: the **thoracic cavity** and the **abdominopelvic cavity.** By using your scalpel, free the diaphragm (do not remove it, however) where it is in contact with the body wall.

6. Carefully peel back flaps A, B, C, and D (see Fig. 32-1), and pin them beneath your pig. It may be necessary to cut through the ventral part of the rib cage with a pair of scissors or bone shears to separate body wall flaps A and B. Do so carefully, so as not to damage the heart and lungs, which lie in this region.

7. When the body cavity is fully exposed, carefully remove any excess red or blue latex that may be present in the abdominal cavity. (This occurs when some veins and arteries burst when injected with latex.)

8. To free the umbilical cord and the flesh immediately surrounding it (flap E), locate the umbilical vein—a dark, tubular structure extending from the umbilical cord forward (anteriorly) to the liver. Tie small pieces of string around the vein in two places (approximately 1.5 cm apart), and with your scissors cut through the vein between the pieces of string. The umbilical cord and attached tissue (flap E) can now be

laid back between the hindlegs and pinned to the body. The pieces of string around the umbilical vein will aid in identifying this structure during the exercise on the circulatory system.

I. DIGESTIVE SYSTEM (Starr, pp. 361–366)

The digestive system of a vertebrate consists of the **digestive tract,** or *alimentary canal* (mouth, oral cavity, pharynx, esophagus, stomach, small intestine, large intestine or colon, rectum, and anus) and associated structures and glands (salivary glands, gall bladder, liver, and pancreas). In this exercise you will locate and identify all of these structures. In addition, you will identify the thymus and thyroid, two endocrine system glands. The thymus is also an important source of one kind of lymphocyte.

MATERIALS

Per student pair:

- one fetal pig preserved in formalin or phenol-based chemical and injected with red and blue latex
- dissecting tray
- paper towels
- dissecting kit
- pins
- bone shears
- 4 large rubber bands *or* 2 pieces of string each 2 feet long
- plastic gloves and/or lanolin hand creme
- goggles if you wear contact lenses
- dissecting microscope

Per lab room:

- several meter rulers
- liquid waste disposal bottle
- sink

PROCEDURE

A. Thymus and Thyroid Glands

1. Work from the ventral side of your pig with the legs secured by string or rubber bands and the body wall flaps pinned to the sides of the body.

Identify the **thymus gland,** a whitish structure that is divided into two lobes and is in the neck and upper thoracic cavity (Fig. 32-2). This gland, which partially covers the anterior portion of the heart and extends along the trachea to the larynx, plays important roles in the development and maintenance of the body's defense system.

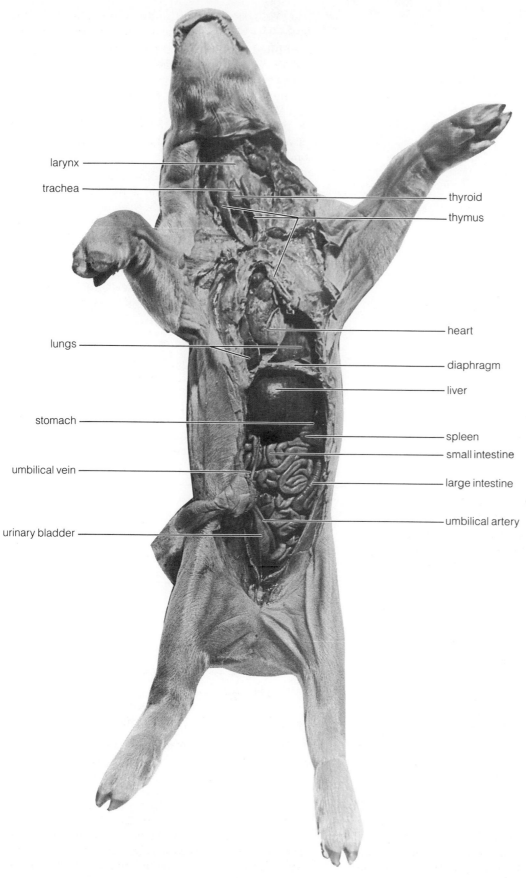

larynx

trachea

thyroid

thymus

heart

lungs

diaphragm

liver

stomach

spleen

small intestine

umbilical vein

large intestine

umbilical artery

urinary bladder

Figure 32-2 Ventral view of the general internal anatomy of the fetal pig. (Courtesy T. Odlaug.)

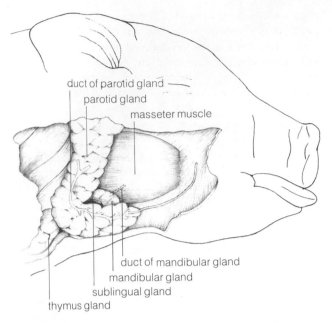

Figure 32-3 Lateral view of the fetal pig head with salivary and thymus glands exposed. (After Gilbert, 1966.)

2. Immediately beneath the thymus in the neck region is the **thyroid gland** (Fig. 32-2). This is a small reddish, oval mass with a consistency more solid than the thymus. Thyroid hormones function in the regulation of metabolism, growth, and development.

B. Salivary Glands

There are three pairs of salivary glands: the *parotid*, the *mandibular* (*submaxillary*), and the *sublingual* (see Fig. 32-3). Together they produce *saliva*. To expose these glands, it will be necessary to remove the skin and muscle tissue from one side of the face and neck of your specimen. Place your pig on its right side and, proceeding from the base of the ear, carefully cut through the skin to the corner of the eye, then ventrally toward the chin, and, finally, continue the incision posteriorly toward the forelimb. Carefully remove the skin as described in the first fetal pig exercise.

1. Parotid gland. After removing the skin, tease away the muscle tissue below the ear to reveal a large, relatively dark, triangular gland, the *parotid*. This gland extends from the edge of the ear posteriorly to the midregion of the neck (Fig. 32-3). Can you distinguish the parotid from the large masseter muscle? The difference is that the parotid appears to be composed of many small nodules, while the muscle tissue is fibrous.

2. Mandibular (*submaxillary*) **gland.** The *mandibular gland* is relatively large and somewhat lobed. It lies just below the parotid (Fig. 32-3). You will need to cut through the middle of the parotid gland to expose the mandibular gland. Do not confuse the mandibular

gland with the small, oval lymph nodes present in the head and neck region.

3. Sublingual gland. The third salivary gland, the *sublingual*, is long and rather slender. It is difficult to locate.

The fluids secreted by the mandibular and sublingual glands are more viscous than that secreted by the parotid. Collectively, the secretions by the three salivary glands maintain the oral cavity in a moist condition, ease the mixing and swallowing of food, and contain enzymes that begin the breakdown of starch to sugars.

C. Mouth

1. The mouth is the beginning of the digestive tract, a long tubelike structure where food digestion and absorption of nutrients occur. Observe the area between the **lips** and **gums;** this is called the **vestibule.** The larger area behind the gums is referred to as the **oral cavity.**

2. With the scissors or bone shears, carefully cut through the corners of the mouth and back toward the ears until the lower jaw can be dropped and the oral cavity exposed (see Fig. 32-4).

3. If teeth are present, carefully extract a **tooth** and examine it. A tooth consists of the *crown*, the *neck* (surrounded by the gum), and the *root* (embedded in the jawbone). If your specimen does not have exposed

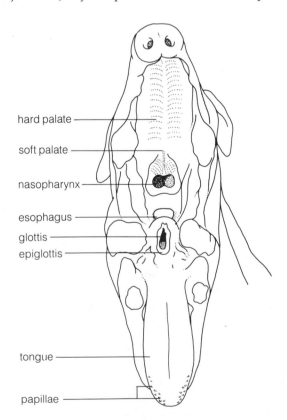

Figure 32-4 Structures of the oral cavity and pharynx

teeth, cut into the gums and determine whether developing teeth are present.

4. Feel the roof of the oral cavity, and determine the position of the **hard palate** and **soft palate** (Fig. 32.4). What is the difference between the two regions?

_____

_____

5. Posterior to the soft palate is the **pharynx** (in humans the portion of the pharynx just posterior to the oral cavity is also referred to as the *throat*). Note that unlike humans, the pig does not have a fingerlike piece of tissue, the **uvula,** projecting from the posterior region of the soft palate. Confirm its presence in humans by looking into the throat of your lab partner.

6. Carefully close the pig's jaws. Now, in the neck locate the **trachea** (Fig. 32-2), a tube that is supported throughout its length by a series of cartilaginous rings. Although the trachea is actually a part of the respiratory system, its identification will aid in finding the **esophagus,** which lies on its dorsal surface. Carefully slit the esophagus and insert a blunt probe into it and run it back toward the mouth. Open the mouth and note where the probe emerges. This is the opening of the esophagus (Fig. 32-4). If you would run your probe posteriorly through the esophagus, where would it emerge?

_____

7. Continue your study of the oral cavity by locating the opening to the *larynx,* the **glottis.** It can be identified by the presence of a small white cartilaginous flap, the **epiglottis,** on its ventral surface (Fig. 32-4). The epiglottis covers the glottis when a mammal swallows.

D. Liver, Gall Bladder, and Pancreas

1. The largest organ of the abdominopelvic cavity is the **liver** (Fig. 32-2), a brownish four-lobed gland that produces *bile.* Count and carefully determine the extent of the four lobes.

In addition to producing bile, the liver plays a very important role in maintaining a stable composition of the blood. When the nutrients from a digested meal are absorbed by the blood capillaries of the small intestine, they contain high concentrations of such compounds as sugars like glucose and amino acids. This blood is transported from the small intestine to the liver (via the hepatic portal vein), and there the excess glucose is converted to glycogen for storage. If the liver has stored a full capacity of glycogen, it converts the glucose into fat, which is stored in other parts of the body. The liver also removes excess amino acids from the blood by converting them to carbohydrates and fats. During this process, an amino group ($-NH_2$) is removed from the amino acid and converted into ammonia (NH_3). Ammonia is a very toxic substance, and the liver combines it with carbon dioxide to form urea. The urea, which is less toxic than ammonia, will be eliminated from the body in urine.

2. By lifting the right central lobe of the liver, you can locate the **gall bladder.** This structure is a sac for storing bile secreted by the liver.

The *cystic duct* from the gall bladder unites with the *hepatic duct* from the liver to form the common *bile duct.* The latter empties into the first portion of the *small intestine* (Fig. 32-2). *If your instructor indicates,* attempt to locate the hepatic and common bile ducts and trace them from the liver to the duodenum. Be careful not to injure the *hepatic portal vein,* which parallels these ducts.

3. Carefully move the small intestine and locate the **pancreas,** an elongated granular mass lying between the *stomach* and small intestine. The *pancreatic duct* carries digestive enzymes and other substances produced by the pancreas to the duodenum. (Do not attempt to find the pancreatic duct, however, as it is too small to be dissected satisfactorily.)

The pancreas is both an **exocrine gland** (secretions are released into a duct) and an **endocrine gland** (hormones are released into the blood). The endocrine portion of the pancreas secretes insulin and other hormones involved with controlling the levels of glucose in the blood of mammals.

E. Structures of the Lower Portion of the Digestive Tract: Stomach, Small Intestine, Large Intestine or Colon, Rectum, and Anus

1. Earlier in the exercise, you made a small slit in the esophagus. Return to this incision and trace the esophagus from the oral cavity to the stomach. You may be aided in this by inserting a blunt probe through the slit in the esophagus and pushing it (posteriorly) toward the stomach, a bean-shaped organ dorsal to the liver. (You will need to carefully push the lobes of the liver to one side to fully expose the stomach.) Note that the esophagus penetrates the diaphragm before joining the *cardiac end* (near the heart) of the stomach. The other end of the stomach, which empties into the small intestine, is called the *pyloric end.*

2. Cut the stomach lengthwise with your scissors. Describe the contents of the stomach.

_____

_____

The contents of the fetal pig's digestive tract are called *meconium* and are composed of a variety of substances including amniotic fluid swallowed by the fetus, epithelial cells sloughed off from the digestive tract, and hair.

3. Clean out the stomach and note the folds or *rugae* on its internal surface. What role might the rugae play in digestion?

_____

_____

_____

Two muscular rings, the *cardiac sphincter* and the *pyloric sphincter,* control the movement of food through the stomach. Unlike skeletal muscles, the cardiac and pyloric sphincters are involuntary smooth muscle tissue.

4. The **small intestine** (Fig. 32-2) is divided into three regions: the *duodenum,* the *jejunum,* and the *ileum.* The first portion, the duodenum leaves the pyloric end of the stomach and runs along the edge of the pancreas. The end of the duodenum and beginning of the ileum cannot easily be distinguished from one another.

5. The coils of the small intestine are held together by thin membranes called **mesenteries.** *At the direction of your instructor,* cut the mesenteries and uncoil the small intestine. How long is it?

_____

_____

A rule of thumb is that the small intestine in both pigs and humans is about 5 times the length of the individual.

6. Using your scissors, cut a 0.5-cm section of the intestine, slit it lengthwise, and place it in a clear, shallow dish filled with water. Now examine it using a dissection microscope. How does the inner surface appear?

_____

_____

_____

Locate the *villi.* Most of the nutrients provided by the digestive process are absorbed by these small projections from the wall of the small intestine.

7. Locate the juncture of the **large intestine** (Fig. 32-2), or *colon,* and the ileum. This may be more difficult in a pig than in a human because in the former there is not such a noticeable difference in the size of the small and large intestines. In the direction opposite to the large intestine is a blind pouch, the *cecum,* which in the pig is relatively large. In humans, the cecum is very short and bears a small fingerlike projection known as the *appendix.*

8. As with other junctures in the alimentary canal, the region where the ileum joins the large intestine is the site of a muscular sphincter, the *ileocecal valve.*

9. The large intestine stretches from the caecum to the **rectum** and opens to the outside at the **anus.** The anus is the site of the final muscle in the alimentary canal, the *anal sphincter.* Locate the rectum, anus, and anal sphincter, but do not dissect these structures at this time. You may, however, *at the direction of your instructor,* remove a piece of the colon and examine it with a dissecting microscope as you did earlier with the small intestine. How do their internal surfaces compare?

_____

_____

_____

II. RESPIRATORY SYSTEM   (Starr, pp. 410, 411)

The respiratory system of a mammal consists of various organs and structures associated with the intake (*inhalation*) of air rich in oxygen, the exchange of oxygen and carbon dioxide between the blood in the lung capillaries and the air sacs of the lungs, and the release (*exhalation*) of gases rich in carbon dioxide.

MATERIALS

Per student pair:

- one fetal pig preserved in formalin or phenol-based chemical and injected with red and blue latex
- dissecting tray
- paper towels
- dissecting kit
- pins
- 4 large rubber bands *or* 2 pieces of string each 2 feet long
- plastic gloves and/or lanolin hand creme
- goggles if you wear contact lenses

PROCEDURE

A. Structures of the Head and Oral Cavity

1. In the pig and other mammals, molecules of air enter the body through the nostrils and pass through a pair of **nasal cavities** dorsal to the hard palate and into the *nasopharynx* (Fig. 32-4). Examine the nostrils and hard and soft palates, and then carefully cut the soft palate longitudinally to examine the nasopharynx of your specimen.

2. From the nasopharynx, air passes through the glottis into the **larynx.** In humans, the front of the larynx is often referred to as the *Adam's apple* or *voice box.* Slit the larynx longitudinally to expose the *vocal cords.*

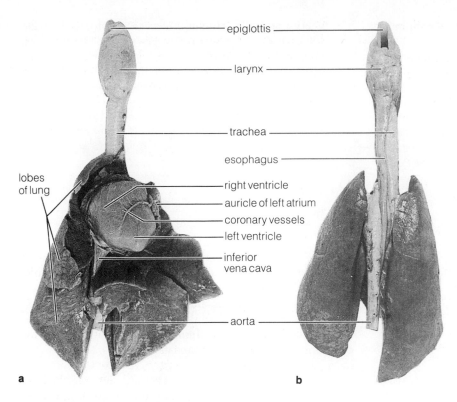

epiglottis

larynx

trachea

esophagus

lobes
of lung

right ventricle
auricle of left atrium
coronary vessels
left ventricle
inferior
vena cava

aorta

a   b

Figure 32-5 Ventral view (**a**) and dorsal view (**b**) of the
respiratory system of the fetal pig. (Courtesy T. Odlaug.)

B. Trachea, Bronchial Tubes, and Lungs

1. Locate the trachea and, as a review, distinguish it
from the esophagus (Fig. 32-5). The trachea extends
from the larynx and divides into two major branches,
the **bronchi** (singular *bronchus*), to the lungs. Note
again the series of cartilaginous rings that prevent the
trachea from collapsing. These rings are actually incomplete on their dorsal side.

2. Direct your attention to the thoracic cavity. It is
divided into two **pleural cavities,** which contain the
lungs, and the **pericardial sac,** which contains the
heart. The latter is located between the pleural cavities. Carefully examine the lungs and note the two
pleural membranes: thin, transparent tissues, one of
which lines the inner surface of the thoracic cavity
and the other the outer surface of the lungs. The right
lung consists of four lobes and the left of two or three.
Are the lungs of the fetal pig filled with air?

(Yes or no) _____

The pericardial sac is likewise lined by a **pericardial
membrane** as is the surface of the heart.

3. Carefully push the heart to one side and gently
tease away some of the lung tissue to expose the bronchi. Attempt to see that the bronchi divide into smaller
and smaller branches. These are called *bronchioles;* they
continue to divide and branch into finer and finer
structures, eventually ending as microscopic air sacs

called **alveoli** (singular *alveolus*). The thin walls of
the alveoli are extensively supplied with blood capillaries, and it is here that the exchange of carbon dioxide and oxygen occurs in an adult. Where does this
exchange occur in the fetus?

_____

4. Now relocate the diaphragm and note the position
of this thin sheet of muscular tissue in relation to the
lungs.

5. Complete your study of the respiratory system of
the fetal pig by tracing the pathway of a carbon dioxide
molecule from an alveolus to the nostrils.

III. CIRCULATORY SYSTEM   (Starr, pp. 378, 379)

The circulatory system of the fetal pig consists of a
vast network of **vessels** (*arteries, arterioles, capillaries,
veins,* and *venules*) that contain blood and transport
water, oxygen, carbon dioxide, nutrients, metabolic
wastes, hormones, and other substances to and from
every living cell in the body. In mammals, blood is
propelled through the arteries, arterioles, and capillaries by a muscular four-chambered **heart.**

Through this extensive system, oxygen is added
to the blood in the capillaries of the alveoli of the
lungs, while carbon dioxide is removed for exhalation
from the body. In the capillaries of the small intestine,

various nutrients (for example, glucose and amino acids) are added to the blood, while in the capillaries of the kidneys, the blood is cleansed of various metabolic wastes and excess ions.

The circulatory system is divided into two circuits: the *pulmonary circuit,* which involves blood flow to and from the lungs, and the *systemic circuit,* which is concerned with the flow of blood to and from the rest of the body. In this exercise, you will study these two circuits and examine how the heart directs the flow of blood through them both in a fetus and in an adult.

MATERIALS

Per student pair:

- one fetal pig preserved in formalin or phenol-based chemical and injected with red and blue latex
- dissecting tray
- paper towels
- dissecting kit
- pins
- 4 large rubber bands *or* 2 pieces of string each 2 feet long
- plastic gloves and/or lanolin hand creme
- goggles if you wear contact lenses

PROCEDURE   For the following dissections, refer to Fig. 32-6 for the major arteries of the fetal pig, and to Fig. 32-7, which is a diagrammatic representation of the heart and the major arteries and veins of a fetal mammal.

A. Pulmonary Circuit and Surface Anatomy of the Heart

1. Continue working from the ventral side of your specimen as you did in the preceding section. Make sure that the legs of the pig are secured with string or rubber bands and that the skin flaps are pinned to the sides, or dorsal portion, of the body.

2. Locate the heart in the thoracic cavity and carefully remove the pericardial sac that surrounds it. Identify the four chambers of the heart: the thin-walled **right atrium** (Fig. 32-6) and **left atrium** (Fig. 32-5), and the larger **right** and **left ventricles** (see Fig. 32-5). You should also be able to locate the **coronary artery** and **coronary vein** lying in the diagonal groove between the two ventricles (Fig. 32-5). The coronary artery and its branches supply blood directly to the heart. (The heart is a muscle and as such has the same requirements as any other organ.) When these vessels become severely occluded, a heart attack may occur. It is the coronary arteries and their branches that are replaced or "bypassed" in coronary bypass surgery.

3. In adult pigs, oxygen-poor (or carbon-dioxide–rich) blood returning to the right atrium of the heart from the systemic circulation does so through the large

veins known as the **superior vena cava** (or anterior vena cava) and the **inferior vena cava** (or posterior vena cava). Gently push the heart to the right and identify these relatively large blue veins. Describe the difference in diameter between the superior and inferior vena cava.

_____

_____

_____

Trace the inferior vena cava a short distance from the heart. Where does it lead?

_____

4. The blood that enters the right atrium passes to the right ventricle and then to the **pulmonary trunk,** a large vessel that branches to the lungs. This vessel, which lies between the left and right atria and extends dorsally and to the pig's left, branches to form the **left** and **right pulmonary arteries.** Do these arteries contain red or blue latex?

_____

In the adult, do they carry oxygen-rich or oxygen-poor blood?

_____

Carefully move the heart aside and follow the pulmonary arteries to the lungs.

5. In the adult, once the blood has been oxygenated (and the carbon dioxide removed) in the lungs, it returns to the left atrium of the heart via the **left** and **right pulmonary veins.** Carefully move the lungs and heart and locate these large vessels that carry the blood from the lungs to the left atrium of the heart.

6. From the left atrium of the adult, the oxygenated blood passes to the left ventricle and from there passes into the **aorta** and into the systemic circulatory system. Locate the aorta, which leads dorsally out of the left ventricle of the heart. Note that its base is partially covered by the pulmonary trunk coming from the right ventricle.

7. Blood circulation in the fetal pig

a. The preceding description of blood flow through the heart and lungs is only representative of a pig or other mammal following birth. In the fetus, most of the pulmonary circuit is bypassed twice. First, most blood from the right ventricle enters into the aorta directly from the pulmonary trunk through the **ductus arteriosus,** a large but short vessel connecting the pulmonary trunk directly to the aorta. With the first breath of the fetus, the ductus arteriosus contracts and circulation is established with the pulmonary system. Then, during the eight weeks following birth, the ductus arteriosus forms a fibrous strand of connective tissue, the *ligamentum arteriosum.* Locate the ductus arteriosus in your fetal pig.

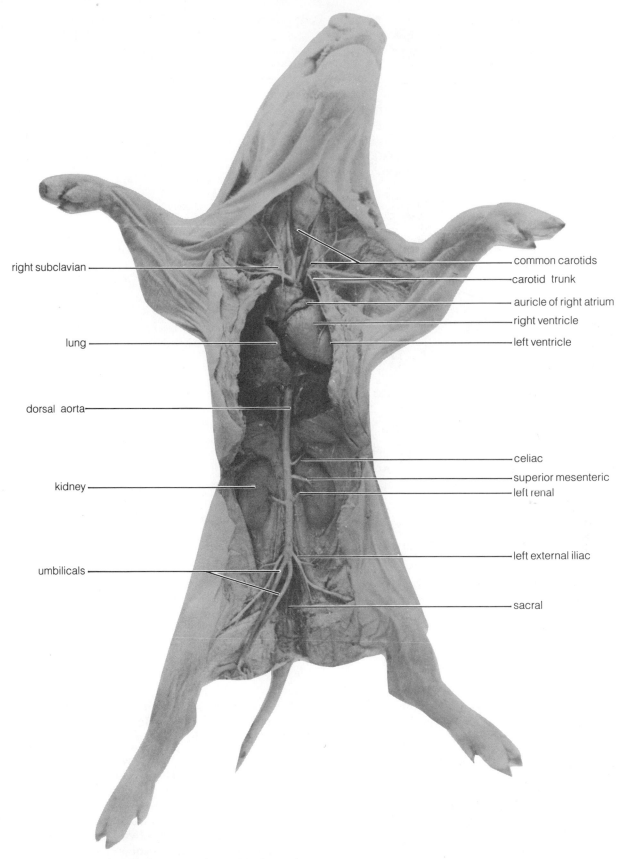

right subclavian ——

lung ——

dorsal aorta ——

kidney ——

umbilicals ——

common carotids

carotid trunk

auricle of right atrium

right ventricle

left ventricle

celiac

superior mesenteric

left renal

left external iliac

sacral

Figure 32-6 Ventral view of major arteries and the heart of a fetal pig. (Courtesy T. Odlaug.)

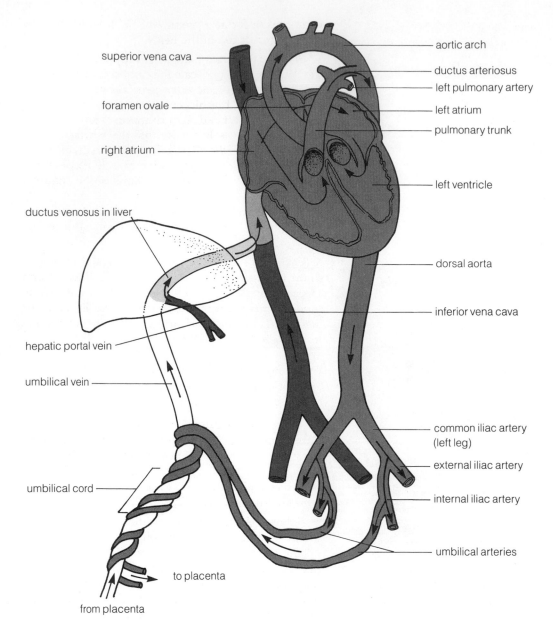

superior vena cava

aortic arch

ductus arteriosus

left pulmonary artery

foramen ovale

left atrium

pulmonary trunk

right atrium

left ventricle

ductus venosus in liver

dorsal aorta

inferior vena cava

hepatic portal vein

umbilical vein

common iliac artery (left leg)

external iliac artery

internal iliac artery

umbilical cord

umbilical arteries

to placenta

from placenta

Figure 32-7 Generalized diagram of the circulatory system of a fetal mammal. Arrows indicate the flow of blood. White represents fully oxygenated blood. The darkest gray indicates oxygen-depleted blood. (After Weller and Wiley, 1985.)

b. The second bypass occurs when most of the blood arriving from the posterior portions of the body via the inferior vena cava and entering the right atrium passes directly into the left atrium via a temporary opening (*foramen ovale*) in the wall separating the right and left atria. Thus, this blood bypasses the pulmonary circuit. As with the ductus arteriosus, the foramen ovale closes at birth and pulmonary circulation is established. Why is it not necessary for large quantities of blood to enter the pulmonary system of a fetus?

_____

_____

B. Systemic Circuit—Major Arteries and Veins Anterior to the Heart (see Fig. 32-6 and Fig. 32-8)

1. The systemic circuit begins with the aorta. This large vessel leads anteriorly out of the left ventricle of the heart and makes a sharp turn to the left (the so-called **aortic arch**) and proceeds posteriorly through the body as the **dorsal aorta.** All of the major arteries of the body arise from these two regions (the aortic arch and dorsal aorta) of the aorta.

2. Locate the first visible vessel, the **brachiocephalic artery,** to branch from the aortic arch. The first vessels to branch from the aorta are the coronary ar-

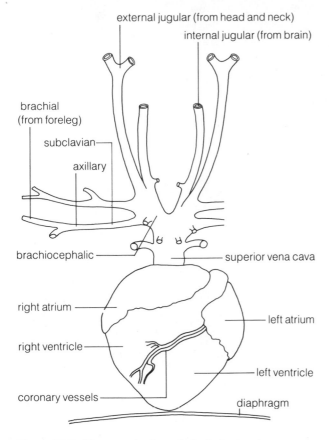

external jugular (from head and neck)

internal jugular (from brain)

brachial
(from foreleg)

subclavian

axillary

brachiocephalic

superior vena cava

right atrium

left atrium

right ventricle

left ventricle

coronary vessels

diaphragm

Figure 32-8 Diagram of veins anterior to the heart

teries, these cannot be seen without dissecting the heart. The brachiocephalic artery branches to give rise to the **right subclavian artery,** going to the right forelimb, and the **carotid trunk,** whose branches course anteriorly through the neck and head. Trace the right subclavian artery and its branches through the shoulder region to the right forelimb. As it passes through the shoulder region, the name of the right subclavian changes to the *axillary artery* and then to the *brachial artery* when it enters the upper forelimb.

3. Return to the aorta and locate the second visible vessel, the **left subclavian artery,** to branch from the aortic arch. The left subclavian artery and its branches pass through the shoulder and left forelimb or arm in the same manner as the right subclavian artery, described above. As you trace the course of the left subclavian, notice that some of its branches feed the muscles of the chest and back.

4. Return to the right forelimb and locate the venous system that passes through this appendage. Because the veins are relatively thin-walled, this may be very difficult. Also, some of them may not be injected with blue latex and will appear a brownish color. If possible, follow the *brachial vein* to the *axillary* and the **subclavian vein** until the latter becomes the **brachiocephalic vein.** It should be relatively easy to follow the brachiocephalic to its juncture with the superior

vena cava (it forms a prominent "V"), which returns to the right atrium of the heart.

5. In order to examine the arterial system that serves the throat and head, locate the carotid trunk (a branch of the brachiocephalic artery; see above). This short branch of the brachiocephalic artery immediately splits into the **left** and **right common carotid arteries.** Each of these vessels divides into the *internal* and *external common carotid arteries.* Remove the thymus and thyroid glands and considerable muscle tissue in the throat to locate the anterior portions of the common carotid arteries.

As you locate and trace the carotid arteries, look for a white "fiber" that parallels them. This is the *vagus nerve.*

6. On either side of the neck are the major veins that drain the head and throat region. The **internal** and **external jugular** *veins* join the subclavian veins (from the forelimbs) to form the **brachiocephalic vein.** The latter leads into the superior vena cava, which returns to the right atrium of the heart.

C. Systemic Circuit—Major Arteries and Veins Posterior to the Heart (see Fig. 32-6 and Fig. 32-9)

1. The posterior extension of the aortic arch is the dorsal aorta. As the name implies, the dorsal aorta lies in a middorsal position along the spine. From this large vessel arise all of the arterial branches that feed the organs, glands, and muscles of the abdominal region and the muscles of the hindlimbs and tail.

2. Follow the dorsal aorta posteriorly. Carefully move the liver and stomach of the pig and use a dissection needle to scrape away the sheet of tissue that connects the dorsal aorta to the pig's back. Locate the **celiac artery,** whose branches deliver oxygenated blood to the stomach, spleen, and liver. Continue to follow the dorsal aorta posteriorly and locate the **superior mesenteric artery.** This vessel, just posterior to the celiac artery, branches to the pancreas and duodenum of the small intestine.

3. Posterior to the superior mesenteric artery are the **renal arteries,** relatively short vessels that connect the dorsal aorta and the kidneys. At this time it is easy to locate the **renal veins,** which drain blood from the kidneys to the inferior vena cava.

4. As you follow the dorsal aorta posteriorly beyond the kidneys, the **external iliac arteries** branch, one into each hindlimb. Each leg is also supplied with a major vein, the **iliac vein,** which joins the inferior vena cava.

5. Follow the branches of the dorsal aorta into the tail region, being careful not to cut the two intervening branches. The small extension toward the tail region is called the *sacral artery* as it leaves the dorsal aorta and the *caudal artery* when it enters the tail.

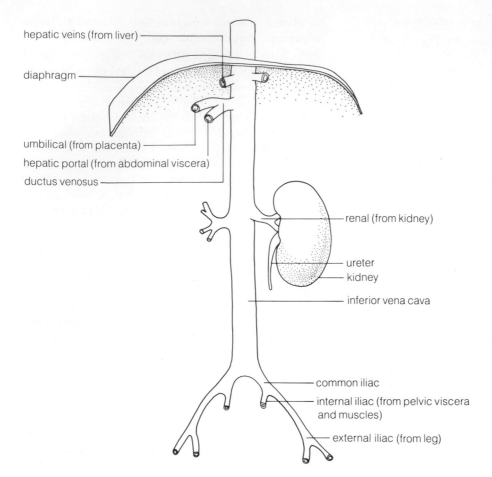

hepatic veins (from liver)

diaphragm

umbilical (from placenta)

hepatic portal (from abdominal viscera)

ductus venosus

renal (from kidney)

ureter

kidney

inferior vena cava

common iliac

internal iliac (from pelvic viscera and muscles)

external iliac (from leg)

Figure 32-9 Diagram of veins posterior to heart

6. Just anterior to the sacral artery, the **internal iliac arteries** branch from the dorsal aorta. These enlarge and form the two **umbilical arteries,** which run through the umbilical cord to the placenta. Cut the umbilical cord transversely and note the arrangement of the umbilical arteries within it. Consider the composition of the blood as it travels through the umbilical arteries to the placenta. Is it oxygenated or rich in carbon dioxide?

—————————————————————

7. Locate the two pieces of vein that you tied with string in Exercise 31. This is the **umbilical vein,** through which blood rich in nutrients and oxygen flows from the placenta of the mother back to the fetus. Locate the umbilical vein in the umbilical cord and follow it anteriorly toward the liver. When the umbilical cord reaches the liver, it becomes the **ductus venosus** (Fig. 32-7), which continues anteriorly within the substance of the liver and joins the inferior vena cava. The umbilical arteries, the umbilical veins, and ductus venosus become modified into ligaments fol-

lowing the birth of the fetus. What is the relationship between the navel and the umbilical cord?

—————————————————————

—————————————————————

—————————————————————

8. The *hepatic portal system* consists of a network of veins that collects blood from the lower digestive tract and associated organs (stomach, small intestine, pancreas, and spleen) and carries it to the liver via the **hepatic portal vein.** In general, a portal vein is one that collects blood from the capillaries of one organ and transfers it to the capillaries of another organ. Locate the hepatic portal vein.

9. In the adult, the blood of the liver is drained by the **hepatic veins.** These join the inferior vena cava just anterior to the point where the ductus venosus joins the inferior vena cava (Fig. 32-9).

10. Complete your dissection of the systemic circulatory system by tracing the inferior vena cava from

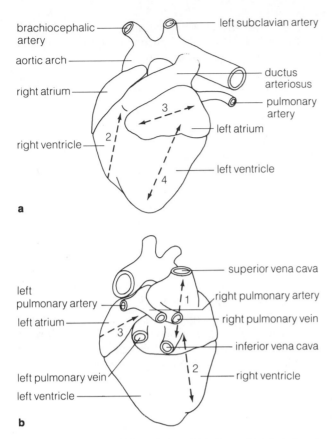

brachiocephalic artery

aortic arch

right atrium

right ventricle

left subclavian artery

ductus arteriosus

pulmonary artery

left atrium

left ventricle

a

left pulmonary artery

left atrium

left pulmonary vein

left ventricle

superior vena cava

right pulmonary artery

right pulmonary vein

inferior vena cava

right ventricle

b

Figure 32-10 Ventral (**a**) and dorsal (**b**) views of fetal pig heart showing numbered incisions for dissection

the abdominal cavity through the diaphragm and to the thoracic cavity where it joins the superior vena cava before entering the right atrium of the heart.

D. Internal Structure of the Heart

Refer to Fig. 32-10 for the dissection of the heart and the study of its internal structure.

1. Carefully free the heart from the fetal body by cutting through the superior and inferior venae cavae, the subclavians, the common carotids, and the dorsal aorta just posterior to the heart. Cut through the left and right pulmonary veins and the pulmonary arteries at their juncture with the lungs. Remove the heart from the fetus and place it on paper towels with its ventral surface facing you, as it was in the thoracic cavity. If any of the whitish membranous pericardium is still present, carefully remove it from around the heart.

2. Review the location of the four heart chambers: the left and right atria and the left and right ventricles. Locate the coronary artery and vein in the longitudinal groove running between the left and right ventricles. Carefully follow these vessels to the dorsal side of the heart. Although you do not need to confirm this, the coronary arteries are the first vessels to branch from the aortic arch.

3. Place the heart dorsal-side up and locate the inferior and superior venae cavae. Cut through these vessels with your scissors and expose the interior chamber of the right atrium (see Fig. 32-10, incision 1). Carefully remove any latex and coagulated blood in the right atrium. Between the atrium and the right ventricular cavity are three membranous cusps attached to the wall of the right atrium. This is the *tricuspid valve*. The open ends of the cusps face downward into the cavity of the right ventricle.

4. Continue working from the dorsal side and cut into the right ventricle as indicated by incision 2 in Fig. 32-10. With your forceps and needle, remove any latex that obstructs your view. You should be able to see the three cusps of the *semilunar valve* at the juncture of the pulmonary artery and the right ventricle. The open ends of the cusps face into the pulmonary trunk and thus prevent a backward flow of blood into the ventricle.

5. Examine the internal walls of the ventricle. If you wish, you may extend incision 2 to the ventral side of the right ventricle. Can you see the muscular ridges? These are the *papillary muscles*, and arising from them are relatively fine fibers, the *chordae tendinae*. The chordae tendinae are attached to the edges of the tricuspid valve.

6. Next, with the heart's ventral surface facing you, locate the ductus arteriosus and the aorta. (Remember, the ductus arteriosus is a shunt between the pulmonary trunk and the aorta.)

Cut open the left atrium (incision 3) and the left ventricle (incision 4). Remove the latex. From the dorsal side of the heart, examine the pulmonary veins from the inside of the left atrium. Next, locate the *bicuspid valve* (consisting of two cusps) between the left atrium and left ventricle. Do the cusps appear similar to the tricuspid valve?

(Yes or no) _____

7. Turn the heart so that the ventral surface is facing you, and examine the cavity of the left ventricle. Note the papillary muscles and the chordae tendinae in the left ventricle. The chordae tendinae are commonly called "heart strings." Do they appear similar to those in the right ventricle?

(Yes or no) _____

8. Insert a probe into the aorta from the exterior of the heart and note where it enters the cavity of the left ventricle. At this point there is another valve, the *aortic semilunar valve*, with three cusps. Is the orientation of the aortic semilunar valve similar to that of the semilunar valve between the pulmonary trunk and the right ventricle?

(Yes or no) _____

9. Recall that in the fetus a temporary opening, the foramen ovale, exists between the right and left atria.

It is not necessary to locate this opening in the fetal heart, but you should review its function.

10. Complete your study by looking for differences in the thickness of the walls of the atria and those of the ventricles. Describe any differences you find.

Explain why the wall of the left ventricle is thicker than the wall of the right ventricle.

_____

_____

_____

_____

_____

When you have completed this exercise, return the body wall flaps to their original positions and place the pig in the plastic bag. Dispose of any paper towels that contain formalin or preservative as directed by your instructor. Clean your dissecting tools and laboratory table.

_____

PRE-LAB QUESTIONS

_____ **1.** The two *major* ventral body cavities of a fetal pig are the (a) thoracic and pleural, (b) thoracic and pericardial, (c) abdominopelvic and thoracic, (d) abdominopelvic and pericardial.

_____ **2.** The diaphragm is a muscular sheet of tissue that separates the (a) thoracic and pleural cavities, (b) thoracic and pericardial cavities, (c) thoracic and abdominopelvic cavities, (d) pleural and pericardial cavities.

_____ **3.** The digestive system is concerned with (a) blood circulation, (b) digestion and the absorption of nutrients, (c) reproduction, (d) excretion of urine.

_____ **4.** One of the most important functions of the liver is to (a) remove urea from the blood, (b) regulate the levels of glucose and amino acids in the blood, (c) both of the above, (d) none of the above.

_____ **5.** A vein is a blood vessel that always carries (a) blood toward the heart, (b) blood away from the heart, (c) oxygen-rich blood, (d) oxygen-poor blood.

_____ **6.** As a general rule, the small intestine of a pig or human is (a) about 2 feet long, (b) about 5 feet long, (c) about as long as the individual is tall (or long in the case of the pig), (d) about five times the height of the individual.

_____ **7.** In humans, the front of the larynx is commonly referred to as the (a) voice box, (b) Adam's apple, (c) food pipe, (d) a and b.

_____ **8.** The microscopic air sacs or alveoli are the sites where (a) oxygen is taken up and carbon dioxide is released by the blood, (b) oxygen is released and carbon dioxide is taken up by the blood, (c) air exchange occurs in the heart, (d) none of the above.

_____ **9.** The hearts of a fetal pig and a human are similar in that they are (a) the primary pump of the circulatory system of the body, (b) both four-chambered, (c) composed of cardiac muscle tissue, (d) all of the above.

_____ **10.** The cardiac, pyloric, anal, and ileocecal sphincters are all part of the (a) digestive tract, (b) heart, (c) kidney, (d) oral cavity.

EXERCISE 33

DISSECTION OF THE FETAL PIG:
UROGENITAL AND NERVOUS SYSTEMS

OBJECTIVES After completing this exercise, you will be able to:

1. define ovulation, meningitis, homologous structures, inguinal hernia, vasectomy, nephron, urea, urine;

2. describe and give the functions of the organs of the excretory, reproductive, and nervous systems;

3. explain the importance of the excretory, reproductive, and nervous systems to a living mammal;

4. locate, name, and describe the function of the internal structures of the kidney.

INTRODUCTION In today's exercise you will complete your study of the internal anatomy of the fetal pig. In previous exercises, you have dissected the digestive system, whose function is to break down the large complex organic compounds present in food to smaller molecules that can be absorbed by the body, and the respiratory system, which brings oxygen into the body and exchanges it for carbon dioxide. In addition, you have examined the circulatory system, which transports nutrients and gases dissolved in the blood throughout the body.

Now you will examine the system that removes metabolic wastes from the bloodstream (the *excretory system*) and the system largely responsible for integration and control in the organism (the *nervous system*). In addition, you will study both male and female specimens in order to examine the system responsible for the production of new individuals or offspring (the *reproductive system*). The excretory or urinary and reproductive systems are traditionally studied together (as the *urogenital system*) because they share several anatomical features.

I. UROGENITAL SYSTEM   (Starr, pp. 420, 467–474)

A. Excretory System

Like humans, the pig is a terrestrial organism and, as such, must conserve body fluids or water. At the same time, metabolic wastes must be continuously removed from the blood. Furthermore, the composition of the blood must be constantly monitored and adjusted so that the cells of the body are bathed in a fluid of constant composition.

Much of the potentially poisonous waste occurs in the form of *urea* and results from the metabolism of amino acids in the liver. Urea is filtered from the bloodstream in the kidneys, which also regulate water and salt balance.

MATERIALS

Per student pair:

- one fetal pig preserved in formalin or phenol-based chemical and injected with red and blue latex
- dissecting tray
- paper towels
- dissecting kit
- pins
- 4 large rubber bands *or* 2 pieces of string each 2 feet long
- plastic gloves and/or lanolin hand creme
- goggles if you wear contact lenses

PROCEDURE

1. Place your pig on its back in the dissecting tray and use string or rubber bands to secure the legs as you did in the preceding exercises. Pin the lateral body wall flaps to the dorsal side of your specimen, and pull the umbilical cord and surrounding tissue back between the hindlimbs.

2. The **kidneys** are situated in the lumbar region of the body cavity against the dorsal body wall (Fig. 33-1). In addition to the various abdominal viscera, they are covered by the **peritoneum,** the smooth, rather shiny membrane that lines the abdominopelvic cavity. (You may have already removed much of this during the dissection of the circulatory system in the previous exercise.) To expose the right kidney and its *ureter,* carefully lift up the abdominal organs and move them anteriorly and to the pig's left. Using a dissecting

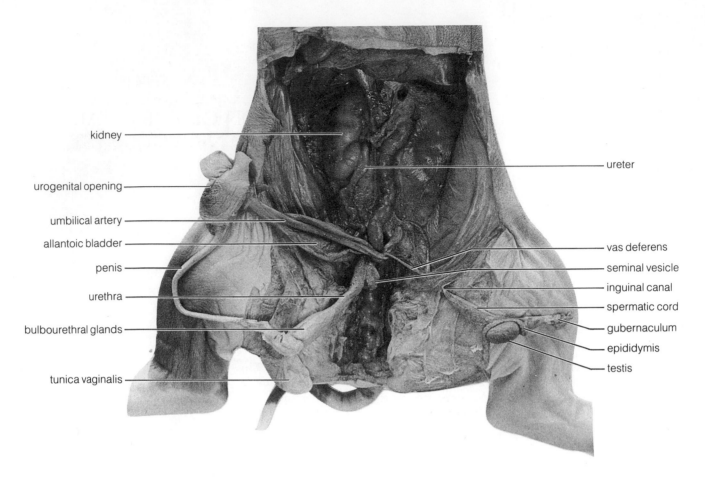

kidney

urogenital opening

umbilical artery

allantoic bladder

penis

urethra

bulbourethral glands

tunica vaginalis

ureter

vas deferens

seminal vesicle

inguinal canal

spermatic cord

gubernaculum

epididymis

testis

Figure 33-1 Ventral view of the male urogenital system of the fetal pig. (Courtesy T. Odlaug.)

needle, carefully scrape away the peritoneum so that the kidney, a bean-shaped structure, and its ureter are easily seen. Note the central depression in the surface of the kidney. This is the *hilus*, the region where the ureters and the *renal vein* leave and the *renal artery* enters the kidney.

3. Carefully follow the ureters from the hilus to the **allantoic bladder.** Then lift the bladder and find the **urethra.** The latter is the structure through which urine passes from the bladder to the outside of the animal. In the male, the urethra is very long and passes through the penis to the outside of the body (Fig. 33-1). Notice that the urethra passes posteriorly for a distance of approximately 2 cm and then turns sharply anteriorly and ventrally before entering the penis. In the female fetal pig, the urethra is short and passes posteriorly to join with the *vagina* to form the **urogenital sinus** (see Fig. 33-2). How does this compare with the structure of the adult female pig?

_____

_____

_____

4. Locate the *allantoic duct*, which leads from the allantoic bladder into the umbilical cord. The allantoic duct is largely a vestigial structure, for most of the wastes produced in the kidneys of the fetus are carried in the bloodstream through the umbilical vein to the placenta, where they are removed in the body of the mother. Following the birth of the fetus, the allantoic duct collapses and the allantoic bladder is incorporated into the **urinary bladder.**

5. We will examine the urinary system and the internal structure of the kidneys more closely following the study of the male and female reproductive systems.

B. Female Reproductive System

In terrestrial organisms, fertilization (the fusion of a male and female gamete) occurs internally, where a relatively stable aquatic environment is maintained. Once fertilization has occurred, the zygote divides to form an embryo and eventually a fetus. In mammals, all of this growth and development occurs within the female's uterus, which nourishes the developing offspring until it can pass through the birth canal and exist on its own in the outside world.

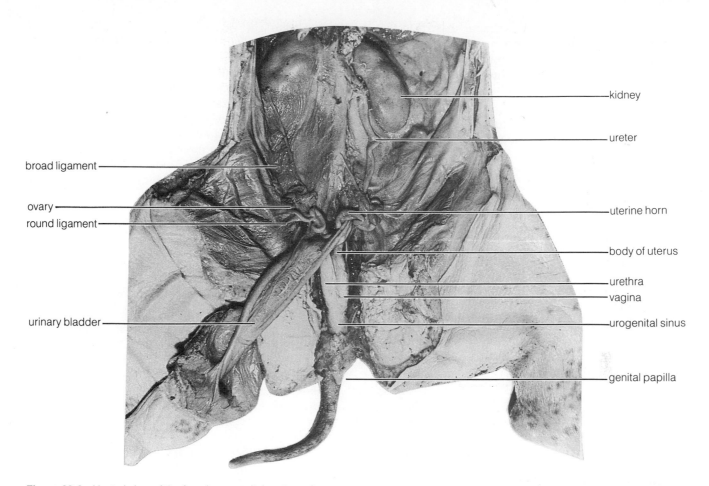

broad ligament

ovary
round ligament

urinary bladder

kidney

ureter

uterine horn

body of uterus

urethra
vagina

urogenital sinus

genital papilla

Figure 33-2 Ventral view of the female urogenital system of the fetal pig. (Courtesy T. Odlaug.)

In this portion of the exercise, you will study the structures (external genitalia, uterus, oviducts, ovaries, and associated ducts) of the mammalian female reproductive system and examine how each contributes to the process of internal fertilization and to the growth and development of a viable fetus.

1. Examine the **vulva,** the collective term for the external genitalia of the female. In the pig, the vulva includes the *genital papilla* on the outside of the body, the **labia** or "lips" found on either side of the urogenital sinus, and the **clitoris,** a small body of erectile tissue on the ventral portion of the urogenital sinus. Also included in the vulva is the opening of the urogenital sinus itself (Fig. 33-2).

2. In the female fetal pig, the urogenital sinus is the common passage for the urethra and the **vagina.** To locate these structures, carefully insert your scissors into the opening of the urogenital sinus and cut this structure from the side. Locate where the ducts of the vagina and urethra enter to form the urogenital sinus.

The urogenital sinus is not present in the adult female pig. During the subsequent development of the fetus, the sinus is reduced in size until the vagina and

the urethra each develop their own, separate opening to the outside. Thus, in the adult female pig, urine exits through the urinary opening. This is the situation in most adult female mammals. How does this compare with the structure of the reproductive system of most male mammals?

_____

_____

_____

3. Again, locate the clitoris. This small, rounded region on the inner ventral surface of the urogenital sinus is **homologous** (i.e., similar in structure and origin) with the male *penis.* In the male, the tissues of the penis develop around and enclose the urethra, while in the female, the urethra opens posteriorly to the clitoris.

4. Follow the urogenital sinus anteriorly and identify the thick-walled muscular vagina that is continuous with the **uterus.** In the pig, the uterus consists of three structures or regions: the **cervix** at the entrance to the uterus, the **uterine body,** and the two **uterine**

horns (Fig. 33-2). Determine in your dissection that the uterine horns unite to form the body of the uterus. The pig has a *bicornuate uterus* in which the fetuses develop in the uterine horns. In the human female there are no uterine horns, and the fetus develops within the body of the *simplex uterus*.

5. From the uterine horns, follow the **oviducts** to the **ovaries** (female gonads). The ovaries are small, yellowish kidney-shaped structures that lie just posterior to the kidneys. They are the sites of egg production and the source of the female sex hormones estrogen and progesterone.

All of the eggs that a female will produce during her lifetime are present in the ovaries at birth. After puberty, eggs will mature, rupture from the surface of the ovaries, and pass through the oviducts after the proper hormonal stimulation. This process is referred to as **ovulation.**

If viable sperm are present in the upper third of an oviduct when it contains eggs, fertilization may occur. In this case, the fertilized egg or zygote will develop into an embryo and pass down the oviduct to become implanted in the wall of the uterine horn. In the human female, however, it will become implanted in the uterine body.

6. The ovaries, oviducts, and uterine horns are supported by a sheet of mesentery or connective tissue, the **broad ligament,** which originates from the dorsal body wall (Fig. 33-2). A second mesentery, the **round ligament,** which also supports the ovaries, extends from the lateral wall and crosses the broad ligament diagonally. Identify the broad and round ligaments.

C. Male Reproductive System

The male reproductive system of a mammal consists of the external genitalia (penis, scrotum, and testes) and various internal structures: the urethra, sex accessory glands, and associated ducts. In this exercise, you will locate and identify these structures and glands in the fetal pig and examine how they contribute to the reproductive process.

1. Begin your dissection of the male reproductive system by locating the **testes** (male gonads), the site of *sperm* production and source of *testosterone*, the male sex hormone. In older fetuses they are located in the *scrotum*, but in younger fetuses they may be found anywhere between the kidneys and the scrotum.

In order for viable sperm to be produced in adult males, the testes must be situated outside of the abdominal cavity, where body temperatures are slightly lower than within. Thus, during normal development, the testes undergo a posterior migration, or descent, into the scrotum.

The following discussion and dissection applies to an older male fetus with testes fully descended into the scrotum.

2. Locate the scrotum. Make a midline incision through this structure, cutting through the muscle tissue. Pull out the two elongated bulbous structures covered with a transparent membrane. This membrane is the *tunica vaginalis* and is actually an outpocketing of the abdominal wall. Notice the tough white cord that connects the posterior end of the testes to the inner face of the sac (see Fig. 33-1). This cord, the **gubernaculum,** is homologous to the round ligament in the female reproductive system. It grows more slowly than the surrounding tissues and thus aids in pulling the testes posteriorly into the scrotal sacs.

3. Cut through the tunica vaginalis to expose a single testis and locate the **epididymis.** This is a tightly coiled tube that lies along one side of the testis. Sperm produced in the testes are stored in the epididymides until ejaculation moves them out of the body in the *seminal fluid*.

4. The slender, elongated structure that emerges from each testis is the **spermatic cord.** Gently pull the cord and note that it moves through an opening, the **inguinal canal,** which is actually an opening in the abdominal wall between the abdominopelvic cavity and the cavity of the scrotum. It is through this opening that the testes descend during their migration into the scrotum.

Some human males develop an **inguinal hernia,** a condition in which part of the intestine drops through the inguinal canal into the scrotum. Pigs and other quadripeds (hint) do not develop inguinal hernias. Explain why.

_____

_____

_____

5. The spermatic cord consists of the **vas deferens** (plural is *vasa deferentia*), the spermatic vein, the spermatic artery, and the spermatic nerve. It is the vas deferens that is severed when a human male has a **vasectomy.** Follow the vas deferens to the base of the bladder, where it loops up and over the ureter and then continues posteriorly to enter the urethra. During ejaculation, sperm stored in the epididymis move through the vas deferens into the urethra for transport out of the body.

6. Before the sex accessory glands can be identified, it is necessary to expose the full length of the **penis** and its juncture with the urethra. Make an incision with a scalpel through the muscles in the midventral line between the hindlegs until they lie flat. Carefully remove the muscle tissue and pubic bone on each side of the cut until the urethra is exposed. With a blunt probe, tear the connective tissue connecting the urethra to the rectum, which lies dorsal to it.

7. Locate a pair of small glands, the **seminal vesicles,** on the dorsal surface of the urethra where the two vasa deferentia enter. Situated between the bases of the seminal vesicles is the **prostate gland.** The other sex accessory glands are the **bulbourethral glands,** two elongate structures lying on either side of the juncture of the penis and urethra.

The seminal vesicles, the prostate gland, and the bulbourethral glands all secrete fluids that, together with sperm, form **seminal fluid,** or *semen,* that is ejaculated during sexual intercourse. In addition to sperm, seminal fluid is mostly water, sugar, and other nutrients that provide the correct aquatic environment for the flagellated sperm.

II. INTERNAL ANATOMY OF THE KIDNEY
(Starr, pp. 420–424)

In the first portion of today's laboratory, you located the kidneys, a pair of bean-shaped structures lying on either side of the spine in the lumbar region of the body. Although the kidneys are situated below the diaphragm, they are actually located outside of the peritoneum, the membrane that lines the abdominal cavity. During this exercise, you will examine in greater detail the structure of the kidney, including its internal anatomy, and the activity of its functional unit, the **nephron.**

MATERIALS

Per student pair:

- one fetal pig preserved in formalin or phenol-based chemical and injected with red and blue latex
- dissecting tray
- paper towels
- dissecting kit
- pins
- 4 large rubber bands *or* 2 pieces of string each 2 feet long
- plastic gloves and/or lanolin hand creme
- goggles if you wear contact lenses

PROCEDURE

A. Gross Internal Anatomy of the Kidney (Fig. 33-3a)

1. Locate one of the kidneys and free it by severing the renal vein, renal artery, and ureter. Remove the kidney from the body cavity and place it on a paper towel with the central depression to the right. Attempt to identify the *adrenal gland,* a yellowish-brown body located on the medial, anterior side of the kidney.

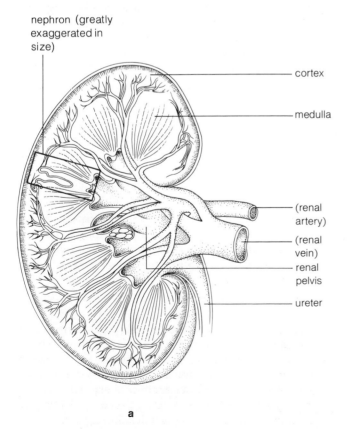

a

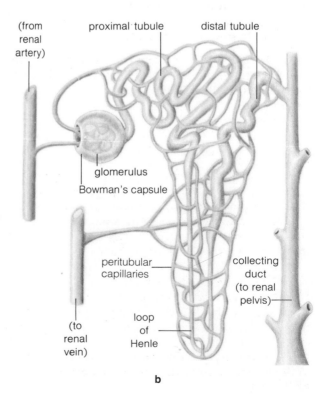

b

Figure 33-3 **(a)** Longitudinal section of kidney; **(b)** a nephron, the functional unit of the kidney

2. With your scalpel, carefully cut the kidney in half lengthwise as you would separate the two halves of a bean or peanut. Examine the cut surface of one of the halves and locate the three major regions of the kidney: the outer *cortex,* the *medulla,* and the *renal pelvis* (see Fig. 33-3). The cortex and medulla contain the functional units of the kidney: the *nephrons* and their associated blood vessels. Nephrons remove **urea** (produced in the liver), excess water, salts, and other wastes from the blood to form **urine.** Urine then collects in the space of the renal pelvis, travels through the ureters to the bladder, and eventually leaves the body through the urethra. Before proceeding, review the location of the ureters, bladder, and urethra (see Figs. 33-1 and 33-2).

B. Microscopic Structure of the Nephrons

1. Although you will not be able to observe a nephron in the laboratory, use Fig. 33-3b and the following narrative to familiarize yourself with the functional unit of the kidney.

2. Blood enters the kidney via branches of the renal artery (from the dorsal aorta) and is filtered out of knots of capillaries called *glomeruli* (singular is *glomerulus*). The filtrate is literally forced out of the capillaries by the relatively high blood pressure and collects in funnel-shaped structures called *Bowman's capsules.* As the filtrate travels through the rest of the nephron (the *proximal tubule,* the *loop of Henle,* and the *distal tubule*), much of the water, ions, sugars, and other useful substances are reabsorbed into the blood in the *peritubular capillaries.* While the reabsorption process carries many useful materials back into the blood, such substances as ammonia, potassium, and hydrogen ions are actually secreted from the blood to the convoluted tubules, where they join urea and other substances, forming urine.

The urine moves out of the nephrons into the *collecting ducts,* where it drains into the renal pelvis. It is in the collecting ducts that the final urine concentration is determined. The blood from the peritubular capillaries returns to vessels that join to form the renal vein, which returns the filtered blood to the inferior vena cava. Thus, the nephron carries out its excretory and osmoregulatory functions in three steps: filtration, reabsorption, and tubular secretion.

3. Although all of the activities of the nephron are extremely vital to the health of a mammal, the importance of the reabsorption function is especially easy to understand. For example, the glomeruli of the kidneys produce approximately 180 L (approximately 180 quarts) of filtrate each day in adult humans. However, about 99% of this filtrate is reabsorbed as water primarily by the nephrons. If they were not so efficient, we would have to drink constantly just to replenish the fluid lost through filtration.

III. NERVOUS SYSTEM   (Starr, pp. 434–438)

The nervous system of the pig and, in fact, all vertebrates can be divided into two major components: the *central nervous system* (CNS) and the *peripheral nervous system* (PNS). The CNS consists of the brain and the spinal cord, which serves as the primary link between the brain and much of the PNS. The PNS includes the large network of nerves outside of the central system. Through the cranial and spinal nerves of the peripheral system, impulses enter and leave the central nervous system.

MATERIALS

Per student pair:

- one fetal pig preserved in formalin or phenol-based chemical and injected with red and blue latex
- dissecting tray
- paper towels
- dissecting kit
- plastic gloves and/or lanolin hand creme
- goggles if you wear contact lenses
- dissection microscope

PROCEDURE   As it is very time-consuming to expose the full length of the spinal cord, it is suggested that the students work in groups of four during the first portion of the exercise (**Section A**). One pair in the group could expose the anterior portion of the spinal cord and the other expose the posterior region. In any case, each pair of students should carry out **Section B,** the dissection of the brain into its juncture with the spinal cord.

A. Spinal Cord

1. Turn the body wall flaps of the fetal pig inward and place your specimen ventral side down on the dissecting tray. Proceed carefully with this portion of the dissection, as the nervous tissues of a fetus are extremely delicate and may be easily destroyed.

2. Carefully remove a strip of muscle, about 1.5 cm wide, from the base of the neck posteriorly along the spinal column to the tail. This will expose the *spines* and the *neural arches* of the vertebrae (see Fig. 33-5). With a sharp scalpel, gradually cut away the spines and the neural arches of several vertebrae until the **spinal cord** is exposed.

3. Note the enlargements of the spinal cord at the level of the forelimbs and hindlimbs (Fig. 33-4). These are the *cervical* and *lumbar enlargements,* respectively; they result from the large number of nerve cells, *neurons,* supplying the appendages in these regions.

4. At the anterior end of the body, the spinal cord widens to become the **medulla oblongata,** the most

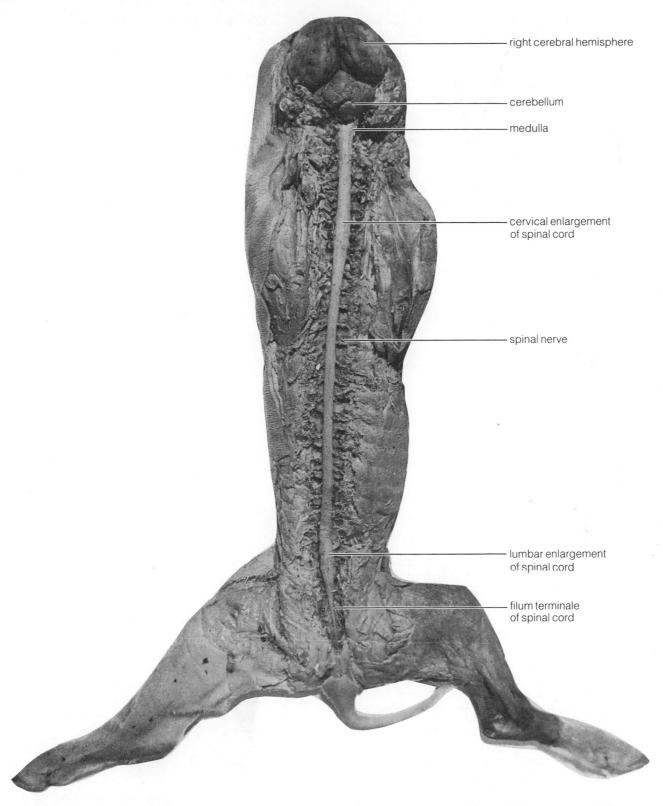

right cerebral hemisphere

cerebellum

medulla

cervical enlargement
of spinal cord

spinal nerve

lumbar enlargement
of spinal cord

filum terminale
of spinal cord

Figure 33-4   Dorsal view of the brain, spinal cord, and spinal
nerves of a fetal pig. (Courtesy T. Odlaug.)

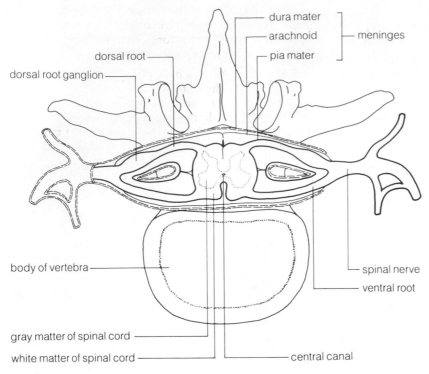

Labels on figure:
dorsal root ganglion
dorsal root
dura mater
arachnoid — meninges
pia mater
body of vertebra
spinal nerve
ventral root
gray matter of spinal cord
white matter of spinal cord
central canal

Figure 33-5 Cross section of a vertebra and spinal cord of a mammal

posterior portion of the brain. At its caudal end, the spinal cord narrows to a relatively thin strand of tissue called the *filum terminale* (Fig. 33-4).

5. Surrounding the spinal cord and the brain are a set of three membranes, the **meninges.** The outermost layer, the **dura mater,** is the most apparent and adheres to the underside of the cranial and spinal bones. The dura mater is a tough, fibrous sheath that must be slit in order to expose the spinal cord. The middle layer, the **arachnoid,** will not be apparent. The innermost layer, the **pia mater,** adheres closely to the surface of the spinal cord and brain. If you cannot identify the outer and inner meninges, attempt to locate them when you dissect the brain in Section **B.3** of this exercise.

6. Note the origin of the **spinal nerves** on either side of the spinal cord (see Fig. 33-5 to determine the relationships of the spinal nerves to the spinal cord). There are 33 pairs of spinal nerves associated with the spinal cord: 8 cervical, 14 thoracic, 7 lumbar, and 4 sacral. (Do not attempt to locate all of these, however.) Determine that a spinal nerve is composed of a **dorsal** and a **ventral root** (Fig. 33-5). The dorsal root, carrying sensory impulses into the spinal cord, can be easily identified by the presence of a distinct swelling called the **dorsal root ganglion** (Fig. 33-5). The ventral root, which carries motor impulses from the cord to some type of effector, a muscle, for example, has no ganglion and is not as easily identified as is the dorsal root.

7. Remove a short cross section (0.5 cm long) of the spinal cord and examine it with a dissection microscope. Notice the tissue that forms the prominent "H" in the transverse plane. This is the **gray matter,** composed of the cells of motor neurons. The **white matter** around the "H" is made up of neuron fibers that conduct messages to and from the brain.

B. Brain

1. Using your scalpel and scissors, make a longitudinal cut through the skin and muscle tissue of the dorsal portion of the head beginning at the base of the snout and ending at the base of the skull. From the anterior portion of this incision, make a transverse cut to the angle of the jaws and another transverse incision from the base of the skull to a level just ventral to the ears. You should now be able to remove the skin and muscle to expose the skull.

2. To remove the skull, make a longitudinal cut along the middorsal line of the skull. Do not cut too deeply and damage the brain, however. Now make two cuts, about 2 cm apart, at right angles to the longitudinal incision. To expose the brain, carefully break off pieces of the skull until the entire dorsal and lateral areas of the brain are exposed.

3. If you did not identify the meninges or membranes covering the spinal cord in Section **A.5,** locate the dura mater and the pia mater on the surface of the brain at this time. As with the spinal cord, the arachnoid layer (middle layer) will not be apparent.

In certain severe viral or bacterial infections, the meninges around the spinal cord and/or brain may become inflamed. This serious condition is known as **meningitis.**

4. The gross features of the brain can be more easily identified by cutting the spinal cord at the base of the brain and carefully removing it from the skull. The brain is composed of the right and left *cerebral hemispheres* (collectively, the largest part of the brain, the **cerebrum**), separated by a prominent *longitudinal groove;* a smaller mass posterior to the cerebrum, the **cerebellum;** and the medulla oblongata, or more simply, the *medulla,* under the cerebellum. The **pons,** which is not easily seen, lies just anterior to the medulla.

In general, most involuntary, unconscious, and mechanical processes are directed by the more posterior portions of the brain (centers in the medulla control breathing, digestion, and heartbeat). The cerebellum unconsciously controls posture and contains motor programs (like computer programs) for many complex movements. The cerebrum is responsible for such activities as reasoning, memory, conscious thought, language, and sensory decoding—activities that are generally associated with intelligence.

PRE-LAB QUESTIONS

_____ **1.** The urogenital system refers to the (a) excretory and reproductive systems, (b) urinary and excretory systems, (c) the reproductive system, (d) the external genitalia.

_____ **2.** Two of the major functions of the kidneys are to (a) remove metabolic wastes and regulate the ionic composition of the blood, (b) remove metabolic wastes and excess nutrients from the blood, (c) control the sugar and amino acid levels in the blood, (d) none of the above.

_____ **3.** The clitoris of the female and the penis of the male are homologous structures. This means they have (a) a similar function, (b) a similar structure, (c) a similar origin, (d) a similar origin and structure.

_____ **4.** The testes of a male differ from the ovaries of a female in that the testes (a) develop in the body cavity and migrate to a position outside of the body cavity, (b) require a slightly lower temperature than that of the body to produce viable gametes, (c) a and b, (d) none of the above.

_____ **5.** When a human male has a vasectomy, the operation involves (a) removal of the male gonads or testes, (b) removal of the urethra, (c) the severing of the vas deferens, (d) removal of the prostate gland.

_____ **6.** Seminal fluid consists mostly of (a) sperm, water, sugar, and other nutrients, (b) male sex hormones and other nutrients, (c) an egg, water, sugar, and other nutrients, (d) none of the above.

_____ **7.** The functional unit of the kidney is the (a) renal pelvis, (b) ureter, (c) cortex, (d) nephron.

_____ **8.** The central nervous system of a mammal includes (a) the brain, (b) the spinal cord, (c) the brain and spinal cord, (d) the brain, spinal cord, and every major nerve in the body.

_____ **9.** The brain is surrounded by a set of membranes called the (a) pleural membranes, (b) thoracic membrane, (c) pericardial membranes, (d) meninges.

_____ **10.** The largest part of the brain of a mammal is the (a) cerebrum, (b) cerebellum, (c) pons, (d) medulla oblongata.

TERMS OF ORIENTATION
IN AND AROUND THE ANIMAL BODY

A. Body Shapes

1. Symmetry. The body can be divided into almost identical halves.

2. Asymmetry. The body cannot be divided into almost identical halves (e.g., many sponges).

3. Radial symmetry. The body is shaped like a cylinder (e.g., sea anemone) or wheel (e.g., sea star).

4. Bilateral symmetry. The body is shaped like ours in that it can be divided into halves by only one symmetrical plane (midsagittal).

B. Directions in the Body

1. Dorsal. At or toward the back surface of the body.

2. Ventral. At or toward the belly surface of the body.

3. Anterior. At or toward the head of the body—ventral surface of humans.

4. Posterior. At or toward the tail or rear end of the body—dorsal surface of humans.

5. Medial. At or near the midline of a body. The prefix *mid-* is often used in combination with other terms (e.g., midventral).

6. Lateral. Away from the midline of a body.

7. Superior. Over or placed above some point of reference—toward the head of humans.

8. Inferior. Under or placed below some point of reference—away from the head of humans.

9. Proximal. Close to some point of reference or close to point of attachment of an appendage to the trunk of the body.

10. Distal. Away from some point of reference or away from point of attachment of an appendage to the trunk of the body.

11. Longitudinal. Parallel to the midline of a body.

12. Axis. An imaginary line around which a body or structure can rotate. The midline or *longitudinal axis* is the central axis of a symmetrical body or structure.

13. Axial. Placed at or along an axis.

14. Radial. Arranged symmetrically around an axis like the spokes of a wheel.

C. Planes of the Body

1. Sagittal. Passes vertically to the ground and divides the body into right and left sides. The *midsagittal* or *median plane* passes through the longitudinal axis and divides the body into right and left halves.

2. Frontal. Passes at right angles to the sagittal plane and divides the body into dorsal and ventral parts.

3. Transverse. Passes from side to side at right angles to both the sagittal and frontal planes.

ILLUSTRATION REFERENCES

Abramoff, P., and R. G. Thomson. 1982. *Laboratory Outlines in Biology III.* New York: W. H. Freeman.

Boolootian, R. A., and K. A. Stiles Trust. 1981. *College Zoology.* Tenth Edition. New York: Macmillan.

Case, C. L., and T. R. Johnson. 1984. *Experiments in Microbiology.* Menlo Park, California: Benjamin/Cummings.

Creager, J. G. 1983. *Human Anatomy and Physiology.* Belmont, California: Wadsworth.

Fowler, I. 1984. *Human Anatomy.* Belmont, California: Wadsworth.

Gilbert, S. G. 1966. *Pictorial Anatomy of the Fetal Pig.* Second Edition. Seattle, Washington: University of Washington Press.

Hickman, C. P., et al. 1978. *Biology of Animals.* Second Edition. St. Louis, Missouri: C. V. Mosby.

Jensen, W. A., et al. 1979. *Biology.* Belmont, California: Wadsworth.

Kent, G. C. 1983. *Comparative Anatomy of the Vertebrates.* Fifth Edition. St. Louis, Missouri: C. V. Mosby.

Kessel, R. G., and R. H. Kardon. 1979. *Tissues and Organs.* New York: W. H. Freeman.

Kessel, R. G., and C. Y. Shih. 1974. *Scanning Electron Microscopy in Biology.* New York: Springer-Verlag.

Lytle, C. F., and J. E. Wodsedalek. 1984. *General Zoology Laboratory Guide.* Complete Version. Ninth Edition. Dubuque, Iowa: Wm. C. Brown.

Mathews, W. W. 1972. *Atlas of Descriptive Embryology.* New York: Macmillan.

Odlaug, T. O. 1975. *Laboratory Anatomy of the Fetal Pig.* Fourth Edition. Dubuque, Iowa: Wm. C. Brown.

Patten, B. M. 1951. *American Scientist* 39: 225–243.

Scagel, R. F., et al. 1982. *Nonvascular Plants.* Belmont, California: Wadsworth.

Sheetz, M., et al. 1976. *The Journal of Cell Biology* 70: 193.

Shih, C. Y., and R. G. Kessel. 1982. *Living Images.* Boston: Science Books International/Jones and Bartlett Publishers.

Stanier, R., et al. 1986. *The Microbial World.* Fifth Edition. Englewood Cliffs, New Jersey: Prentice-Hall.

Starr, C., and R. Taggart. 1984. *Biology.* Third Edition. Belmont, California: Wadsworth.

Starr, C., and R. Taggart. 1987. *Biology.* Fourth Edition. Belmont, California: Wadsworth.

Steucek, G. L., et al. 1985. *American Biology Teacher* 471: 96–99.

Storer, T., et al. 1979. *General Zoology.* New York: McGraw-Hill.

Villee, C. A., et al. 1973. *General Zoology.* Fourth Edition. Philadelphia: W. B. Saunders.

Weller, H., and R. Wiley. 1985. *Basic Human Physiology.* Boston: PWS Publishers.

Wischnitzer, S. 1979. *Atlas and Dissection Guide for Comparative Anatomy.* Third Edition. New York: W. H. Freeman.

Wolfe, S. L. 1985. *Cell Ultrastructure.* Belmont, California: Wadsworth.

Wynn, C. M., and G. A. Joppich. 1984. *Laboratory Experiments for Chemistry.* Third Edition. Belmont, California: Wadsworth.